59.95

OTHER TITLES OF INTEREST FROM ST. LUCIE PRESS

Ecosystem Management for Sustainability: Illustrated by an International Reserve Cooperative

Ecological Integrity and the Management of Ecosystems

Everglades: The Ecosystem and Its Restoration

The Everglades Handbook: Understanding the Ecosystem

The Everglades: A Threatened Wilderness

Development, Environment, and Global Dysfunction: Toward Sustainable Recovery

Resolving Environmental Conflict: Toward Sustainable Community Development

Economic Theory for Environmentalists

From the Forest to the Sea: The Ecology of Wood in Streams, Estuaries and Oceans

Sustainable Forestry: Philosophy, Science and Economics

Environmental Fate and Effects of Pulp and Paper Mill Effluents

Environmental Effects of Mining

For more information about these titles call, fax or write:

St. Lucie Press
100 E. Linton Blvd., Suite 403B
Delray Beach, FL 33483
TEL (407) 274-9906 • FAX (407) 274-9927

$\mathrm{S^{t}_{L}}$

ECOLOGY and MANAGEMENT of TIDAL MARSHES

A Model from the Gulf of Mexico

ECOLOGY and MANAGEMENT of TIDAL MARSHES

A Model from the Gulf of Mexico

Editors

CHARLES L. COULTAS
YUCH-PING HSIEH

St. Lucie Press
Delray Beach, Florida

Phone: (407) 274-9906
Fax: (407) 274-9927

S^t_L

Published by
St. Lucie Press
100 E. Linton Blvd., Suite 403B
Delray Beach, FL 33483

Dedication

This book is dedicated to Mr. Olin C. Lewis, former state soil scientist with the Soil Conservation Service (USDA). In the mid-1960s, when few people were concerned with wetlands, Mr. Lewis encouraged scientists at Florida A&M University to study these systems. It is unlikely that this book would have been written without his suggestions. Mr. Lewis is presently retired and living in Gainesville, Florida.

Table of Contents

Foreword

Lt. Governor Buddy MacKay

All Floridians, even those whose families go back generations like mine, are only recent residents of Florida. Thousands of years, millions of creatures, and billions of events carved and shaped this peninsula long before we arrived.

As recent immigrants to this peninsula, we have found a natural treasury that we value but little understand. We have settled on the edge of great natural resources that are marvels and mysteries. The Everglades, the Keys, the Gulf of Mexico, seagrassed dunes, needlerush tidal marshes, the Atlantic Ocean, the Great Green Swamp, Tampa Bay, Barrier Islands and the Intercoastal, the St. Johns River, Apalachicola Bay, bird-filled wetlands, cypress-lined lakes, and mangroved rivers are imprinted on the memories of millions of Floridians.

These natural marvels have fueled an economic marvel in the last fifty years, as Florida has become a national leader. Those of us who have grown up with these changes have witnessed an astonishing transformation. Florida is no longer the sleepy edge of a growing nation; it has become the center of an emerging hemisphere—a major state in the only global superpower.

Growth in population and the development that it brings have fueled the state's economy throughout the last half century of expansion and prosperity. Growth has played the same historic role in Florida's economy as oil in Texas, autos in Michigan, and steel in Pennsylvania. But single-track economies provide a roller-coaster ride with highs and lows. When the dominant industry exhausts itself, the state's economy and job picture crash. Florida's roller-coaster ride is coming to the end of its track.

Each day our newspapers signal warnings that our natural resources are in trouble: water shortages in Pasco and Pinellas counties, fish kills in Florida Bay, flooding in southeast Florida, storm devastation of homes perched on beachfronts, insurers anxious to leave Florida behind, net bans to save the dwindling fishery.

The trends are even more unsettling than the headlines:

- Intertidal marshes are among the most productive ecosystems on earth, and Florida has more intertidal wetlands than Georgia and the Carolinas combined.
- The economic value of tidal marshes is harvested by each region's economy, not the owners of the marsh. This drives the owners to seek alternative uses for their property.
- Sixty-five percent of these wetlands have already been lost to agriculture and commercial and residential use.

It is becoming clear that we must slow down and unburden our increasingly overburdened environment in order to save its economic value. But how is that done?

We must inventory our treasury. We must take stock of the natural resources that drew us to Florida and that we all cherish.

We must enlist our best minds and our best science to unravel these natural mysteries. We, too often, damage them unknowingly.

We must sleuth out the complex linkages within these resources and understand how they collectively impact our lives.

We must translate the value of these natural assets into denominations that citizens and elected officials can understand.

We must fashion new public and private strategies to steward these natural systems that underpin our communities.

We must act before development overgrows the natural treasures that have triggered our state's growth.

Ecology and Management of Tidal Marshes follows this prescription. This collection assembles the complex inventory of tidal marshes: their soil, mineral, animal, and plant life. Each of these assets is explored and explained. The complex equation of these tidal marshes is then detailed. The real dollar value to the public is calculated. The public systems to protect and manage these ecosystems are then assessed, analyzed, and critiqued. A strategy to sustain these natural/economic resources is outlined.

The resulting blueprint is clear and enactable. With it, we have a benchmarking of these resources. What makes them work? We have an equation to compute their worth. How much value do they provide us? We have a formula to sustain them. What saves them? This collection details the way that we should go. But do we have the will to follow it? If we are to maintain the life and the economic prosperity that we all appreciate today, we surely must.

Preface

This book is targeted toward scientists and students interested in the complex world of coastal marshes. It will be helpful to government officials who must propose and enforce laws and regulations which affect these crucial natural areas. Significant and widespread research on wetlands in the United States is a fairly recent phenomenon. The large majority of published work has occurred within the past 25 years. This is due to the fact that the importance of wetlands has only recently been realized. This is especially true of coastal wetlands, since a large proportion of the world's population lives and works near the sea/land interface. We have slowly come to realize that intertidal marshes provide many functions for humankind.

This book is intended as a compendium of our knowledge of the intertidal marshes using data primarily from the Gulf coast states. Some information is rather site specific, but much discussion has a wider applicability than the Gulf. The authors discuss the marsh as a part of a larger system, specifically the plants, soils, and fisheries. They also present different methodologies in valuing the marshes and discuss human impact on this system. They suggest management alternatives and legal options and their effect.

CLC
YPH
Tallahassee

Acknowledgments

We wish to thank the many reviewers who gave of their time to make this book as well written and scientifically sound as possible. We especially want to recognize Dr. Elizabeth D. Purdum and Dr. Edward A. Fernald of the Institute of Science and Public Affairs at Florida State University for assistance with final editing.

The editors are grateful for the assistance of Drs. H.T. Odum and Clay Montague and Ms. Joan Breeze of the Department of Environmental Engineering, University of Florida for their assistance in obtaining copyright releases.

The editors acknowledge the support of the Cooperative State Research, Education, and Extension Service of the USDA and the encouragement and assistance of Dr. Sunil K. Pancholy, Research Director in the College of Engineering Sciences, Technology, and Agriculture at Florida A&M University.

The Editors

Charles L. Coultas has performed research in the wetlands of Florida since 1967. His first efforts were studies on the soil–plant relationship of intertidal areas from the Perdido River (Pensacola) to Rookery Bay below Naples. He also worked with Soil Conservation Service and University of Florida scientists on the extensive marshes of Duval County. More recently, he has centered his studies on freshwater wetlands.

Dr. Coultas received his training in soil and plant sciences in the Midwest and earned his Ph.D. degree from the University of Minnesota. He has authored over 36 scientific papers and book chapters and has received numerous grants. Much of his professional work has centered around land use, soil survey, and sustainable agriculture. He was employed by Florida A&M University in 1966, where he taught courses in soil science and horticulture. Although presently retired, he maintains a part-time relationship with that university and is involved in numerous consultancies, most relative to wetlands. He has had extensive professional experience in Central America, the Caribbean, and Africa.

Yuch-Ping Hsieh is a professor of wetland ecology at Florida A&M University. He received his B.S. degree in agricultural chemistry from National Chung-Hsing University, M.S. degree in soil mineralogy from Montana State University, and Ph.D. degree in soil and environmental science from Rutgers University.

His research interests include soil organic carbon dynamics, sulfate reduction in wetlands, and landscape-scale soil erosion assessment. He was instrumental in the development of several innovative methods for ecological research, including a bomb ^{14}C and ^{13}C method to determine sizes and turnover rates of active and stable soil organic carbon pools, diffusion methods to determine reduced inorganic sulfur species in sediments, a simple method to determine aboveground net primary production in marshes, a rapid method to quantify living roots in marshes, a simple mesh-bag method to determine soil erosion in field conditions, and the use of clay mineralogy to trace the source of erosion in a watershed.

Contributing Authors

Jonathan D. Arthur
Florida Geological Survey
Department of Natural Resources
Tallahassee, Florida

Frederick W. Bell
Department of Economics
Florida State University
Tallahassee, Florida

Andre F. Clewell
A.F. Clewell, Inc.
Quincy, Florida

Charles L. Coultas
Wetland Ecology Program (retired)
Florida A&M University
Tallahassee, Florida

C.S. Gidden
St. Marks National Wildlife Refuge
(retired)
U.S. Department of the Interior
St. Marks, Florida

Kenneth D. Haddad
Division of Marine Resources
Department of Environmental Protection
Tampa, Florida

Richard Hamann
Center for Governmental Responsibility
School of Law
University of Florida
Gainesville, Florida

Douglas A. Hornbeck
Gainesville Regional Utilities
Gainesville, Florida

Yuch-Ping Hsieh
Wetland Ecology Program
Florida A&M University
Tallahassee, Florida

Michael D. Hubbard
Entomology Research
Florida A&M University
Tallahassee, Florida

Edwin A. Joyce, Jr.
Division of Marine Resources
Florida Department of Environmental
   Protection
Tampa, Florida

Clyde F. Kiker
Department of Food and Resource
   Economics, IFAS
University of Florida
Gainesville, Florida

William L. Kruczynski
U.S. Environmental Protection Agency
Pensacola, Florida

Gary D. Lynne
Department of Agricultural Economics
University of Nebraska
Lincoln, Nebraska

Earl D. McCoy
Department of Biology
University of South Florida
Tampa, Florida

Clay L. Montague
Department of Environmental
   Engineering Sciences
University of Florida
Gainesville, Florida

Howard T. Odum
Department of Environmental
   Engineering Sciences
University of Florida
Gainesville, Florida

Jorge R. Rey
Florida Medical Entomology Lab, IFAS
University of Florida
Vero Beach, Florida

Frank R. Rupert
Florida Geological Survey
Department of Natural Resources
Tallahassee, Florida

Barbara F. Ruth
Department of Environmental Protection
Pensacola, Florida

John Tucker
Center for Governmental Responsibility
School of Law
University of Florida
Gainesville, Florida

Introduction:
The Intertidal Marshes
of Florida's Gulf Coast

Clay L. Montague and Howard T. Odum

The intertidal marshes of Florida's Gulf coast are exhilarating to view and exciting to explore. They are among Florida's most productive ecosystems, annually giving rise to vast quantities of both plants and animals. They exist at the water-logged fringe of land over which coastal waters inundate and recede from the force of the wind and tide. Abundant food and cover alternately attract aquatic and terrestrial animals. The marsh vegetation is highly productive, but not very diverse. A single species may dominate, forming nearly monotypic stands that from a distance look like a vast mowed playing field or pasture. Fields of corn and sugarcane are similarly productive, but their high production and low diversity depend on artificial subsidies of fuel, human labor, and pesticides. Damaging epidemics of disease and pests, common in agricultural monocultures, are virtually unknown in intertidal marshes.

Residential property in Florida with a view of intertidal marsh often sells for a far greater price than nearby property without such a view. Access to tidal creeks adds even more value. Although these benefits of intertidal marshes accrue to the fortunate owners of marsh-front property, several ecological functions hidden within the spectacular view benefit the public at large (Table I.1). Intertidal marshes, which include salt-, brackish, and freshwater marshes in fluctuating coastal waters, are one of a set of coastal ecosystems (including seagrasses, mud flats, rocky outcroppings, mangroves, plankton, and others) that together account for Florida's productive commercial and recreational fish-

1

Table I.1   Known or Suspected Public Values of Intertidal Marshes

1   Shoreline protection from storm surge, winds, and waves

2   Habitat: simultaneous food and cover for numerous fishes, crustaceans, mollusks, birds, and other vertebrates and invertebrates, many of commercial or recreational importance, some threatened or endangered

3   Part of a *set* of estuarine and nearshore habitats of varying scale and location that ensure a diversity of productive fish and wildlife at all life history stages

4   Buffer of estuarine food supply

5   Buffer of estuarine water quality

6   Contributor to attractive coastal image that encourages development

eries. They provide food and cover for juvenile stages of most estuarine fish and invertebrates, including commercially valuable shrimp and blue crabs as well as sportfish such as red drum and spotted seatrout. Intertidal marshes reduce damage from storms by attenuating the energy of winds, waves, and storm surges. Furthermore, as coastal waters alternately fill and drain, materials used and produced by the marsh are exchanged, which results in a stabilizing effect both on estuarine water quality and on the food supply to the web of estuarine biota.

Intertidal marshes help trap and stabilize estuarine sediments, increasing water clarity in coastal waters. If water were to become more turbid because of the loss of intertidal marshes, seagrass beds would die because of lack of light. Seagrass also stabilizes sediments, reduces damage from storms, and is habitat for numerous aquatic animals. Intertidal marshes form part of the image of coastal Florida, which stimulates many to tour and live on Florida's coast.

Intertidal marshes are a key part of the coastal ecosystem and the coastal economy. Florida's population continues to grow by more than 1000 people per day, more than three-quarters of whom settle in Florida's coastal counties. Coastal management is required to ensure the continuation of the functions and public values of intertidal marshes. The best management practices arise from management principles derived by integrating scientific knowledge from a variety of scales and perspectives, including models of ecological and economic theoreticians, tests of hypotheses by experimentalists, and large-scale field trials by marsh managers aimed at fish and wildlife enhancement and mosquito control.

To develop sound management principles, at least three scales of ecological study must be integrated: studies of the living components of the marsh, studies that link the components to one another and with the overall environment, and studies that link the marsh ecosystem to other estuarine subsystems and the coastal economy. Information needs at larger scales depend upon knowledge at smaller scales. However, from the perspective of coastal zone management, the whole is more than (or at least different from) the sum of the parts.

Controversies often arise concerning the use and management of marshes. Scientific, economic, and legal studies have addressed such public policy questions as the economic value of marshes, the impact and need of mosquito control, the maintenance or enhancement of coastal and estuarine fisheries, enhancement of wildlife habitat, waste disposal practices, land building on marshes, canal construction, restoration to pre-Columbian conditions, and preservation of present conditions. Unfortunately, study of such management issues is often on a crisis-by-crisis basis, initiated by those with an immediate information need and ended often prematurely when a new and different crisis begins. Usually, the entire collection of management issues is not simultaneously considered when decisions are made about one.

Effective management of the expansive intertidal marshes of Florida requires an integration of scientific knowledge with legal and economic perspectives and the development of management principles. This book is a first step in that direction. Its purpose is to summarize what is known about the structure and function of intertidal marshes as well as to present varying economic, legal, and management perspectives. It is hoped that this synthesis will lead to a more informed public and more effective management of Florida's remaining intertidal marshes. Studies of management questions at several scales are included, together with various efforts to analyze and synthesize them. This collection of information forms a basis for predicting the consequences of management decisions.

Although intertidal marshes are found along both the Atlantic and Gulf coasts, conditions for their occurrence (low wave energy and relief, high tidal range) are most favorable on the Gulf coast, particularly north of Tampa, where it is too cold for mangroves. The most extensive intact intertidal marshes are in the Big Bend region of Florida where—because of their distance from urban centers—they have been relatively protected from development pressures, though perhaps not for long given the huge growth of Florida's population and the value of coastal property. Because of their extent and attractiveness, and because there is still time to manage them properly, the intertidal marshes of the Gulf coast, particularly the Big Bend coast of Florida, are the subject of this book.

An Energy Systems Diagram of the Intertidal Marsh

To begin to synthesize knowledge of intertidal marshes and as a framework for much of what is included in this book, we have constructed an energy systems diagram (Figure I.1). A full understanding of the functions of marshes in the larger coastal ecosystem and coastal economy requires integration of scales of study, from the small to the larger. Systems methods for integrating knowledge into a fuller understanding usually involve the construction of diagrams to represent those parts and the causal connections among them that investigators believe to exist from their measurements and intuitions. These diagrams represent large-scale, complex hypotheses that reflect current knowledge and ideas. Quantifying these connections allows computer simulation of the dynamics among components that result from the hypothesis. Quantified diagrams allow other analyses as well (e.g., the "EMERGY" analysis of Odum and Odum, 1987).

Simulation modeling and analysis techniques explore the system-level consequences of the hypothesis and allow experimentation with alternative ways of connecting the parts. Often the output of a computer model is surprisingly different from what was expected or known to be true. Such a result may require a change of ideas, new hypotheses, new measurements, and revised models. In the interim, the limits of our understanding at the large scale have been learned in a rigorous manner. Old ideas, some perhaps often used in management, may have been rejected. From successful simulations come "what if" experimental manipulations that can be explored. The management consequences of allowing specific alterations to some areas in an estuary and restoring others may be predictable.

The energy systems diagram is drawn according to a protocol in which small, quickly replaced items and processes are shown on the left and slower processes with components that occupy larger territories on the right (Odum, 1983; Odum and Odum, 1987). Thus, phytoplankton are on the left and larger animals are toward the right. Products from the small components flow from left to right. Material recycling and human services form feedback loops from the right over the top of the diagram. Through these feedbacks, the high-transformity items on the right side of the diagram amplify and control the lower-transformity items on the left side.

The amount of energy of one kind (e.g., plant biomass) that is transformed in the process of producing a unit of energy of another type (e.g., herbivore biomass) is defined as the "transformity" of that product. It may take 10 calories of plankton, for example, to produce each calorie of small fish. In terms of plankton, the transformity of the fish is 10:1. All components in Figure I.1 are forms of energy and each can be expressed in units of one kind of energy,

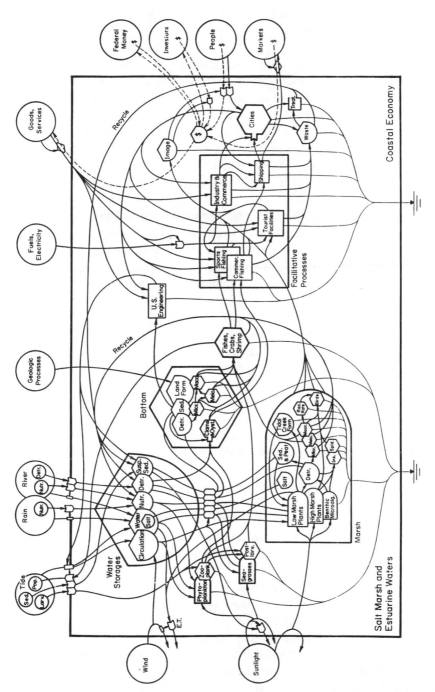

Figure I.1 The salt marsh as a component of a larger coastal ecosystem and the coastal economy as illustrated by an energy-flow network.

solar insolation, which is required both directly and indirectly to produce a unit of each component. The resulting ratios are known as "solar transformities" (Odum and Odum, 1987).

Components in the energy systems diagram are arranged from left to right in order of increasing solar transformity. Hence, basic energy sources (such as sunlight, wind, and tide) enter the diagram to the left and those which enter the system already with higher solar transformities (such as fuels, electricity, investments, and people) enter to the right. These energy sources combine at various points in the system. The flow is from left to right as the components of the system transform these energy sources into other components of higher and higher transformity. Thus, the diagram represents the natural energy hierarchy: a process in which many units of energy on the left are required to produce those to the right, which, in turn, augment or control those on the left.

Although the diagram includes many components and processes, these can be grouped into two main divisions: a natural coastal ecosystem on the left and a coastal economy on the right. Each division has a few major components and several subcomponents. The major components of the natural coastal ecosystem are the water column, the subtidal bottom, and the intertidal marsh. The main producers in these systems (indicated by bullet-shaped symbols) are phytoplankton, seagrasses, marsh plants, and microalgae. Consumers (the hexagons) include zooplankton, postlarval fish and invertebrates, microbes, meiofauna, and larger animals such as shrimp, crabs, fish, birds, and raccoons. Energy sources for the natural ecosystem (circles) include sunlight, wind, tide, rain, rivers, and long-term geologic processes that determine such things as coastal topography. Nonliving storages (the tank-shaped symbols) include circulating water (and accompanying nutrients, salts, detritus, and other dissolved and suspended materials), sediments, tidal creeks, and landforms (e.g., dunes and lowlands).

The coastal economy occupies the right third of the diagram. Cities are the main consumer. The main exogenous energy sources for this economy are fuels, electricity, goods and services, federal money (mostly defense spending), investors (many in real estate), people, and markets. Facilitative processes transform natural resources into the coastal economy. The natural ecosystem to the left produces fishes, crabs, and shrimps for commercial and recreational fisheries; contributes to an attractive coastal image that encourages tourism and development; and recycles nutrients from domestic and industrial wastes.

The diagram is the beginning of a synthesis process that will be improved by the participation of all those with specialized knowledge of its various aspects. It is hoped that this book will stimulate such a synthesis—that it will inspire research in areas where more knowledge is needed and that it will help us all, regardless of our special skills and interests, maintain a sense of the whole. Management depends on whole-system understanding.

References

Odum, H. T. 1983. *Systems Ecology: An Introduction*. John Wiley and Sons, New York. 644 pp.

Odum, H. T. and E. C. Odum. 1987. *Ecology and Economy: "EMERGY" Analysis and Public Policy in Texas*. Policy Research Report No. 78. Lyndon B. Johnson School of Public Affairs, University of Texas, Austin. 178 pp.

Setting and Functions

1

Clay L. Montague and Howard T. Odum

Florida has more intertidal wetlands (salt marshes and mangroves) than Georgia, South Carolina, and North Carolina combined (Table 1.1). Over three-
quarters of Florida's intertidal wetlands are on the Gulf coast, with the greatest
acreage of salt marsh in the Big Bend between Tarpon Springs and Lighthouse
Point at the mouth of the Ochlockonee River (Table 1.2, Figure 1.1). Extensive
salt marshes are also found landward of wide bands of mangroves along the
southwestern coast south of Naples.

Variations in lunar tides, topographic slope, and winter temperature account
for the distribution of intertidal marshes in Florida. Sediment supply, so important to the extensive marshes of Georgia and Louisiana, is relatively insignificant along Florida's Gulf coast. Intertidal wetlands occur where coastal topographic relief is low relative to tidal range. Where coastal relief is especially
low and tidal ranges high, intertidal wetlands are most expansive. A crude
estimate of the inland reach of Florida's Gulf coast intertidal wetlands (in
kilometers) can be obtained by dividing the predicted lunar tidal range (meters)
by the change in elevation (meters/kilometers) for each region (Table 1.2).

The Gulf coast of Florida consists of a very broad continental shelf gradually rising toward land and ancient dune ridges. A little more than half of
Florida's intertidal wetlands (marshes and mangroves) occur on the southwestern Gulf coast. Here, coastal topographic relief is extremely low and tidal range
is higher than anywhere else along the Florida Gulf coast. In the Big Bend area,
ancient dune ridges are generally well inland, nearshore topographic slope is
very gradual, and predicted tidal ranges are intermediate. At the limits of the
Big Bend region (Pasco County to the south and western Wakulla County to
the north), geological scarps lie very near the coast and the band of marshes

Table 1.1 Intertidal Wetland Area (Marshes and Mangroves) in the Southeastern United States

State	Marsh area (ha)	Coastline length (km)[a]
North Carolina	64,300	500
South Carolina	204,200	350
Georgia	159,000	160
Total	427,500	1,010
Florida		
Atlantic	77,800	660
Gulf	359,700	1,340
Total	437,500	2,000

[a] Measured as a boat would most likely travel if 5 km offshore.

From Montague et al., 1987a; U.S. Fish and Wildlife Service, 1984.

Table 1.2 Physical Description and Population Density Estimates for Four Regions of the Gulf Coast of Florida

Region	Area of marsh[a] (ha)	Area of mangrove[a] (ha)	Tidal range[b] (cm)	Topographic slope[c] (m/km)	Coastline length[d] (km)	1990 population density[e] (per km^2)
Panhandle	16,754	97	35	10	383	50
Big Bend	66,537	6,091	100	0.4	345	37
Pinellas to Lee County	4,501	35,201	80	5	320	216
Southwest	25,345	205,142	120	0.1	290	29
Florida total	163,121	274,347	—	—	2,000	93

[a] U.S. Fish and Wildlife Service (1984).
[b] National Oceanic and Atmospheric Administration (1989).
[c] Estimate measured on 1:150,000 USGS bathymetric-topographic maps of Florida.
[d] Estimate measured on 1:500,000 USGS map of the state of Florida.
[e] Shoemyen et al. (1989).

Figure 1.1 Four regions of the Gulf coast of Florida based on environmental differences: Panhandle, Big Bend, Pinellas to Lee County coastal zone, and the southwestern coast.

narrows. The band also narrows near Cedar Key (Levy County) and between Stake Point and Rock Point in Taylor County, where ancient dune fields lie just inland (Brooks, 1981; U.S. Fish and Wildlife Service, 1985). Both southward and westward of the Big Bend region, dune ridges are close to shore and tidal ranges are lower. In the Panhandle, predicted tidal ranges are very low and coastal topography is very steep.

On the Gulf coast south of Tampa Bay, dense mangroves occupy most of the intertidal zone. Shorter marsh plants cannot survive under the shade of these trees. North of Tampa Bay, killing freezes are too frequent to allow extensive mangrove development (Table 1.3), although pockets of mangrove occur on the outer islands near Cedar Key, where the intensity of freezes is reduced by the surrounding water.

The northern extent of mangroves fluctuates with the occurrence of occasional freezes. Hard freezes in December 1983 and January 1985, for example, killed many mangroves not only at Cedar Key but also much farther south on both coasts (e.g., Indian River and Tampa Bay). At Cedar Key, where over 95% of the mangroves were killed, nonwoody salt marsh plants (mostly *Spartina alterniflora* and *Batis maritima*) have rapidly replaced mangroves in many areas. Mangrove seeds sprouted, however, during the summer of 1984 and

Table 1.3   Climate Normals for Florida Gulf Coast

Temperature normals (°C)							Days/year below freezing
	January		July		Annual		
Region	Max	Min	Max	Min	Max	Min	
Panhandle	15.9	5.5	32.0	23.0	24.8	14.5	5–39
Big Bend	18.7	7.9	32.1	23.1	26.1	15.9	4–5
Pinellas to Lee County	22.7	11.2	32.6	23.3	28.3	17.7	0–1
Southwest	24.5	13.6	32.5	23.5	28.9	19.0	0–0.5

Precipitation normals (cm)					
Region	Annual mean	Percent from May to October	Monthly Max	Min	Annual range as % of mean
Panhandle	152	58.1	20.2	8.3	7.8
Big Bend	134	63.8	20.5	6.1	10.8
Pinellas to Lee County	130	74.7	21.4	4.1	13.3
Southwest	131	79.3	22.7	3.3	14.8

Note: Computed from National Oceanic and Atmospheric Administration (1983, 1985).

young trees began to bear viable seeds by the autumn of 1987 (Montague, unpublished data).

Marshes may initially grow in a relatively narrow band of intertidal zone, but if suspended sediments are in great supply, they may expand seaward (Davis, 1940; Meade, 1982; Frey and Basan, 1985). Coasts that receive a lot of silt and clay from rivers often have extensive marsh development. In Georgia, two large rivers supply 2 million tonnes of sediment per year along only 160 km of coastline. Coastal marshes there are sediment saturated and have perhaps expanded as much as hydrodynamically possible (Meade, 1982; Meade and Parker, 1985). Additional sediment forms steep levees at the edges of marsh creeks (Frey and Basan, 1985). Likewise, in the Mississippi River Delta, the huge intertidal marsh development is attributable in large part to the (formerly) vast supply of sediments from that river, which are now largely prevented from entering marshes by dikes and levees built for channel stabilization and flood control (Turner, 1987).

The four rivers of Florida's Big Bend (Suwannee, Steinhatchee, Sopchoppy, and Ochlockonee) supply only about 109,000 tonnes of sediment per year

Table 1.4 Mean Total Sediment Discharge (×1000 tonnes per year) in Four Regions of the Gulf Coast of Florida

Region	Sediment discharge[a]	No. of years in mean	Average standard deviation	Ratio of standard deviation to mean	No. of NASQUAN stations
Panhandle	1,459	6	36	0.41	5
Big Bend	109	6	10	0.69	8
Pinellas to Lee County	97	7	33	1.34	5
Southwest	3	7	3	1.04	1

Major sources of sediment by region:

Percent from Apalachicola in Panhandle total	78
Percent from Suwannee in Big Bend total	67
Percent from Peace and Caloosahatchee in Pinellas to Lee County total	80
Percent from Tamiami canals in Southwest total	100

[a] Means of several years of discharge reported for all Gulf NASQUAN stations reporting in USGS Water Data for Florida, Water Years 1975–1988.

(Table 1.4). The Apalachicola River just to the west of marsh band supplies about 1 million tonnes per year. How much of the Apalachicola River sediment ever reaches the Big Bend coastal marshes is questionable. Sediments from the Apalachicola River tend to accumulate in shoals near the mouth and slowly drift *westward* into the very deep nearshore waters off the Panhandle (Tanner, 1960). The Big Bend is considered sediment starved, which explains why it has no barrier islands (Tanner, 1960). Because the supply of fine, marsh-building sediments from the Apalachicola River is probably also very low, these marshes have not greatly expanded seaward. Studies of the geological history of Big Bend marshes (performed by direct examination of cores) indicate little or no seaward expansion of intertidal marshes in this part of Florida over the past several thousand years (Kurz and Wagner, 1957).

Sea level along the Florida coast has been rising at a rate of about 25 cm per century and has apparently accelerated considerably in the last century, presumably due to global warming (Marmer, 1954; Wanless, 1989). As sea level rises, intertidal marshes may be expected to move inland. If nearshore conditions change so that a large supply of suspended sediment becomes available, marshes grow vertically as well, and thus the aerial extent of the intertidal zone may increase. This seems unlikely in the Big Bend area since the supply of sediment is currently so low. Furthermore, if topographic slope is steeper inland, the width of the intertidal marsh band will decrease (Mehta et al., 1989). The intertidal zone of the Big Bend area may decline considerably as the shoreline approaches the scarps and dune fields of Taylor and Levy counties. Increasing wave energy may occur with rising sea levels in some areas (Mehta et al., 1989). Whether increased wave energy will significantly reduce the loss of intertidal marsh in the Big Bend as sea level rises requires detailed hydrodynamic study.

Environment, Production, and Diversity

Abundant energy and mineral resources accumulate at the coast: rivers discharge fresh water (Table 1.5) and deposit sediments (Table 1.4) and nutrients, rain (Table 1.3) and wind (Table 1.6) are abundant, and tidal energy is discharged (Table 1.2). Together with sunlight (Table 1.7), these resources stimulate ecological production (see End-of-Chapter Note) in some intertidal marshes that is among the highest per unit area in the world. The dominant primary producer is determined largely by soil salinity and the frequency of inundation. These variables also influence the production and composition of the community of resident marsh animals and the timing of use by transient animals.

Freshwater discharges from land influence estuarine circulation and intertidal soil salinity. Surface water discharge along the Gulf coast of Florida is relatively low and decreases toward the south (Table 1.5). Low discharge of surface water usually means that saline water will extend inland for many kilometers. This is true in most regions of the Gulf coast of Florida (Table 1.8), including the large bays of the Panhandle and those of the coast from Pinellas to Lee counties. South of Lee County, very low and diffuse discharge of surface water from the Everglades allows considerable inland penetration of saline water. The Big Bend coast, however, has a surprisingly low salinity, given the low surface water discharge and the tendency of Apalachicola River water to flow westward. A significant freshwater input to this coast is groundwater. The Floridan aquifer, deep in many parts of the state, is at or near the surface throughout the Big Bend area, and flowing springs are common (Fernald and Patton, 1984). Coastal and submerged springs in the Big Bend region reduce salinity near shore.

Wind and lunar forces combine to inundate the marshes of the Gulf coast of Florida. In Gulf coast intertidal marshes, water levels are often considerably

Table 1.5 Freshwater Discharge (m³/s) from Rivers, Streams, and Canals in Four Regions of the Gulf Coast of Florida

Region	Mean regional discharge	1988[a] discharge	1988[a] max/mean	1988[a] min/max (%)	No. of USGS stations	Years of record
Panhandle	1262	899	10	7.87	14	32
Big Bend	507	423	10	5.35	14	24
Pinellas to Lee County	163	150	29	0.21	28	19
Southwest	34	41	4	0.03	4	40

Major discharges by region	Period of record (year)	Water year 1988
Percent from Apalachicola in Panhandle total	59	56
Percent from Suwanee in Big Bend total	60	65
Percent from Peace and Caloosahatchee in Pinellas to Lee County total	50	52
Percent from Tamiami canals in southwest total	92	92

[a] Average of all stations in region.

Note: Compiled from U.S. Geological Survey Water Data for Florida, Water Year 1988.

Table 1.6 Average Wind Speed at 10 m from the Surface and Annual Storm Frequency for Four Regions of the Gulf Coast of Florida

Region	Wind speed (m/s)[a]	Standard deviation[b] (12 months)	Probability of tropical storm or hurricane[c] (%)	Probability of destruction by hurricane[c] (%)	Mean annual days with thunderstorms[d]
Panhandle	3.5	0.38	11–21	6–13	60–80
Big Bend	3.3	0.34	12–20	4–6	80–100
Pinellas to Lee County	3.5	0.35	9–12	4–9	100–110
Southwest	4.0	0.42	19–21	9–13	85–100

[a] Reed (1979).
[b] Computed from data in state of Florida (1983).
[c] Bradley (1972).
[d] Bryson and Hare (1974).

Table 1.7 **Insolation and Cloud Cover for Four Regions of the Gulf Coast of Florida**

Region	Insolation (Langleys)			Cloud cover[c] (days/year)	
	Annual[a]	December[b]	May[b]	Cloudy	Partly cloudy
Panhandle	408–445	225–265	565–615	120–150	110–115
Big Bend	445–453	265–300	600–615	120–125	110–140
Pinellas to Lee County	428–450	300–320	550–600	105–125	140–155
Southwest	450	300–320	550–580	100–120	155–170

[a] Bryson and Hare (1974).
[b] Bennett (1965).
[c] Conway and Liston (1974).

different than those predicted in lunar tide charts (Figure 1.2). Winds from the north and northeast, more common in the winter, may push water out of the marsh and keep it out. Winds from the south and west, which occur more often in the summer and are usually relatively gentle, may hold water on the marsh for the entire tidal cycle.

Table 1.8 **Inland Extent of Saline Water in Four Regions of the Gulf Coast of Florida**

Region	Average distance (km)[a]				
	Gulf to river mouth[b]	Gulf to 20 ppt	20–5 ppt	5–0.5 ppt	No. of rivers in average
Panhandle	32.3	15.5	18.9	12.6	7
Big Bend	1.1	0.8	4.1	10.2	12
Pinellas to Lee County	26.5	26.5	14.2	24.0	7
Southwest	1.8	21.3	8.8	4.6	4

[a] Data measured from U.S. Fish and Wildlife Service Gulf Coast Ecological Inventory Maps (see Beccasio et al., 1982).
[b] Rivers sometimes discharge into bays rather than directly into the Gulf. The number in this column reflects the prevalence of bays in each region.

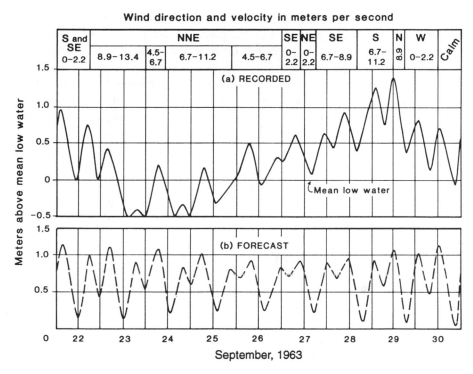

Wind direction and velocity in meters per second

Figure 1.2 Influence of wind on the predicted tides. (Redrawn from Stelzenmuller, 1965.)

The frequency and amplitude of environmental change influence the types of animals and plants found in a coastal ecosystem. The many energies and minerals that stimulate production in coastal marshes are neither continuous nor in phase with one another. Among the most highly variable and influential factors is freshwater discharge. The most variable discharges are in the region from Pinellas to Lee County.

Intertidal marsh organisms are exposed to wide variation in oxygen, moisture, and salt. To survive, they must either avoid the many environmental extremes of the intertidal marsh or possess adaptations for withstanding them. Numerous species of rapidly growing but short-lived organisms, such as microbes, grow and die in coastal marshes in response to the high frequency of unpredictable environmental changes. Microbial "opportunists" take advantage of the abundant resources during brief periods of tolerable or favorable conditions. Patches of blue-green algae, for example, are sometimes noticeably abundant soon after rains. Few long-lived species, however, are well-adapted for withstanding all of the environmental variation in intertidal marshes. Vagile animals (i.e., animals able to move from place to place) can avoid physiologically damaging changes, but resident or sedentary animals and plants must withstand these changes or die. Fiddler crabs, mussels, salt marsh cordgrass, and other such organisms possess unique behavioral and physiological adaptations for survival. They tolerate considerable fluctuation in salinity, water level, temperature, and oxygen. Few such species exist, but resources are abundant for those few, and they are very productive.

Short-lived or vagile species occur in greater diversity. If their movement is unimpeded and changes do not occur too quickly (compared to their rate of movement), vagile animals can survive a long time in estuaries and will repeatedly revisit fringing marshes whenever suitable conditions return.

Intertidal Marsh Ecosystem Processes

Vast, complex networks of tidal creeks fill and drain the marsh except where tidal ranges are very low. The intricacies of these networks are best revealed when viewed from above (Figure 1.3). The tidal channels of Figure 1.3 are an example of "hierarchical self-organization" in which many smaller units support fewer larger ones that return some control or subsidy to the smaller supporting units. The hierarchy of tidal channels results from the mutual interaction of tidal energy, sediments, and marsh vegetation. As tidal or river waters spread into the marsh, energy dissipates and channels of various sizes are formed. Nearer to open water, the greater energy results in larger channels, but as the waters lose energy, smaller channels occur until finally the water spreads out among the grasses and very little physical energy remains. The location of the larger channels determines that of the smaller channels.

A channel hierarchy results in more productivity and diversity than would likely occur without these channels. The hydrodynamic action through the tidal channels facilitates nutrient exchange and removal of excess sulfides and salts. The energies of the wind and tide subsidize the growth of marsh vegetation in the same sense that mechanical tilling and applications of fertilizers and pesticides subsidize agricultural crops (Schelske and Odum, 1961; Odum, 1971). However, it takes some time for the sediment to accumulate, the vegetative structure to develop, and the channels to organize. The result of this process of self-organization can be viewed as the "capital" of the ecosystem, analogous to capital investment in farm buildings and machinery.

The channel hierarchy also facilitates a living hierarchy. On the outgoing tide, the products of the broad area of grass (organic matter, young animals, and rich nutrition) converge from the smaller creeks to the larger and ultimately

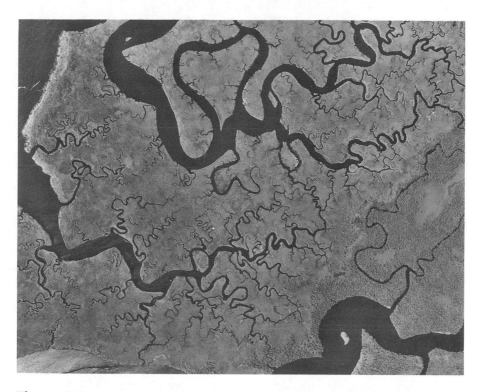

Figure 1.3 Aerial view of an intertidal marsh near the mouth of the Suwannee River. Note the network of tidal creeks. (From Montague and Wiegert, 1990.)

enhance the supply of food to the larger fish and other biota of the larger channels and nearby open coastal waters. Small animals are food for larger ones that in turn regulate the populations of the smaller ones. Since the smaller tidal creeks serve as partial refuges for the smaller fish, the channels of various sizes control the spatial patterns of biota.

Vegetation often grows taller near the edges of the creeks, perhaps due to a combination of greater tidal energy, nutrient availability, and soil–water exchange (Kruczynski et al., 1978; Coultas and Weber, 1980; Montague and Wiegert, 1990). Fiddler crab burrows also stimulate marsh plant growth, and they are more abundant near creeks (Montague, 1982).

Tidal creeks are perhaps the key to some of the greatest values of intertidal marshland to estuarine animal life (Horlick and Subrahmanyam, 1983; Montague and Wiegert, 1990). They are access points for the ingress and egress of fish and invertebrates (Subrahmanyam and Drake, 1975; Subrahmanyam et al., 1976; Subrahmanyam and Coultas, 1980) and feeding sites of wading birds.

Salt Marsh Biota, Food Chains, and Export

The dominant intertidal marsh vegetation is black needlerush (*Juncus roemerianus*). Lesser amounts of the greener smooth cordgrass (*Spartina alterniflora*) occur nearer the water's edge, where tidal flooding and draining are more frequent and consistent. Cordgrass often forms a narrow border at the edges of creeks. Several other species of grasses and a few succulent plants occur in marsh on sediments above mean high water (high marsh) (Kurz and Wagner, 1957; Montague and Wiegert, 1990). The landward edge of the marsh may grade into maritime forest: wax myrtles, junipers (red cedars), cabbage palms, and the characteristic densely branched live oaks, or it may consist of flatwoods with pines, palmettos, gallberry, and other scrub vegetation.

Most of the plant biomass within Florida's Gulf intertidal marshes is produced by a few species of salt-tolerant vascular plants and a large number of species of soil microalgae (mostly microscopic diatoms and blue-greens). Only about 10% of the vascular plant production is eaten alive, mostly by herbivorous insects. Nevertheless, this small amount gives rise to a productive and diverse arthropod community (insects and spiders) that feeds several species of resident passerine birds as well as a variety of avian transients. Although based on only a fraction of total production, secondary (animal) production from this insect-dominated, terrestrial-like grazing food chain is comparatively high (Pfeiffer and Wiegert, 1981).

The arthropod community of the marsh is very diverse (McMahan et al., 1972), especially compared with that in other monotypic ecosystems, such as agricultural crops (Pfeiffer and Wiegert, 1981). This diversity occurs perhaps in part because the aerial portion of the plants, which is the primary habitat of this community, is less subject to the alternating levels of water, salts, temperature, and oxygen. The lack of epidemic disease in intertidal marshes may result from the complexity of the arthropod community.

Most of the vascular plants die and decay in the marsh, decompose by a productive community of bacteria and fungi, and form detritus. Along with microalgae, detritus is food for a web of consumers ranging in size from microscopic animals to snails, fiddler crabs, shrimps, minnows, oysters, and juvenile stages of some larger fish and invertebrates (Montague et al., 1981). These animals are food for a variety of mostly transient fish and birds and even a few mammals (Montague and Wiegert, 1990).

Soil microalgae contribute only about 10% of the total marsh primary production because they grow in the shade of the vascular plants or on the narrow strip of bare mud at the water's edge (Pomeroy, 1959). These algae are very nutritious, however, and contribute as much to the food web as the greater quantity of detritus (Peterson and Peterson, 1979; Montague et al., 1987a).

This detritus–algae food chain is the energy basis for early life stages of many estuarine animals of commercial and recreational importance. Only a small fraction of intertidal production is converted into this form (Montague et al., 1987a), but this portion is highly valued. Factors that control the quantity and type of aquatic animals produced in marshes are still poorly understood, but probably relate to the density of marsh creeks (Montague et al., 1987a; Montague and Wiegert, 1990).

The abundant detritus produced in intertidal marshes accumulates in sediment and decomposes slowly, while being incorporated into the soil and soil microbes and on into the detritus food chain. This accumulation creates a great oxygen demand, which results in anaerobic sediment. Anaerobic microbial processes control the supply of nitrogen and the mobility of phosphorus in marsh soils. Thus, in a dynamic coastal environment in which primary production is potentially variable from year to year, the accumulation of detritus buffers or smooths fluctuations in the supplies of food and nutrients to animals and plants and the levels of nutrients and oxygen in sediment and water (Kalber, 1959; Nixon, 1980; Simpson et al., 1983). Certain anaerobic microbial processes reduce sulfate and nitrate to gases and thus complete the cycles of these elements back to the atmosphere, thereby preventing excessive accumulation in sediment and water and contributing to global atmospheric balance (Lovelock, 1979).

Exchange Between Marshes and Coastal Waters

Intertidal marshes may buffer changes in estuarine water quality by transforming and exchanging materials with estuarine water. As the marsh becomes inundated, particulate and dissolved materials will settle or be actively extracted from water by plants, suspension-feeding animals, and microbes. Simultaneously, living organisms and advective–diffusive processes add particles and dissolved materials to the water. When the water recedes, any materials added while on the marsh become part of the pool of similar materials in adjacent waters. Whether more material leaves than enters depends upon several hydrological and biological factors, recently reviewed elsewhere (Montague et al., 1987a), including tidal range and frequency and the amount of surface and groundwater flow. As groundwater flow from the Floridan aquifer is considerable in the Big Bend marshes, an overall net export of some materials to the surrounding estuarine water may be expected. Although net export is also likely during intense rains at low tide, it is an open question whether the quantity thus exported exceeds possible imports at other times.

The ecological significance of any materials added to estuarine waters depends upon several factors. For detritus, these factors include ease of decomposition, the nutrient and oxygen demand exerted on the system when detritus decomposes, and the relative quantity of similar materials available from other sources (Nixon, 1980; Montague et al., 1987a). Whether any marshes in coastal Florida export significant quantities of useful nonliving materials is unknown. Net transport measurements are difficult and have been accomplished for fewer than 20 intertidal marsh sites worldwide with variable results (Montague et al., 1987a).

Marsh vegetation nearer to tidal creeks (e.g., *Spartina alterniflora*) regularly drops dying tissues into the tidal waters, whereas vegetative production further away from channels (e.g., *Juncus roemerianus* and high-marsh plants) tends to accumulate and may decompose for months or even years before any is washed into estuaries during a hurricane or torrential rain. This greater retention time results in less net community production and less export from such areas. As the biomass of the plants decomposes, however, it fuels a productive web of resident marsh animals; hence, secondary production should be greater within such areas. With sufficient access, transient animals from the estuary at large can still feast on the perhaps greater production of marsh residents than might occur in marshes with greater export.

Stimulation of greater production of resident animals can result from greater food supply only if the already very productive animals are food limited. Most feed on detritus and microalgae. Because detritus is especially abundant, it may be difficult to imagine how detritivores in the marsh could be food limited. Detritus, however, is not particularly nutritious and the more nutritious microalgae are not very productive. Since the consumers are so plentiful, nutrition per consumer may be relatively low (Montague, 1980a, 1980b; Montague et al., 1981; Montague et al., 1987a; Montague and Wiegert, 1990). To test whether fiddler crabs are food limited, Genoni (1985) added *S. alterniflora* detritus to a salt marsh. Significantly greater recruitment of fiddler crabs was found in plots with added detritus.

Intertidal Marshes and Coastal Development

Coastal zones in Florida are intensively used for a variety of purposes (Table 1.9). In the past, interest in the largely unquantified public values of intertidal marshes has paled within the broader context of coastal development. As development of the coastal zone continues, public policy questions involving intertidal marshes will intensify.

In 1993, over three-quarters of Florida's estimated 13.6 million people lived in coastal counties (Bureau of Economic and Business Research). Although they comprise only 57% of Florida's land area, these counties produce 80% of the state's personal income and 87% of its municipal wastes and spend 85% of

Table 1.9 Selected Economic and Environmental Statistics for Florida's Coastal Counties

Statistic	Units	Coastal counties total	Coastal % of state total	Gulf % of coastal total
Economic				
Total personal income[a]	$\times 10^9$ $/year	85	80	36
Value added by manufacturing[a]	$\times 10^9$ $/year	14	77	32
Defense Department expenditures[a]	$\times 10^9$ $/year	7.8	73	43[c]
Recreation and tourism at beaches[b]	$\times 10^9$ $/year	2.3	100	41
Fishery landings value[a,d]	$\times 10^9$ $/year	0.18	100	71
Environmental				
Land area[a]	$\times 10^3$ km^2	79.3	57	63
1990 estimated population[a]	$\times 10^6$	10.2	78	41
Waste treatment discharges[a]				
Municipal	$\times 10^6$ m^3/day	3.7	87	34
Commercial or industrial	$\times 10^6$ m^3/day	0.8	74	62
State mosquito control funds[a]	$\times 10^6$ $/year	2.2	85	58

[a] Shoemyen et al. (1988, 1989).
[b] Personal expenditures by residents and visitors (Bell and Leeworthy, 1986).
[c] Estimate based on distribution of number of employees in tourism-related businesses.
[d] Fishery landings values are the dockside selling price and do not include the stimulation to the economy after the dockside sale. This may be two to seven times greater than the landings value.

the states mosquito control funds (Table 1.9). Population density along Florida's Atlantic coast (estimated to be about 300 people/km^2 in 1990) is much higher than the state average of 93 people/km^2, but along the Gulf, high densities are found only between Tampa Bay and Charlotte Harbor (Pinellas to Lee counties, see Table 1.2).

Florida has several major air force and naval bases along its coast. These military bases make considerable contributions to the state's economy (Table 1.9). Expenditures in coastal counties by the federal Department of Defense account for 73% of total Department of Defense expenditures in Florida. Coastal zones have also attracted heavy industry, both for waterborne shipping of raw materials and finished goods and for the abundant water for manufacturing processes, cooling of power plants, and disposal of wastes. Florida's coastal counties contribute 77% of the value added to the Florida economy by manufacturing and 74% of the commercial and industrial waste discharges (Table 1.9).

Co-occurring with heavy industrial and military uses of coastal zones are recreation, tourism, and commercial fishing. These activities depend more heavily than defense and manufacturing on a healthy coastal ecosystem that includes marshes. The contribution of these industries to the total economy is, however, considerably less than that from heavy industry and defense (see Table 1.9). Salt marshes near these more economically significant activities have been destroyed for airfields, shipping terminals, and ports.

Yet nearly three-quarters of the total personal income generated in coastal counties is not explained by manufacturing and defense expenditures. Some of this income is undoubtedly derived from the image created by the natural coastal environment, of which marshes are an integral part. On Florida's Gulf coast, this image is created by a variety of natural ecosystems including, and mutually interdependent with, intertidal marshes. The coastal image attracts not only tourists but also permanent residents who work in the heavy industries, buy residential real estate, and provide services.

Alteration of Intertidal Marshes

In Florida, the most extensive intertidal marshes are in areas of lowest population density (see Table 1.2). Lack of nearshore deep water and high ground has caused development to lag in areas where intertidal marshes are naturally most expansive. Nevertheless, activities associated with coastal development in Florida have locally eliminated a large fraction of marsh area.

Intertidal marshes have been eliminated by reducing tidal levels, increasing topographic slope, or reducing sediment supply. Restoration requires reversing these processes. Filling and bulkheading (to develop shipping terminals and residential areas) and disposal of various materials eliminated approximately 40% of the intertidal marshes around Tampa Bay between 1948 and 1978 (Estevez and Mosura, 1985).

In their natural state, marshes have considerable value to the public. They can be altered, however, to enhance habitat for certain fish and wildlife or to eliminate nuisance mosquitoes, and they can easily be eliminated by filling to create very valuable real estate. They have also been used as dump sites for trash, sediments dredged from shipping channels, wastes from phosphate ore processing, and wastewater (Montague and Wiegert, 1990).

Mosquito Control

At times, enormous populations of biting salt marsh mosquitoes make human habitation near coastal marshes almost unbearable. Salt marsh mosquitoes (*Aedes taeniorhynchus* and *A. sollicitans*) breed in temporary pools above mean high water where fishes cannot reach the mosquito larvae. Whenever heavy rains or exceptionally high tides occur, pools can remain long enough in these areas to produce another crop of mosquitoes. Any disturbance that results in greater temporary ponding of water above mean high water exacerbates the mosquito problem. Examples include roads, dikes to try to keep out tidal water, and hoof depressions of grazing livestock.

All measures to eliminate mosquitoes are controversial. Pesticides affect much of the aquatic food chain and are self-defeating because they result in selection of resistant strains of pests. By knocking out the complex, controlling, arthropod web, insecticides may also eliminate natural population controls on pest insects, thereby increasing the potential for epidemics of plant disease in salt marshes.

Ditching to drain pools at low tide and increase access of fishes at high tide has left canals with hills of spoil that interrupt marsh circulation. This can reduce the beneficial effects of the physical energy of the tide and can even *enhance* mosquito production behind the spoil piles. New efforts to improve the effectiveness of ditching and reduce negative side effects (open marsh water management) have not yet been perfected in Florida, but are being tested and should improve with sufficient trials. These techniques use very small ditches that are carefully located in mosquito-breeding hot-spots. The spoil from the ditching is spread widely over the marsh. Such ditches could even enhance marsh production if they also increase the distribution of hydraulic energy in the marsh, although too many ditches may create degradation. Quantitative determination of optimal ditch density is an open area for research.

Impounding water on the marsh is another method of mosquito control. Unlike many mosquitoes, salt marsh mosquitoes will not lay eggs on standing water. Hence, this method effectively breaks the life cycle of salt marsh mosquitoes. Impoundments also provide habitat for certain waterfowl, wading birds, alligators, and other wildlife, though perhaps at the expense of other organisms. For example, unless there are special arrangements for larvae to enter and larger animals to go back and forth between the estuary and the shallow impounded waters, the estuarine nursery functions of the marsh are lost with impoundment (Montague et al., 1985, 1987a, 1987b; Percival et al., 1987; Zale et al., 1987).

Impounding marshes for mosquito control has resulted in the alteration of 45% of the Indian River Lagoon's marshes (Atlantic coast). This mosquito control practice has not been prevalent along the Gulf coast not only because development has lagged here, but also because tidal ranges are much greater along the Gulf coast than in the Indian River Lagoon. Greater tidal range significantly reduces salt marsh mosquito-breeding area (Provost, 1973, 1974, 1976, 1977).

Other Alterations

Shallow impoundments have also been constructed in intertidal marshes solely to benefit fish and wildlife, notably ducks. Impounded marshes are successfully managed for wildlife at the Ding Darling (Sanibel Island) and the St. Marks National Wildlife Refuges.

Dams on rivers trap sediments and may reduce the sediment available for salt marsh development (Meade and Parker, 1985). In the Mississippi River delta, levees constructed to stabilize river channels and control flooding have removed the supply of sediment to intertidal marshes. These sediment-starved marshes can no longer grow vertically to keep pace with rising sea level, and large expanses of marsh are disappearing (Turner, 1987; Browder et al., 1989). Dams on the Apalachicola River may affect the future extent of coastal marshes in Apalachicola Bay and perhaps elsewhere, depending upon the future direction of transport of fine sediments as sea level rises.

In some areas where the general appearance of intertidal marshes seems unchanged, use by fish and wildlife may still be reduced by nearby disturbances (noise, distractive movement) or toxic materials (street drainage, industrial effluent, and solid wastes). These more insidious reductions in use by fish and wildlife are difficult to assess without carefully controlled studies (Odum, 1970).

Engineering Uses of Intertidal Marsh

Intertidal marshes have been successfully constructed by adjusting the elevation and slope of sediments and replanting marsh grasses or by simply awaiting natural vegetative growth on the sediments (Krone, 1982). Where the hydrological regime was suitable, marsh grasses and some associated organisms have grown in substrates of human origin such as gravel and broken pavement. One of us (HTO) has observed marsh grasses growing through road asphalt. Mitigation projects in hydraulically less suitable areas have not been as successful. The pressures of legally required mitigation (addressed in Chapter 11) require producing fully functional ecosystems in a short time, which has been difficult in many cases. Proactive ecological engineering, however, may obviate some of these concerns by providing some marsh functions, though incomplete for a time, in areas where they are needed.

Salt marsh ecosystems may be used to treat waste. Nutrient-rich waters from treated sewage stimulate and are absorbed partly by intertidal marshes. The nutrients accelerate productivity and can enhance estuarine nursery functions of marshes (Marshall, 1970). Although adding sewage to natural marshes is controversial, new intertidal marshes may be constructed as a buffer between sewage outfalls and estuaries.

A serendipitous example one of us (HTO) has observed over some years is the sewage waste outflow from a small treatment plant at Port Aransas, Texas. Wastes were released to a bare sand flat starting about 1950. As the population grew, wastes increased. Now there is an expansive marsh with a zonation of species outward from the outfall. Freshwater cattail marsh occurs immediately around the outfall. Beyond that is a salt marsh of *Spartina* and *Juncus* through which the wastewaters drain before reaching adjacent coastal waters.

Economic activities on land generate discharges. Storm runoff from streets, sewage in various degrees of treatment, and even some industrial wastes now flow directly into many coastal waters without first passing through coastal marshes. By first flowing through bands of intertidal marsh, substances in these discharges may be transformed into more innocuous and even useful forms by being bound into organic matter, buried in sediments, or converted into gases and vapors. Heavy metals can even be removed by marshes (Wolverton and Bounds, 1988).

In managing and restoring marshes and building interfaces between human settlements and estuaries, we should remember the hierarchical, branching, and tapering geometry of a system of tidal creeks, which is built by the self-organizing marsh as its own functional performance improves. For example, rather than square-cornered, bulkheaded "finger canals" for boats or for mosquito prevention, hierarchical, tapering canals with gently sloped banks provide self-flushing and an intertidal surface for the growth of some salt marsh. In this way, access is supplied for small boats, and better tidal flushing reduces the accumulation of toxic bottom-paint leachate and the depletion of dissolved oxygen. Better flushing is also a key to salt marsh mosquito control. The more geometrically natural canals have banks that are inexpensive to construct and are self-maintaining; wildlife and nursery roles are retained, and vertical bulkhead walls, which erode and are dangerous to children, are avoided.

Design and implementation of tidal creeks in constructed marshes is a new engineering challenge that can be met with appropriate consideration of the amount of hydraulic tidal energy at a site and the resistance of the marsh to erosion. Adding tidal creeks allows coastal waters to circulate through the marsh and provides essential access for fish, wading birds, and other aquatic animals.

Determining the total value of all the natural functions of intertidal marshes and effectively mitigating these values remain scientific and economic imperatives for research. Scientific as well as social controversy remains, however, over ways of evaluating nature. Several ways to quantify intertidal marsh values are presented in Chapters 8, 9, and 10. Each way yields a different value because each differs in what is measured. One method, "EMERGY synthesis" (spelled with an m, for embodied energy), estimates the public value of intertidal marsh by evaluating the work of nature and its percentage in the total work of the coastal economy (Odum and Odum, 1987). This approach estimates the marsh's indirect support of the economy through functions such as receiving and treating wastes or protecting the coast from storms. Natural processes thus contribute to the public welfare without requiring taxes.

Alternative approaches compute the economic market value, the human willingness to pay for marsh (or to pay for being assured that healthy marsh still exists), and the marginal value of marsh as it relates to the market value of a marsh-dependent product such as blue crabs. As the issue of marsh values is addressed more and more in the courts (see Chapter 11), a clear distinction in each case will be necessary between individual human rights (market value) and the welfare of the public economy (nature's direct and indirect work for the public, of which the public may not be aware).

The processes that generate each of these kinds of value are shown in the energy systems diagram in the Introduction (see Figure I.1). Market value is measured by the flow of money at the interface of the ecosystem with the coastal economy, as with the sale of fishery products. Public valuation involves the extensive environmental basis for production of wealth (on the left side of the diagram) for which no money is ever paid (money is only paid to people, but the work of the marsh is considered to be free).

Much of the history of Florida's economic development has involved intertidal marshes and their values. Perceptions on the one hand have ranged from worthless, mosquito-ridden, briny wastelands that needed to be "reclaimed" to a more modern view of marshes as a place of beauty, the basis of fisheries, real estate scenic vistas, and part of our life-support system to be protected.

Development pressures are increasing around many Gulf towns near expansive marshes as newcomers discover the natural beauty and serenity of the area (Crystal River, Cedar Key, and Steinhatchee, for example). An appreciation of the history of the effects of uncontrolled coastal development elsewhere in Florida will, it is hoped, encourage all Floridians to more effectively preserve the remaining intertidal marshes (and their investment) for the future.

Attempting a publicly acceptable balance among marsh preservation, use, and development within the economic diversity of the coastal zone is an ongoing political process that involves management agencies, businesses, citizen groups, and the courts. Teaching the naturally subsidized public values of intertidal marshes will create a more informed public as future decisions are made. It is hoped that this book will become an integral part of this educational process.

End-of-Chapter Note

Gross primary production is the formation of organic matter from raw materials (carbon dioxide, water, nutrients) using visible sunlight and aided by other energy inputs, i.e., water circulation and infrared insolation (which may aid transpiration by salt marsh plants). The net production within the marsh, however, is of special interest to those considering the organic food matter that goes into the estuary to support all aquatic food chains—ultimately the growth of many commercial species of shrimp, crabs, and fish. Because much of the gross photosynthesis is being used by the respiration of plant tissues, animals, and some microorganisms at the same time as the production is occurring, what is often measured is the difference between production and concurrent consumption. The difference between gross production and respiration is "net production." Although there are many uses of this term in this book and elsewhere, the term "net *primary* production" is only properly used when only *plant* respiration is subtracted from gross production. If respiration by unknown animal and microbial components is included, the interpretation of net production becomes difficult. Reported measurements of net production in intertidal marshes are highly variable (Montague and Wiegert, 1990). Great caution is required when comparing net production among various conditions of measurement. If net production in the same system is measured over an hour, a day, a month, a season, a year, or 10 years, entirely different results are obtained, because different amounts of consumption are included. The shorter the time, the greater the ratio of production to consumption, and thus the greater the net community production appears. Likewise, inclusion of different amounts of area also results in different results. Greater areas are more likely to include more of the larger (but rarer) consumers. Although measurements of net production may differ for other technical reasons, the time and spatial scale of measurement must be considered first in comparing measurements of net production.

References

Beccasio, A. D., N. Fotheringham, A. E. Redfield, et al. 1982. *Gulf Coast Ecological Inventory: User's Guide and Information Base.* FWS/OBS-82/55, U.S. Fish and Wildlife Service, Washington, D.C. 191 pp.
Bell, F. W. and V. R. Leeworthy. 1986. *An Economic Analysis of the Importance of*

Saltwater Beaches in Florida. Report No. 82. Florida Sea Grant College, University of Florida, Gainesville. 166 pp.

Bennett, I. 1965. Monthly maps of mean daily insolation for the United States. *Solar Energy* 9:145–158.

Bradley, J. T. 1972. *Climates of the States. Climate of Florida*. National Oceanic and Atmospheric Administration, U.S. Department of Commerce, Washington, D.C. 31 pp.

Brooks, H. K. 1981. Physiographic divisions (Florida). Map scale 1:500,000. Center for Environmental and Natural Resources, Institute of Food and Agricultural Sciences, University of Florida, Gainesville. Single sheet.

Browder, J. A., L. N. May, Jr., A. Rosenthal, J. G. Gosselink, and R. H. Baumann. 1989. Modeling future trends in wetland loss and brown shrimp production in Louisiana using thematic mapper imagery. *Remote Sens. Environ.* 28:45–59.

Bryson, R. A. and F. K. Hare, Eds. 1974. *Climates of North America*. Volume 11. Elsevier, New York. pp. 226 and 238.

Conway, H. M. and L. L. Liston. 1974. *The Weather Handbook*. Conway Research, Inc., Atlanta, Georgia. pp. 58–63.

Coultas, C. L. and O. J. Weber. 1980. Soil characteristics and their relationship to growth of needlerush. *Proc. Soil Crop Soc. Fla.* 39:73–77.

Davis, J. H., Jr. 1940. The ecology and geologic role of mangroves in Florida. *Papers from the Tortugas Laboratory* 32:304–412.

Estevez, E. D. and E. L. Mosura. 1985. Emergent vegetation. Pages 248–278 in S. F. Treat, J. L. Simon, R. R. Lewis, III, and R. L. Whitman, Jr., Eds. *Proceedings, Tampa Bay Area Scientific Information Symposium, May 1982*. Bellwether Press, Tampa, Florida.

Fernald, E. A. and D. J. Patton, Eds. 1984. *Water Resources Atlas of Florida*. Institute of Science and Public Affairs, Florida State University, Tallahassee. 291 pp.

Frey, R. W. and P. B. Basan. 1985. Coastal salt marshes. Pages 225–301 in R. A. Davis, Jr., Ed. *Coastal Sedimentary Environments*. Springer-Verlag, New York.

Genoni, G. P. 1985. Food limitation in salt marsh fiddler crabs *Uca rapax* (Smith, 1870) (Decapoda, Ocypodidae). *J. Exp. Mar. Biol. Ecol.* 87:97–110.

Horlick, R. G. and C. B. Subrahmanyam. 1983. Macroinvertebrate infauna of a salt marsh tidal creek. *Northeast Gulf Sci.* 6:79–89.

Kalber, F. A., Jr. 1959. A hypothesis of the role of tide-marshes in estuarine productivity. *Estuarine Bull.* 4:3.

Krone, R. B. 1982. Engineering wetlands: circulation, sedimentation, and water quality. Pages 53–58 in M. Josselyn, Ed. *Wetlands Restoration and Enhancement in California* (Report No. T-CSGCP-007). California Sea Grant College Program, California State University, Hayward.

Kruczynski, W. L., C. B. Subrahmanyam, and S. H. Drake. 1978. Studies on the plant community of a north Florida salt marsh. Part 1. Primary production. *Bull. Mar. Sci.* 28:316–334.

Kurz, H. and K. Wagner. 1957. *Tidal Marshes of the Gulf and Atlantic Coasts of Northern Florida and Charleston, South Carolina*. Florida State University Studies No. 24. Florida State University, Tallahassee. 168 pp.

Lovelock, J. E. 1979. *Gaia: A New Look at Life on the Earth.* Oxford University Press, New York. 157 pp.

Marmer, H. A. 1954. Tides and sea level in the Gulf of Mexico. Pages 101–118 in P. S. Galtsoff, Ed. *Gulf of Mexico: Its Origin, Waters, and Marine Life.* U.S. Department of the Interior, Washington, D.C.

Marshall, D. E. 1970. Characteristics of *Spartina* marsh receiving treated municipal sewage wastes. Work towards Master's degree. Department of Zoology, University of North Carolina, Chapel Hill. 51 pp.

McMahan, E. A., R. L. Knight, and A. R. Camp. 1972. A comparison of microarthropod populations in sewage-exposed and sewage-free *Spartina* salt marshes. *Environ. Entomol.* 1:244–252.

Meade, R. H. 1982. Sources, sinks, and storage of river sediment in the Atlantic drainage of the United States. *J. Geol.* 90:235–252.

Meade, R. H. and R. S. Parker. 1985. Sediment in rivers of the United States. Pages 49–60 in *National Water Summary 1984: Hydrologic Events, Selected Water-Quality Trends, and Ground-Water Resources* (Water-supply Paper 2275). U.S. Geological Survey, Reston, Virginia.

Mehta, A. J., R. M. Cushman, R. G. Dean, C. L. Montague, and W. R. Dally. 1989. *Workshop on Sea Level Rise and Coastal Processes, Palm Coast, Florida. March 9–11, 1988.* DOE/NBB-0086. U.S. Department of Energy, Washington, D.C. 289 pp.

Montague, C. L. 1980a. A natural history of temperate western Atlantic fiddler crabs (genus *Uca*) with reference to their impact on the salt marsh. *Contrib. Mar. Sci.* 23:25–55.

Montague, C. L. 1980b. The net influence of the mud fiddler crab, *Uca pugnax*, on carbon flow through a Georgia salt marsh: the importance of work by macroorganisms to the metabolism of ecosystems. Ph.D. dissertation. University of Georgia, Athens. 157 pp.

Montague, C. L. 1982. The influence of fiddler crab burrows and burrowing on metabolic processes in salt marsh sediments. Pages 283–301 in V. S. Kennedy, Ed. *Estuarine Comparisons.* Academic Press, New York.

Montague, C. L. and R. G. Wiegert. 1990. Salt marshes. Pages 481–516 in R. L. Myers and J. J. Ewel, Eds. *Ecosystems of Florida.* University of Central Florida Press, Orlando.

Montague, C. L., S. M. Bunker, E. B. Haines, M. L. Pace, and R. L. Wetzel. 1981. Aquatic macroconsumers. Pages 69–85 in L. R. Pomeroy and R. G. Wiegert, Eds. *The Ecology of a Salt Marsh.* Springer-Verlag, New York.

Montague, C. L., A. V. Zale, and H. F. Percival. 1985. *A Conceptual Model of Salt Marsh Management of Merritt Island National Wildlife Refuge, Florida.* Technical Report 17. Florida Cooperative Fish and Wildlife Research Unit, University of Florida, Gainesville. 92 pp.

Montague, C. L., A. V. Zale, and H. F. Percival. 1987a. Ecological effects of coastal marsh impoundments: a review. *Environ. Manage.* 11:743–756.

Montague, C. L., A. V. Zale, and H. F. Percival. 1987b. The nature of export from fringing marshes, with reference to the production of estuarine animals and the effect of impoundments. Pages 437–450 in W. R. Whitman and W. H. Meredith, Eds. *Waterfowl and Wetlands Symposium*. Delaware Department of Natural Resources and Environmental Control, Dover.

National Oceanic and Atmospheric Administration. 1983. *Climate Normals for the U.S. (Base 1951–80)*. Gale Research Company, Detroit, Michigan. pp. 125–138.

National Oceanic and Atmospheric Administration. 1985. *Climatic Summaries for Selected Sites, 1951–1980, Florida*. National Climatic Data Center, Asheville, North Carolina.

National Oceanic and Atmospheric Administration. 1989. *Tide Tables 1990: East Coast of North and South America*. U.S. Department of Commerce, Washington, D.C.

Nixon, S. W. 1980. Between coastal marshes and coastal waters—a review of twenty years of speculation and research on the role of salt marshes in estuarine productivity and water chemistry. Pages 437–525 in P. Hamilton and K. B. MacDonald, Eds. *Estuarine and Wetland Processes*. Plenum Press, New York.

Odum, E. P. 1971. *Fundamentals of Ecology*. W. B. Saunders, Philadelphia. 574 pp.

Odum, H. T. and E. C. Odum. 1987. *Ecology and Economy: "Emergy" Analysis and Public Policy in Texas*. Policy Research Report No. 78. Lyndon B. Johnson School of Public Affairs, University of Texas, Austin. 178 pp.

Odum, W. E. 1970. Insidious alteration of the estuarine environment. *Trans. Am. Fish. Soc.* 99:836–847.

Percival, H. F., C. L. Montague, and A. V. Zale. 1987. A summary of positive and negative aspects of coastal wetland impoundments as habitat for waterfowl. Pages 223–230 in W. R. Whitman and W. H. Meredith, Eds. *Waterfowl and Wetlands Symposium*. Delaware Department of Natural Resources and Environmental Control, Dover.

Peterson, C. H. and N. M. Peterson. 1979. *The Ecology of Intertidal Flats of North Carolina: A Community Profile*. FWS/OBS-79/39. Office of Biological Services, U.S. Fish and Wildlife Service, Washington, D.C. 73 pp.

Pfeiffer, W. J. and R. G. Wiegert. 1981. Grazers on *Spartina* and their predators. Pages 87–112 in L. R. Pomeroy and R. G. Wiegert, Eds. *The Ecology of a Salt Marsh*. Springer-Verlag, New York.

Pomeroy, L. R. 1959. Algal productivity in the salt marshes of Georgia. *Limnol. Oceanogr.* 4:386–397.

Provost, M. W. 1973. Mean high water mark and use of tidelands in Florida. *Fla. Scientist* 36:50–66.

Provost, M. W. 1974. Salt marsh management in Florida. *Proc. Tall Timbers Conf. Ecol. Anim. Contr. Habitat Manage.* 5:5–17.

Provost, M. W. 1976. Tidal datum planes circumscribing salt marshes. *Bull. Mar. Sci.* 26:558–563.

Provost, M. W. 1977. Source reduction in salt-marsh mosquito control: past and future. *Mosquito News* 37:689–698.

Reed, J. W. 1979. *Wind Power Climatology of the United States—Supplement*. Sandia Laboratories Energy Report. SAND78-1620. Sandia Laboratories, Albuquerque, New Mexico.

Schelske, C. L. and E. P. Odum. 1961. Mechanisms maintaining high productivity in Georgia estuaries. *Proc. Gulf Caribb. Fish. Inst.* 14:75–80.

Shoemyen, A. H., S. S. Floyd, and L. L. Drexel. 1988. *1988 Florida Statistical Abstract.* University Presses of Florida, Gainesville.

Shoemyen, A. H., S. S. Floyd, L. D. Rayburn, and D. A. Evans. 1989. *1989 Florida Statistical Abstract.* University Presses of Florida, Gainesville. 718 pp.

Simpson, R. L., R. E. Good, M. A. Leck, and D. F. Whigham. 1983. The ecology of freshwater tidal wetlands. *BioScience* 33:255–259.

State of Florida. 1983. *Renewable Energy: Energy from Wind.* Governor's Energy Office, Tallahassee. 20 pp.

Stelzenmuller, W. B. 1965. Tidal characteristics of two estuaries in Florida. *J. Waterways and Harbors Div.* pp. 25–36.

Subrahmanyam, C. B. and C. L. Coultas. 1980. Studies on the animal communities in two north Florida salt marshes. Part III. Seasonal fluctuations of fish and macroinvertebrate communities. *Bull. Mar. Sci.* 30:790–818.

Subrahmanyam, C. B. and S. H. Drake. 1975. Studies on the animal communities in two north Florida salt marshes. Part I. Fish communities. *Bull. Mar. Sci.* 25:445–465.

Subrahmanyam, C. B., W. L. Kruczynski, and S. H. Drake. 1976. Studies on the animal communities in two north Florida salt marshes. Part II. Macroinvertebrate communities. *Bull. Mar. Sci.* 26:172–195.

Tanner, W. F. 1960. Florida coastal classification. *Trans. Gulf Coast Assoc. Geol. Soc.* 10:259–266.

Turner, R. E. 1987. *Relationship Between Canal and Levee Density and Coastal Land Loss in Louisiana.* Biological Report 85(14). U.S. Fish and Wildlife Service, Washington, D.C. 58 pp.

U.S. Fish and Wildlife Service. 1984. *Draft Highlights of Reconnaissance Level Mapping and National Trend Analysis: Results for the State of Florida.* Regional Wetlands Coordinator, Atlanta, Georgia.

U.S. Fish and Wildlife Service. 1985. *Wetlands and Deepwater Habitats of Florida. Map Scale 1:500,000.* National Wetlands Inventory, Washington, D.C. Single sheet.

Wanless, H. R. 1989. The inundation of our coastlines: past, present, and future with a focus on south Florida. *Sea Frontiers* 35:264–271.

Wolverton, B. C. and B. K. Bounds. 1988. Aquatic plants for pH adjustment and removal of toxic chemicals and dissolved minerals from water supplies. *J. Mississippi Acad. Sci.* 33:71–80.

Zale, A. V., C. L. Montague, and H. F. Percival. 1987. A synthesis of potential effects of coastal impoundments on the production of estuarine fish and shellfish and some management options. Pages 424–436a in W. R. Whitman and W. H. Meredith, Eds. *Waterfowl and Wetlands Symposium.* Delaware Department of Natural Resources and Environmental Control, Dover.

Geology and Geomorphology

2

Frank R. Rupert and Jonathan D. Arthur

Geology

Florida, along with southern Alabama, southern Georgia, Cuba, and the Bahamas, is in the eastern Gulf of Mexico sedimentary basin (Puri and Vernon, 1964). Below the state's land surface are thousands of meters of Cenozoic (67 million years ago to present) and Mesozoic (250 to 67 million years ago) marine limestones, dolomites, sandstones, and claystones, resting on Paleozoic (600 to 250 million years ago) and Precambrian basement rocks, lying at depths in excess of 1200 m.

The emergent portion of Florida and the offshore continental shelf and slope are principally Cenozoic marine sedimentary rocks. Most of these rocks were deposited in the shallow seas that covered the Floridian Platform sporadically during the last 66 million years. Over the millennia, younger rock layers, or formations, were successively deposited "layer-cake" style on top of the older layers. Many of the layered formations vary locally in thickness or have been tilted, downwarped, or modified by erosion since deposition. Figure 2.1, a generalized geologic cross-section of the west coast from Pensacola to St. Petersburg, illustrates the stratigraphy of the near-surface formations underlying the coastal marshes. Data used in the construction of the cross-section were obtained from lithologic logs of well cores and samples on file at the Florida Geological Survey (FGS). The "W-" numbers shown on the cross-section represent FGS well accession numbers.

The oldest rocks underlying Florida are known only from samples brought to the surface during the drilling of deep oil test wells. These basement rocks

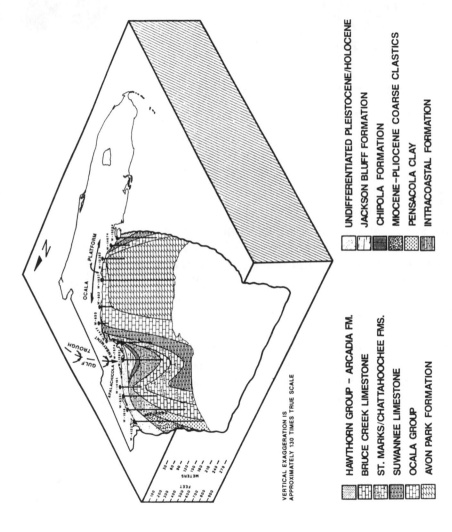

Figure 2.1 Generalized geologic cross-section along Florida's west coast.

range in age from late Precambrian (about 700 million years ago) to middle Mesozoic (about 150 million years ago). In the eastern Florida Panhandle and northern peninsula, the basement rocks are igneous and sedimentary and lie below 1200 m (4000 ft) in depth; in southern Florida, the basement rocks are generally more than 4500 m (15,000 ft) deep and are comprised primarily of middle Mesozoic igneous rocks (Arthur, 1988).

Overlying the basement rocks are Mesozoic sedimentary rocks, including sandstones, claystones, and limestones. Most are marine in origin. Florida's two oil-producing regions, the Jay trend in northwestern Florida and the Sunniland trend in south Florida, produce oil from horizons within these Mesozoic rocks.

Cenozoic (predominantly marine) sediments in turn overlie the Mesozoic rocks. These rocks contain Florida's drinking water aquifers and industrial mineral deposits and comprise the visible portions of the state today. Figure 2.2 is a generalized stratigraphic chart summarizing the Eocene and younger Cenozoic formations under west coastal Florida. The Paleocene rocks are marine limestones and dolomites. These older rocks are not important freshwater aquifers and lie deeper than the depths attained by most water wells. For the purposes of this chapter, discussion will be limited to middle Eocene and younger sediments. Many of these sediments crop out at the surface and most directly affect the geology and hydrology of Florida's Gulf coastal marsh region. Data on the lithology, depth, thickness, and occurrence of the formations are derived from well logs on file at the Florida Geological Survey.

Middle Eocene Series

Avon Park Formation

The Avon Park Formation (Applin and Applin, 1944; Miller, 1986) is typically a cream to brown to tan, fossiliferous, marine dolomite. It commonly contains pasty limestone beds and peat flecks and seams. The Avon Park Formation was deposited in a shallow sea which covered the area of present-day Florida about 54 million years ago. Numerous shallow-water fossils are found in Avon Park Formation sediments, including heart urchins, sand dollars, mollusks, foraminifera, and a "turtle grass-like" marine plant, very similar to the species living off the Big Bend coastline today.

The oldest rock exposed at the surface in Florida, the Avon Park Formation occurs in west central Florida along the crest of a gentle, northwest-southeast trending anticlinal feature variously called the Ocala Uplift, Ocala Arch, Ocala High, or Ocala Platform (Puri and Vernon, 1964; Scott, 1988), the origin of which is uncertain. Eocene and younger rocks have been removed by erosion.

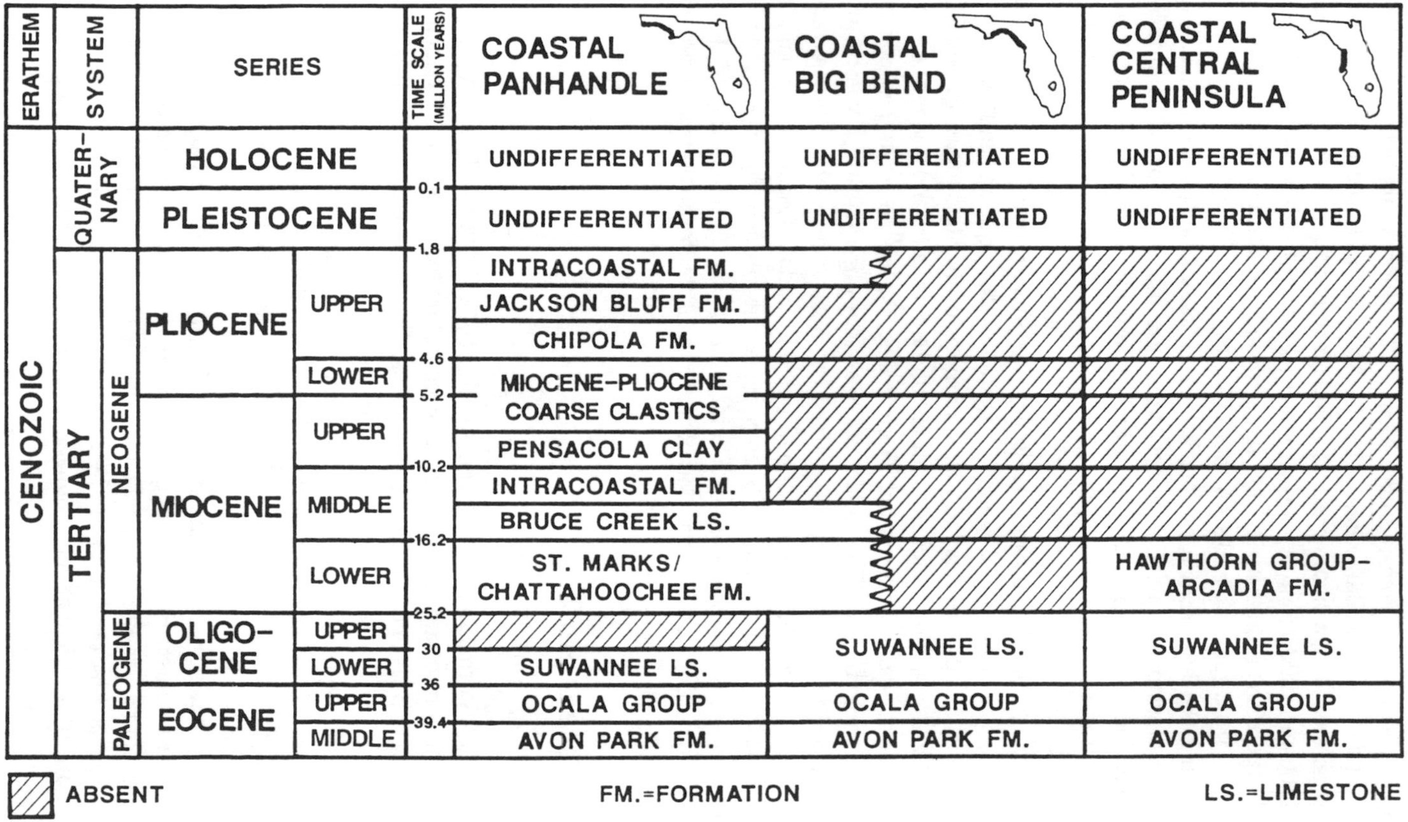

ERATHEM	SYSTEM	SERIES		TIME SCALE (MILLION YEARS)	COASTAL PANHANDLE	COASTAL BIG BEND	COASTAL CENTRAL PENINSULA
CENOZOIC	QUATERNARY	HOLOCENE		0.1	UNDIFFERENTIATED	UNDIFFERENTIATED	UNDIFFERENTIATED
		PLEISTOCENE		1.8	UNDIFFERENTIATED	UNDIFFERENTIATED	UNDIFFERENTIATED
	TERTIARY (NEOGENE)	PLIOCENE	UPPER	4.6	INTRACOASTAL FM. / JACKSON BLUFF FM. / CHIPOLA FM.		
			LOWER	5.2	MIOCENE–PLIOCENE COARSE CLASTICS		
		MIOCENE	UPPER	10.2	PENSACOLA CLAY / INTRACOASTAL FM.		
			MIDDLE	16.2	BRUCE CREEK LS.		
			LOWER	25.2	ST. MARKS/ CHATTAHOOCHEE FM.		HAWTHORN GROUP– ARCADIA FM.
	TERTIARY (PALEOGENE)	OLIGOCENE	UPPER	30	SUWANNEE LS.	SUWANNEE LS.	SUWANNEE LS.
			LOWER	36			
		EOCENE	UPPER	39.4	OCALA GROUP	OCALA GROUP	OCALA GROUP
			MIDDLE		AVON PARK FM.	AVON PARK FM.	AVON PARK FM.

ABSENT   FM.=FORMATION   LS.=LIMESTONE

Figure 2.2   Stratigraphic chart of the Eocene and younger formations underlying Florida's Gulf coastal marshes.

The Avon Park Formation is exposed only in small areas of central and southern Levy County and northernmost Citrus County, corresponding to the crest of the Ocala Platform (see Figure 2.1). It dips westward, where it interfingers with the Lisbon Formation (of the same age) under the western Florida Panhandle. Thickness of the unit in west central Florida is variable, but generally ranges from 0 to 240 m (0 to 800 ft) statewide (Chen, 1965).

Dolomite from the Avon Park is quarried in Citrus and Levy counties for construction materials. The Avon Park is also a unit of the Floridan aquifer, one of Florida's primary drinking water aquifers. It is unconformably overlain by marine limestone of the upper Eocene Ocala Group.

Upper Eocene Series

Ocala Group

The late Eocene (38 to 41 million years ago) Ocala Group (Puri, 1957) is composed of three marine limestone formations. In ascending order, these are the Inglis, Williston, and Crystal River Formations, all named for towns in the local outcrop areas of each formation. The formations have traditionally been differentiated on the basis of fossil content and, to a lesser extent, lithology. These three units will be referred to collectively as the Ocala Group.

The Ocala Group, a white to cream, abundantly fossiliferous, chalky to coquinoid limestone, was deposited in a shallow, late Eocene sea. Microfossils are generally the most common fossil forms present, especially the large species of foraminifera *Lepidocyclina, Nummulites,* and *Heterostegina.* Mollusks, bryozoans, and echinoids are also common in these sediments. The Ocala Group also crops out in west central Florida, along the crest of the Ocala Platform, and dips generally westward and southwestward (see Figure 2.1). The thickness of the Ocala Group in west central Florida averages about 60 m (200 ft), but is locally variable and generally thins over the crest of the Ocala Platform.

Limestone of the Ocala Group is exposed sporadically or is blanketed only by a thin veneer of Pleistocene and Holocene sands in all of Dixie County and portions of Lafayette, Gilchrist, Levy, Citrus, Marion, and Hernando counties. Throughout this area it is quarried as construction material and road base.

The exposed surface of the Ocala Group limestone forms a relatively flat-lying, gently seaward-sloping, solution depression, and sinkhole-pocked plain over most of its outcrop area. This plain continues offshore onto the broad western Florida continental shelf. Boulders, pinnacles, and small islands of Ocala Group limestone are common along the Gulf coastline of Dixie, Levy, and Citrus counties (Vernon, 1951). Many of the coastal marshes in this area developed in the thin calcitic muds, silt, organics, and relict marine sands deposited in solution depressions in the underlying limestone (see Figure 2.3).

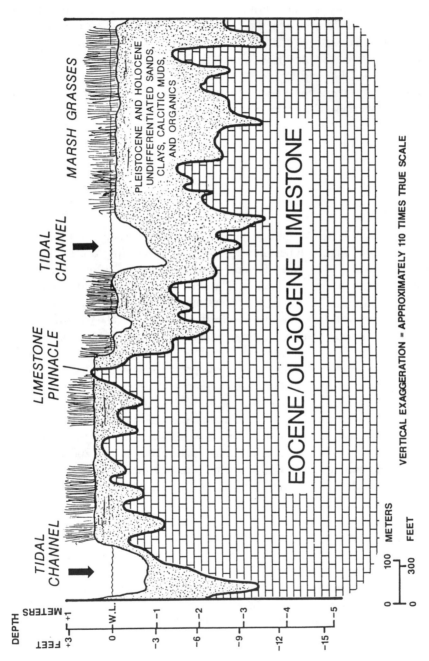

Figure 2.3 Generalized near-surface cross-section in the coastal marshes, west central Florida. (Modified from Hine and Belknap, 1986.)

The Ocala Group is an important unit of the Floridan aquifer system. Along much of the central west coast and within many of the coastal marsh areas, the potentiometric surface of the Floridan aquifer is at or above land surface. As a result, numerous seeps and small springs feed fresh water into the marshes. Flow from some of these springs is sufficient to form small tidal creeks and tributaries through the otherwise dense marsh grasses (Nettles, 1976).

Oligocene Series

Suwannee Limestone

The Oligocene (33 to 25 million years old) Suwannee Limestone (Cooke and Mansfield, 1936) is a white to tan, fossiliferous, commonly dolomitic marine limestone, named after exposures along the Suwannee River in northern Florida. As with the underlying Eocene formations, the Suwannee was probably deposited in a shallow temperate sea. It contains abundant microfossils, mollusks, echinoids, bryozoans, and rare corals. Early Indians of the region used the abundant chert occurring in both the Ocala Group and Suwannee Limestone to fashion tools and weapons. The Suwannee Limestone has been truncated against the flanks of the Ocala Platform (see Figure 2.1) and is exposed at the surface at the northern and southern ends of this feature. It occurs as the surficial unit in eastern Jefferson County, most of Taylor County, western Hernando County, and northern Pasco County. In these areas, the Suwannee Limestone forms a flat-lying, karstic plain which extends seaward onto the continental shelf. Boulders and pinnacles of Suwannee Limestone are common along the low-energy coasts of Jefferson and Taylor counties. The Suwannee Limestone is the upper unit of the Floridan aquifer in its outcrop area. Small freshwater springs and seeps commonly occur in the marshes of southern Jefferson and Taylor counties, some forming creeks or contributing to the flow of rivers such as the Aucilla and Econfina.

Miocene Series

The Miocene epoch marked a significant change in the depositional regime of the Florida peninsula. Previously a shallow carbonate bank, Florida experienced an influx of continental siliciclastic sediments during the Miocene from the continental mainland to the north. This was due in large part to the closing of a feature called the Suwannee Straits or Gulf Trough, which trended southwestward across the eastern Florida Panhandle (Figure 2.1). The Gulf Trough is thought possibly to have been an ancient channel or seaway, existing from the Cretaceous period through the end of the Oligocene epoch (Dall and Harris, 1892; Chen, 1965). It probably connected two major depositional basins, the

Southeast Georgia Embayment and the Apalachicola Embayment. Currents within this paleostrait may have functioned as both a zoological and a sedimentological barrier between the carbonate banks and islands of peninsular Florida and the continental mainland.

Phosphogenesis also occurred on a large scale beginning in the Miocene. Many Miocene formations contain phosphate grains, ranging from silt to pebble size. Florida's economic phosphate deposits are concentrated in the Miocene Hawthorn Group sediments in the Central Florida Phosphate District of Polk County and surrounding counties and in the North Florida Phosphate District in Hamilton County.

The Miocene formations in Florida are primarily marine in origin. In contrast to the nearly pure calcium carbonate limestones of the Eocene, the Miocene and younger sediments, including the carbonates, typically contain terrigenous quartz and heavy mineral sands and clays, and phosphate. Starting in the lower Miocene, a massive influx of river-borne continental sediments poured into the seas covering present-day peninsular Florida (Scott, 1988). The Miocene seas reworked and deposited these sediments in a broad blanket over the carbonates of earlier epochs. In the western Panhandle, a series of marine limestones and shelly sands were laid down. The Miocene seas supported a rich marine fauna, as shown by the numerous fossils found in these sediments. Based on fossils, paleoenvironments ranged from shallow nearshore and lagoon to deepwater continental shelf. Many of Florida's commercial mineral deposits, including phosphate and fuller's earth, were formed in these ancient sea floor sediments.

Lower Miocene Series

St. Marks Formation

The St. Marks Formation (Puri and Vernon, 1964), exposed along the northern Big Bend coastline in Wakulla County, is the only Miocene carbonate unit exposed in the coastal marsh zone. This formation is a white to very pale orange to light gray, quartz sandy, fossiliferous marine limestone. Foraminifera and mollusks are the dominant fossil forms, generally present as molds. Along the coast the St. Marks Formation contains abundant chert, which probably supplied the Paleo-Indians of the region with tool and spear point material. The St. Marks Formation occurs near or at the surface in most of south central and southeastern Wakulla County. Here it forms a slightly seaward-sloping, karstic plain called the Woodville Karst Plain. Thickness varies from 0 m at the Wakulla-Jefferson County line to over 30 m (100 ft) along the eastern flank of the Apalachicola Embayment. Extensive salt marshes, developed in the thin sand

and mud veneer overlying the St. Marks Formation limestone, border the south-ern edge of Wakulla County.

The St. Marks Formation is the uppermost unit of the Floridan aquifer system in the eastern Panhandle. In southern Wakulla County, the potentiomet-ric surface of this aquifer is at or above the land surface. Many small freshwater springs are scattered through the coastal marshes, and several submarine springs are on the offshore extension of the karst plain. The St. Marks pinches out against the underlying Suwannee Limestone to the east in Jefferson County (see Figure 2.1) and dips generally to the west-southwest. To the west under Franklin and Gulf counties it becomes indistinguishable from the Bruce Creek Limestone. It interfingers to the north and possibly under the western Pan-handle with the age-equivalent Chattahoochee Formation.

Hawthorn Group—Arcadia Formation

The lower Miocene Arcadia Formation of the Hawthorn Group (Scott, 1988) underlies the west central Gulf coast from approximately New Port Richey to Sarasota (see Figure 2.1). This unit is comprised primarily of white to yellow-ish-gray to light olive-gray limestone and dolomite containing variable amounts of quartz sand, clay, and phosphate grains. It unconformably overlies the Suwannee Limestone and is overlain by undifferentiated Pleistocene quartz sands. Hawthorn Group sediments are absent along the Panhandle coast.

Hawthorn Group sediments serve as an intermediate confining unit to the underlying Floridan aquifer system. Locally, carbonates within the Hawthorn Group may also function as an intermediate freshwater aquifer system or in some areas are in hydrologic continuity with the underlying Floridan aquifer system and considered part of the Floridan aquifer.

The Hawthorn Group does not crop out in the vicinity of the coastal marshes. It is probably eroding offshore however, as pebbles and cobbles of Hawthorn Group carbonate are often found washed up on Pinellas County beaches.

Middle Miocene Series

West of Wakulla County in the Panhandle, the Miocene units dip and thicken southwestward into the Apalachicola Embayment (see Figure 2.1). Both the lower Miocene St. Marks Formation and the overlying Middle Miocene Bruce Creek Limestone dip into the trough of the embayment, reaching a maximum local depth and thickness under the axis of the basin. These units shallow again on the western side of the embayment and attain a depth and thickness similar to that on the eastern edge of the embayment.

Bruce Creek Limestone

Bruce Creek Limestone was the name given by Huddlestun (1976) to a white to light yellowish-gray, middle Miocene marine limestone underlying part of Panhandle Florida. It extends from the eastern edge of the Apalachicola Embayment westward, with generally southwest dip. The Bruce Creek Limestone is commonly quartz sandy, phosphoritic, and both micro- and macrofossiliferous. Fossils are generally preserved as molds. This formation is primarily a subsurface unit and does not crop out in the vicinity of the coastal marshes. The Bruce Creek Limestone thickens to over 60 m (200 ft) in the Apalachicola Embayment. Down-dip, it is indistinguishable from the underlying St. Marks Formation. In the central and western Panhandle, the Bruce Creek Limestone comprises the uppermost unit of the Floridan aquifer system.

Middle Miocene Through Pliocene Series

Intracoastal Formation

Huddlestun (1976) applied the name Intracoastal Formation to a soft, yellowish-gray to olive green, sandy, highly microfossiliferous, argillaceous marine limestone underlying the coastal area of west Florida. This formation extends from easternmost Franklin County westward and continues to approximately the Santa Rosa County line (Schmidt, 1984). The Intracoastal Formation is predominantly a subsurface unit, dipping and thickening to the west-southwest into the Apalachicola Embayment. It approaches 100 m (300 ft) in thickness in the trough of the Apalachicola Embayment. The unit shallows and thins west of the embayment, then thickens and dips westward toward the Pensacola area. Analysis of the microfossils present indicates an age range for the Intracoastal Formation of middle Miocene (down-dip) to late Pliocene (in the up-dip portions), with a hiatus separating the age extremes. Locally it occurs close to the surface in the central Panhandle, but does not crop out in the coastal marsh region.

Pensacola Clay

The Pensacola Clay (Marsh, 1966) is a pale, yellowish-brown to olive-gray, dense, silty, commonly quartz-rich sandy clay. This unit is restricted to the subsurface, occurring under the Panhandle from approximately Okaloosa County westward into Alabama. It thickens rapidly to the south and west (see Figure 2.1), reaching a maximum thickness of about 150 m (500 ft) under Pensacola Bay (Clark and Schmidt, 1982). The Pensacola Clay is considered late Miocene in age, overlying the Bruce Creek Limestone and interfingering locally with the Intracoastal Formation or the Miocene Coarse Clastics (Clark and Schmidt, 1982).

Miocene–Pliocene Coarse Clastics

Marsh (1966) proposed the name Miocene Coarse Clastics for a series of sands, gravel, clays, and shell beds underlying the western Panhandle. These sediments are largely comprised of light-gray to pale yellowish-brown quartz sand and gravel, with minor clay and marine mollusk shells. The Miocene–Pliocene Coarse Clastics occur from western Okaloosa County westward through Santa Rosa and Escambia counties. Thickness of this unit reaches nearly 150 m (500 ft) under Santa Rosa County. In the westernmost Panhandle, these units comprise an important groundwater aquifer known as the sand and gravel aquifer. The coarse clastics overlie the Pensacola Clay and Intracoastal Formations and locally may be contemporaneous (late Miocene to early Pliocene in age) with portions of both formations; they are overlain by undifferentiated Pleistocene sand (Clark and Schmidt, 1982).

Chipola Formation

The Chipola Formation (Puri and Vernon, 1964) is present at depths under the coastal marshes in the vicinity of Gulf County, along the axis of the Apalachicola Embayment. Lithologically, it is comprised of a yellowish-gray to light-gray, quartz sandy marine limestone. Foraminifera and mollusks are typically the most abundant fossils. Thickness of this unit reaches about 15 m (50 ft) in the central Apalachicola Embayment, thinning and pinching out to the west and east of the embayment. In the down-dip coastal and offshore areas, the Chipola Formation overlies the Intracoastal Formation and is considered to be late Pliocene in age (Schmidt, 1984). It is unconformably overlain by the Jackson Bluff Formation.

Jackson Bluff Formation

The late Pliocene Jackson Bluff Formation (Puri and Vernon, 1964) consists of tan to orangish-brown to grayish-green sandy, clayey shell beds and is restricted to the central Apalachicola Embayment, where the unit reaches about 30 m (100 ft) in thickness. The Jackson Bluff Formation overlies the Chipola and Intracoastal Formations and is in turn overlain by undifferentiated Pleistocene and Holocene sands. The Jackson Bluff Formation does not crop out in the coastal marshes.

Pleistocene and Holocene Series

A series of undifferentiated Pleistocene (1.8 million to 10,000 years old) and Holocene (10,000 years and younger) quartz sands, clayey sands, and sandy clays blanket the older formations along much of Florida's west coast. These sediments are composed largely of reworked relict Pleistocene marine sands and Holocene alluvium, calcitic muds, and organics.

The Pleistocene epoch, or "Ice Age," was characterized by four great glacial periods. During this epoch, global temperatures cooled and huge ice sheets grew southward from Canada. Large quantities of sea water were consumed to build the glaciers, and sea level dropped during the cold intervals. Each glacial period was punctuated by warmer interglacial periods during which the sea level rose, sometimes to levels higher than present-day sea level. As the Pleistocene seas transgressed inland, waves and currents eroded, reworked, and redeposited the sands of earlier formations. At the same time, rivers and an active southward-moving littoral drift system brought new clastic sediments into Florida. The Big Bend area was a drowned karst coastline during the Pleistocene sea level highstands. Planing by wave action and in-filling of the karst features with sand resulted in the flat, seaward-sloping plain characterizing this region today. Figure 2.3 illustrates a typical near-surface cross-section in the coastal marsh zone along the west central Florida coast. The karstic, irregular surface of the underlying limestone is in-filled with Pleistocene and Holocene sediments, with a few limestone pinnacles exposed at the surface. The younger sands, muds, and silts form a substrate for many of the coastal marshes in this area.

The Pleistocene deposits are thinnest in the Big Bend, where limestone is near or at the surface. West of Wakulla County, the undifferentiated Pleistocene sediments thicken to nearly 60 m (200 ft) near the Alabama state line.

Holocene sand deposition continues today. Accumulation of these deposits is primarily concentrated along the banks, bottoms, and mouths of the major Gulf coast rivers, such as the Apalachicola, Ochlockonee, Aucilla, Suwannee, and Withlacoochee. In portions of the central west coast and Big Bend areas, a Holocene intertidal calcitic mud commonly overlies the Pleistocene sand (Nettles, 1976). Organics derived from decaying marsh grasses, intermixed with sand, typically form the surface layer in the coastal marshes (Hine and Belknap, 1986).

Geomorphology

The geomorphology of Florida's Gulf coast has been shaped primarily by coastal processes and sea level changes during the past 5 million years. These sea level changes have caused shifts in groundwater levels, which in turn have enhanced the development of karst landforms during this time. The karst features have thus altered some of the original shoreline erosional and depositional features within the Gulf coast region. Discussion of these two interrelated geomorphic features is limited to an inland distance of 16 km (10 mi), from Pensacola to St. Petersburg. Modern offshore landforms such as spits, barrier islands, and bars are not discussed.

Up to six marine terraces are reported to occur within the region of interest (Cooke, 1945; Healy, 1975). A marine terrace is a surface of erosion or deposition formed along a coast by wave action (Garner, 1974). These terraces represent the floors of ancient shallow seas and are formed step-like, roughly parallel to the present-day coast. Each "step" is a topographic break ranging from a gentle slope to a sharp incline. Where sharp, these changes in topography that separate two terraces are well-preserved seaward-facing wave cut scarps, some of which are regional in extent and up to 15 m (50 ft) high (e.g., the Cody Scarp, Puri and Vernon, 1964, p. 11). Although there are few of this magnitude proximal to the Gulf coast, coastal terraces and scarps appear to control surface drainage in localized areas. For example, Healy's (1975) Silver Bluff terrace generally coincides with the location of several Gulf coastal swamps.

Physiographic and geomorphic evidence (White, 1970), aerial photography (Vernon, 1951), sedimentology/stratigraphy (Altschuler and Young, 1960; Pirkle et al., 1970), field mapping (Parker et al., 1955), and fossil evidence (Alt and Brooks, 1965) have all been used to study early marine terraces. Although conclusions drawn for localized areas are well constrained by these data, studies based on elevation correlations over large areas (e.g., MacNeil, 1950; Healy, 1975) should be considered with reservation. Given the possibility of regional Pliocene–Pleistocene warping (Winker and Howard, 1977a, 1977b) and the predominance of karst development in the region (White, 1970), as well as other erosional processes, some of these long-distance correlations may not be valid. For example, a present-day low-lying area may appear to correlate with a certain low-elevation terrace, whereas in fact it may have been an upland or higher elevation terrace which was subsequently lowered by subsurface limestone dissolution. On the other hand, a "young," low-elevation terrace may have undergone epeirogenic uplift and now topographically correlates with a higher elevation, older terrace. Thus, the correlation between swamps/marshlands and the Silver Bluff terrace may be an artifact of the manner in which some terraces have been delineated. Ancient marine scarps and terraces do exist in the Gulf coast region; however, further study is required in order to more accurately define their occurrence and extent.

Pliocene–Pleistocene sea level fluctuations are the primary reason for the presence of these terraces (Winker and Howard, 1977a, 1977b). During the

Pleistocene "Ice Age" when the fluctuations became more prevalent, four major glacial cycles occurred due to major climatic changes. As glaciers spread over the continents, sea level dropped. During periods of melting or interglacial periods, sea level rose. Healy (1975) and MacNeil (1950) suggested that during subsequent (younger) interglacial periods, the seas stood at levels below that of the prior event, thus preserving the earlier formed terraces and scarps. Several authors have recognized that glacial melting alone cannot account for the present-day anomalously high elevations of the older marine terraces. Tanner (1968, 1985) and Winker and Howard (1977a, 1977b) document evidence for regional uplift during this time.

In a study entitled "The Geomorphology of the Florida Peninsula," White (1970) subdivided the region into a series of uplands and lowlands. Most of the present study area lies within a single, major physiographic province: the Gulf Coastal Lowlands (Figure 2.4). This province stretches from the western Florida Panhandle to southern peninsular Florida and averages 40 km (25 mi) in width. Various highlands, ridges, or scarps constitute the inland limits of the Gulf Coastal Lowlands, whereas the present-day shoreline marks the seaward boundary. Unlike eastern coastal areas of Florida, this geomorphic province does not correspond to any specific marine terrace delineated by Healy (1975). The Gulf Coastal Lowlands contain various erosional and depositional landforms that occurred in response to Pliocene–Pleistocene sea level fluctuations. Several relict bars, spits, and terraces are superimposed on the modern topography of the region. Subdivisions within the Gulf Coast Lowlands include gulf barrier chains, coastal swamps, coastal lagoons, and estuaries.

Using White's (1970) terminology, the Gulf coastal wetlands include areas where a deficiency exists in the sand budget for building beaches. These low-lying areas correspond to areas on topographic maps where the wetlands are immediately adjacent to the coast. No distinction is made between salt marshes and freshwater swamps. Where isolated patches of wetlands are separated from the coast by dry land, the term "coastal lagoons" is applied (White, 1970). This generalized terminology is being revised based on vegetation, soil type, and topography in current investigations (C.L. Coultas, personal communication, 1990).

White (1970) presented a twofold classification for the Gulf coast of peninsular Florida. Within the study area, this includes a coastal salient that extends from Tampa Bay north to the southern part of Pasco County and a coastal reentrant spanning from this point north toward Apalachee Bay. The salient is characterized by a relatively steep offshore profile, which allows wave energy to transport, deposit, and erode sand along the shoreline. This coastal area contains many relict barrier bars, beach ridges, and lagoons. In contrast, the reentrant offshore profiles have gentle slopes and are floored by limestone. White (1970) suggested that these broad shallow profiles sufficiently

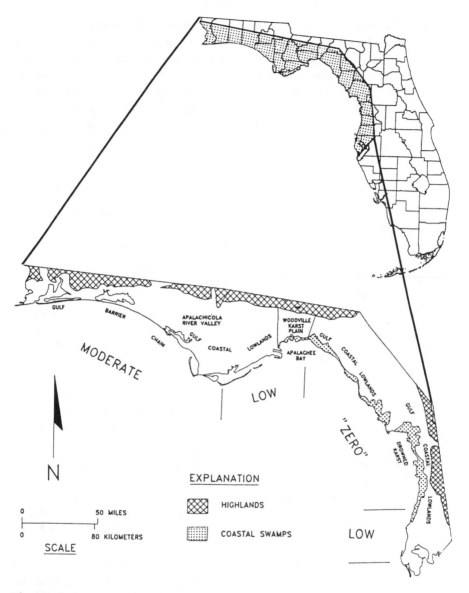

Figure 2.4 Geomorphic features along study area modified from White (1970) and Puri and Vernon (1964). Wave energy zones are from Tanner (1960).

dissipate wave energies to account for the sand-starved marshy coasts. The sand deficiency is also due to a lack of sediment transported by the Suwannee River. The very low-energy re-entrant coast is virtually free of relict shoreline features.

Along the Florida Panhandle coastline, White's (1970) generalization concerning the peninsula's offshore profiles is also applicable. Westward from the west end of Apalachee Bay, the profile is either as steep as or steeper than in the Tampa Bay area. The entire Panhandle coastline consists of spits, lagoons, beach ridges, and offshore barrier islands. An even steeper offshore slope (ramp) in the central Panhandle region has precluded barrier island formation offshore of Walton County and the west half of Bay County (Tanner, 1960). Puri and Vernon's (1964) geomorphic map of the region, which uses White's criteria for landform definition, shows no coastal swamps along the central and eastern Panhandle.

Tanner (1960) presented data that quantify energy levels within the study area. Rather than focusing on shelf slope, he reported average annual breaker heights to estimate wave energy (Figure 2.4). The "zero energy" coast corresponds to the coastal swamps and marshes in the Big Bend area. The lack of both wave energy and sediment transport and deposition are the two most significant factors that have allowed swamp and marsh development in the region.

In addition to the regional geomorphic characteristics of the study area, two localized features are noteworthy: the Ozello Marsh Archipelago and the Woodville Karst Plain. The first of these can be seen on Tanner's (1960) map, which classifies a shoreline region of Citrus County as "drowned karst" (Figure 2.4). This area, discussed in detail by Hine and Belknap (1986), consists of several limestone-cored marsh islands. The outer islands are separated either by creeks or salt marsh vegetation, whereas the interior is pocked with circular ponds. Various karstification processes have controlled the shape and orientation of these water bodies (Hine and Belknap, 1986). Figure 2.3 illustrates the undulatory nature of bedrock limestone in this area due to dissolution.

The Woodville Karst Plain (Hendry and Sproul, 1966) is a subdivision of the Gulf Coastal Lowlands and is located north of the coastal marsh belt in Wakulla County. Portions of the Woodville Karst Plain extend into Jefferson and Leon counties, where it is bounded to the north by the Northern Highlands (Puri and Vernon, 1964). Permeable sands form a veneer over a shallow, southward-dipping limestone bedrock. Dissolution of the underlying carbonate bedrock has caused subsidence and has probably been occurring ever since the area has been above sea level. The prevalence of shallow sand-filled sinkholes geomorphically characterizes the Woodville Karst Plain. Although surrounding areas are also underlain by limestone, dissolution has not been prevalent because the overlying sediments contain clays which both buffer the acidic rainwater and reduce the amount of percolation. Geologic variables such as those characterizing the Woodville Karst Plain (i.e., bedrock type, sediment composition, and sea level history) are significant in the development of Florida's Gulf coast geomorphology.

Acknowledgments

The authors would like to thank Ken Campbell, Dr. Walt Schmidt, and Dr. Thomas Scott of the Florida Geological Survey and Dr. William F. Tanner of Florida State University for critically reviewing the manuscript. Their comments and suggestions were most helpful in the preparation of the final draft of this chapter.

References

Alt, D. and H. K. Brooks. 1965. Age of the Florida marine terraces. *J. Geol.* 73: 406–411.

Altschuler, Z. S. and E. J. Young. 1960. Residual origin of the "Pleistocene" sand mantle in central Florida uplands and its bearing on marine terraces and Cenozoic uplift. *U.S. Geol. Surv. Prof. Pap.* 400-B:202–207.

Applin, P. L. and E. R. Applin. 1944. Regional subsurface stratigraphy and structure of Florida and southern Georgia. *Am. Assoc. Petroleum Geologist Bull.* 28: 1673–1753.

Arthur, J. D. 1988. *Petrogenesis of Early Mesozoic Tholeiite in the Florida Basement and an Overview of Florida Basement Geology.* Florida Geological Survey Report of Investigation 97.

Chen, C. S. 1965. *The Regional Lithostratigraphic Analysis of Paleocene and Eocene Rocks of Florida.* Florida Geological Survey Bulletin 45.

Clark, M. and W. Schmidt. 1982. *Shallow Stratigraphy of Okaloosa County and Vicinity, Florida.* Florida Bureau of Geology Report of Investigation 92.

Cooke, C. W. 1945. *Geology of Florida.* Florida Geological Survey Bulletin 29.

Cooke, C. W. and W. C. Mansfield. 1936. *Suwannee Limestone of Florida* [Abstr.]. Geological Society of America Proceedings for 1935.

Dall, W. H. and G. D. Harris. 1892. *Correlation Papers—Neocene.* U.S. Geological Survey Bulletin 84.

Garner, H. F. 1974. *The Origin of Landscapes.* Oxford University Press, New York.

Healy, H. G. 1975. *Terraces and Shorelines of Florida.* Florida Geological Survey Map Series 71.

Hendry, C. W. and C. Sproul. 1966. *Geology and Ground-Water Resources of Leon County, Florida.* Florida Geological Survey Bulletin 47.

Hine, A. C. and D. F. Belknap. 1986. *Recent Geological History and Modern Sedimentary Processes of the Pasco, Hernando, and Citrus County Coastline: West Central Florida.* Florida Sea Grant College Report No. 79. University of Florida, Gainesville.

Huddlestun, P. F. 1976. The Neogene stratigraphy of the central Florida Panhandle. Ph.D. dissertation. Florida State University, Tallahassee.

MacNeil, F. S. 1950. *Pleistocene Terraces and Shorelines in Florida.* U.S. Geological Survey Professional Paper 221-F.

Marsh, O. T. 1966. *Geology of Escambia and Santa Rosa Counties, Western Florida Panhandle.* Florida Geological Survey Bulletin 46.

Miller, J. A. 1986. *Hydrogeologic Framework of the Floridan Aquifer System in Florida and in Parts of Georgia, Alabama, and South Carolina.* U.S. Geological Survey Professional Paper 1403-B.

Nettles, S. 1976. Intertidal calcitic muds along the west coast of Florida. Thesis. Department of Geology, University of Florida, Gainesville.

Parker, G. G., G. E. Ferguson, and S. K. Love. 1955. *Water Resources of Southeastern Florida with Special Reference to the Geology and Ground Water of the Miami Area.* U.S. Geological Survey Water-Supply Paper 1255.

Pirkle, E. C., W. H. Yoho, and C. W. Hendry. 1970. *Ancient Sea Level Stands in Florida.* Florida Bureau of Geology Bulletin 52.

Puri, H. S. 1957. *Stratigraphy and Zonation of the Ocala Group.* Florida Geological Survey Bulletin 38.

Puri, H. S. and R. O. Vernon. 1964. *Geology of Florida and a Guidebook to the Classic Exposures.* Florida Geological Survey Special Publication 5 (revised).

Schmidt, W. and M. Clark. 1980. *Geology of Bay County, Florida.* Florida Bureau of Geology Bulletin 57.

Schmidt, W. 1984. *Neogene Stratigraphy and Geologic History of the Apalachicola Embayment, Florida.* Florida Geological Survey Bulletin 58.

Scott, T. M. 1988. *The Lithostratigraphy of the Hawthorn Group (Miocene) of Florida.* Florida Geological Survey Bulletin 59.

Tanner, W. F. 1960. Florida coastal classification. *Trans. Gulf Coast Assoc. Geol. Soc.* 10:259–266.

Tanner, W. F. 1968. Tertiary sea level symposium—introduction. *Paleogeogr. Paleoclimatol. Paleoecol.* 5:7–14.

Tanner, W. F. 1985. Late Cenozoic sea level history in the southeastern United States. *Inst. Tertiary-Quaternary Studies—TER-QUA Symp. Ser.* 1:3–8.

Vernon, R. O. 1951. *Geology of Citrus and Levy Counties, Florida.* Florida Geological Survey Bulletin 33.

White, W. A. 1970. *The Geomorphology of the Florida Peninsula.* Florida Bureau of Geology Bulletin 51.

Winker, C. D. and J. D. Howard. 1977a. Correlation of tectonically deformed shorelines on the southern Atlantic coastal plain. *Geology* 5:123–127.

Winker, C. D. and J. D. Howard. 1977b. Plio-Pleistocene paleogeography of the Florida Gulf coast interpreted from relict shorelines. *Trans. Gulf Coast Assoc. Geol. Soc.* 27:409–420.

Soils

3

Charles L. Coultas

Water—its chemistry, energy, and upward trend (rising sea level)—has a profound effect on intertidal soils. Salinity and extreme wetness preclude growth of most upland plants. Erosion takes place at higher elevations in the marsh and deposition at lower elevations. Soils developed in an upland position are being "drowned" by rising sea level as saline Gulf waters cause these acid, low-base soils to become near neutral in reaction and saturated with bases. Where an adequate energy source (i.e., organic carbon) and reducing conditions occur, sulfur is reduced from the sulfate in the sea water to sulfides and sulfur Halomorphic plants, as *Juncus roemerianus* (black needlerush), *Spartina alterniflora*, and *Distichlis spicata* occur in abundance.

In estuaries, which receive significant amounts of fresh water, the environment is much different. Tidal amplitude and wave energy are less and erosion is not significant. Particulate matter is deposited and organic matter is accreted *in situ*. The lower levels of salts in the soil permit growth of less salt-tolerant species as sawgrass (*Cladium jamaicense*), sedge or bulrush (*Scirpus* spp.), and cattail (*Typha*).

Intertidal marshes are dissected by abundant sinuous channels, which permit the rapid ingress and egress of sea water and export and import of materials in solution and suspension.

Tidal amplitudes are low in the Gulf coast marshes, ranging from 0.7 to 1.1 m except in estuaries, where they are less. Marshes range in width from 0.5 km to over 3 km (Citrus County) and in elevation from 0.8 to 1.5 m. Slopes are less than 2% except at the interface between the sea and the marsh and the marsh and the upland, where they may exceed 3%.

Intertidal marsh soils are very diverse, but they have many common properties. Most soils are saline, near neutral in reaction, saturated, and have a low load-bearing capacity. Soils at the lowest elevations tend to contain more clay (especially in the surface layers), more nitrogen and organic matter, and support higher primary production (Kruczynski et al., 1978) than soils at higher elevations. Reducing conditions prevail in all parts of the marsh but are greatest in the lower areas. Sulfur content is also highest at lower elevations. Most of the intertidal marsh soils are shallow over limestone. In many areas, the soils are shallower at higher elevations and become thicker nearer the sea due to erosion and deposition. In the major estuaries, the limestone lies at a much greater depth especially in the Panhandle.

Animals, particularly fiddler crabs, and wave action have caused considerable mixing of the soil horizons. This is observed by the mixture of soil colors, other than that due to mottling caused by oxidation-reducing changes. Two other conditions in tidal wetlands that have not been researched, but deserve some consideration, are the sand barrens (or sand flats, pannes, salt barrens) and the thin (<2 cm) veneer of nearly gelatinous mud high in clays and organic matter and deposited over soils in lower elevations.

The sand barrens are considered to be too saline to support significant vegetation, and usually *Salicornia* spp. and *Batis* spp. are found on the fringes. These barrens are usually found near the highest elevations and are being eroded as evidenced by the lack of a surface horizon high in organic matter. Salinity may fluctuate very widely in this zone. Salinity measurements made at the edge of barrens soon after heavy rains were relatively low (C. L. Coultas, unpublished data). Fiddler crabs are more in evidence in this area, perhaps only because they are more easily seen.

Recently deposited muds are ubiquitous at lower elevations. Particle-size analysis of middle and lower marsh soils, except for those in estuaries, indicates that increasing amounts of clay have been deposited over a long period of time. The source of these clays is unknown. Many of Florida's rivers are clear and carry little sediment, except for the Apalachicola, Ochlockonee, and a few other rivers in the Panhandle. Does this widespread depositional feature indicate a greater sediment (clay) load in recent times in a large area of the Gulf? More likely this phenomenon can be explained by an advancing sea: finer particles are being deposited in the low-energy waters of the lower marsh as the sea advances over the sandier uplands.

Soil Geography, Classification, and Genesis

The Soil Conservation Service (USDA) has described, mapped, and classified many of the soils of the intertidal marsh (Table 3.1). It has delineated seven Great Groups of soils: Sulfaquents, Sulfihemists, Psammaquents, Haplaquolls,

Table 3.1 **Intertidal Marsh Soils as Reported in Soil Conservation Service (USDA) Soil Survey Reports (or Personal Communication)**

County and year of publication	Soil series	Classification	Acreage	Percentage of county
Escambia 1960	Described as	"tidal marsh" (probable Sulfaquents)	1,400	0.3
Santa Rosa 1980	Bohicket Hansboro	Typic Sulfaquent Typic Sulfihemits	8,500	1.3
Okaloosa[a]	Duckston	Typic Psammaquents	723	0.1
Walton[a]	Dirego Duckston	Terric Sulfihemists Typic Psammaquents	2,256 973	0.3 0.1
Bay 1984	Bayvi Dirego	Cumulic Haplaquolls Terric Sulfihemists	8,702 2,326	1.8 0.5
Gulf	Survey in progress, 1996			
Franklin[b]	Dirego and Bayvi	Sulfihemists and Haplaquolls	4,847	1.4
	Bohicket and Tisonia	Sulfaquents and Sulfihemists	14,509	4.2
	Maurepas	Typic Medisaprist	5,007	1.4
Wakulla 1991	Bayvi Estero Isles Maurepas	Cumulic Haplaquolls Typic Haplaquods Arenic Ochraqualfs Typic Medisaprists (Sulfihemists probable)	20,467	5.3
Jefferson 1989	Bayvi	Cumulic Haplaquolls	4,009	1.0
Taylor	Survey in progress, 1996			
Levy	Survey in progress, 1996			
Dixie	Survey in progress, 1996			

**Table 3.1 (continued)   Intertidal Marsh Soils as Reported in Soil
Conservation Service (USDA) Soil Survey Reports (or Personal Communication)**

County and year of publication	Soil series	Classification	Acreage	Percentage of county
Citrus 1988	Weekiwachee-Durban	Terric and Typic Sulfihemists	9,842	2.3
	Rock outcrop-Homosassa	Typic Sulfaquents		
	Lacoochee Complex	Spodic Psammaquents	12,775	3.0
	Homosassa	Typic Sulfaquents	7,858	1.8
Hernando 1977	Homosassa	Typic Sulfaquents	4,862	1.6
	Weekiwachee	Typic Sulfihemists	3,747	1.2
	Lacoochee	Spodic Psammaquents	1,238	0.4
	Weekiwachee-Homosassa	Sulfihemists-Sulfaquents	1,179	0.4
Pasco 1982	Homosassa	Typic Sulfaquents	4,373	0.9
	Lacoochee	Spodic Psammaquents	641	0.1
	Weekiwachee	Typic Sulfihemists	249	0.1

[a] Personal communication from J. D. Overing, May 1988.
[b] Personal communication from G. W. Hurt, August 1988.

Medisaprists, Haplaquods, and Ochraqualfs (Soil Conservation Service, USDA Soil Surveys 1960, 1977, 1980, 1982, 1984, 1988, 1989). All these soils are saturated for the greater part of the year. The Sulfihemists and Medisaprists are organic soils, with the former containing ($\geq$0.75%) S in the upper 50 cm. The organic matter is dark-colored and well decomposed in both soils. The other five soils are mineral soils. The Sulfaquents have a dark-colored surface with a sandy loam to clay texture. Sandier material or rock sometimes occurs below 1 m. Sulfur levels in the upper 0.5 m are high ($\geq$0.75%) and may exceed 5%.

Psammaquents are sandy soils with little morphological development. These soils usually have the lowest cation exchange capacity (CEC) and the least clay and organic matter of any of the seven Great Groups. The Haplaquolls have a thick ($\geq$25 cm) dark-colored surface layer, which is well supplied with bases (Ca, Mg, Na, and K). Surface textures are usually sandy loams grading to loamy sands and sands with depth. Limestone may occur within 1 m. Ochraqualfs have a thin, light-colored surface layer and a subsurface accumulation of silicate clays (argillic horizon). These soils are often shallow over limestone. The Haplaquods are sandy soils with a subsurface accumulation of organic matter, Fe, and Al (spodic horizon).

The largest acreage of intertidal marsh soils mapped by the Soil Conservation Service occurs in Citrus County (30, 475a, 12,342 ha). In the Panhandle, the greatest acreage is mainly in the Apalachicola River delta in Franklin County (24,363a, 7406 ha). The Escambia County Soil Survey was published before the adoption of the modern U.S. system of soil classification. In this publication, wetlands were not examined and intertidal marshes were labeled "tidal marsh." These areas occur principally along the Perdido River, Perdido Bay, Big Lagoon, and a small area along the Escambia River. From a brief examination along Perdido Bay (R31W, T2S, sec 6 and Land Grant area 36 immediately to the east of sec 6), soils classified as Sulfaquents (probably Bohicket series) were noted. These soils consisted of a thin histic epipedon (surface organic layer) underlain by soft, very dark gray clays to 2 m. In some areas, loamy sand and sandy loam material were encountered at about 0.8 m. The soil emitted a strong smell of H_2S. Sawgrass, *Spartina patens*, needlerush, and unidentified grasses were common.

In the adjoining marshes of Santa Rosa County, the soils were classified as Bohicket and Hansboro series. The Bohicket soils have a very dark gray clay surface horizon grading to a dark olive gray silty clay to about 1.2 m. This is underlain by sands and sandy loams. Reaction (pH) is near neutral under field conditions and S levels are high. The Hansboro soils are dark-colored organic soils containing from 2.0 to 4.5% S. The organic material is underlain by clays below 1 m. The soils are developing in deltaic deposits at the mouth of the Escambia and Yellow rivers. Vegetation is a mixture of hydrophilic plants adapted to brackish water as *Scirpus* spp., sawgrass, cattail, and some needlerush.

Small areas of a sandy intertidal marsh soil (Duckston) occur in Okaloosa and Walton counties. In Walton County, 2256 acres (668 ha) of an organic soil high in S (Dirego) occurs.

Most of the intertidal marsh soils in Bay County occur in West Bay. These are sandy soils with a thick, dark-colored surface horizon (loamy sand and sandy loam textures) and a high base saturation (Bayvi series). A shallow organic soil high in S (Dirego) comprises about 20% of the total marsh area. Although much of the intertidal marsh has been destroyed in the development of Panama City and its suburbs, some 11,028 acres still remained in 1981 when field work was completed for the survey.

The soil survey of Gulf County is near completion (1996). The Bureau of Coastal Zone Planning estimated that there were 633 acres (256 ha) of intertidal marsh, however. Most of the extensive area of intertidal marsh in Franklin

County developed in deltaic deposits of the Apalachicola River, Florida's largest river, with a mean flow of 6131 hl/sec (21,650 ft^3/sec) draining an area extending into the southern Appalachian Mountains. Most of these soils are organic, containing high S (Dirego, Tisonia, and Maurepas), and clayey mineral soil, also high in S. Sandy soils with a thick dark-colored surface horizon (Bayvi) also occur.

In Wakulla County, Bayvi, Estero, and Isles soils were found. The Bayvi soils occur in Bay and Franklin counties and have been briefly described. The Estero soils have a dark-colored subsurface horizon where organic matter and Fe/Al have accumulated (spodic horizon). These are "drowned" upland soils formed at a higher elevation but now covered by the sea, at least during high tides. The Isles soil also indicates the occurrence of a rising sea level. These soils have a subsurface accumulation of silicate clays (argillic horizon with relatively high base saturation). This phenomenon (illuviation) can only occur under conditions of better internal drainage than now exists.

Soils at the mouth of the Oklockonee River (both in Franklin and Wakulla counties) are classed as Maurepas muck, a nonsulfidic organic soil. From brief examinations in this area, clayey Sulfaquents occur (probably Bohicket). A significant portion of the organic soils are probably sulfidic. Vegetation in the lower portion of this delta is a mixture of plants tolerant to brackish water (*Scirpus*, sawgrass, and needlerush).

Jefferson County intertidal marshes are mapped as Bayvi (Haplaquolls). The adjoining upland soils are Tooles (Albaqualf) and Chaires (Haplaquods). Extensive areas of soils similar to Tooles and Chaires occur in the intertidal position and again indicate rising sea level. Vegetation is predominantly black needlerush. The soil mapped as Bayvi, a Haplaquoll, is probably misclassified since estimations of "n" values of similar soils are high, indicating a poor load-bearing capacity.

Taylor and Levy County soil surveys are presently underway (1996) and complete information is unavailable. This area has been studied by the Wetland Ecology Program (Florida A&M University) and the findings will be discussed in a later section.

Field work has been completed for the Levy County survey, and three soil series were delineated in the saline marshes: Cracker (similar to Bayvi except shallow over limestone), Wulfert (a shallow organic soil high in S), and Tidewater (a medium-textured soil high in S). James D. Slabaugh, former soil scientist with the Soil Conservation Service, had the following to say about the Levy County intertidal marshes (personal communication dated June 8, 1989). Figure 3.1 shows the location of areas he discussed.

> Areas of the tidal marsh in Levy County between Cedar Key and
> Jones Creek, near Yankeetown, are dominated by loamy textured

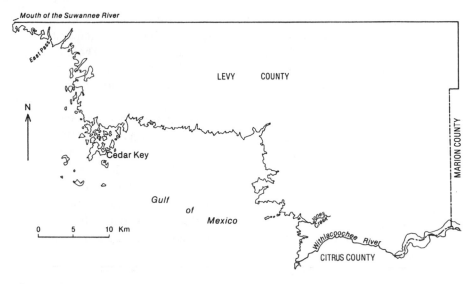

Figure 3.1   The coastline of Levy County, Florida.

Haplaquolls and Sulfaquents that are underlain by limestone bed-rock. The depth to limestone generally is less than 2 feet on the inland edges of the tidal marsh, but the soils become gradually deeper near the seaward edge. The vegetation consists of thick stands of black needlerush on the soils that are deeper to bedrock, but it is usually a mixture of black needlerush, marshy cordgrass and sea-shore saltgrass, with scattered small clumps of cabbage palms on the soils that are shallow to bedrock. The upland soils that are adjacent to or surrounded by this area of tidal marsh generally are loamy textured Haplaquepts or Haplaquolls that have limestone bedrock within a depth of 2 feet.

Areas of the tidal marsh in Levy County between Cedar Key and East Pass are dominated by Sulfaquents that are loamy textured in the upper 2 to 4 feet, and sandy textured below that. Depth to bedrock ranges from 3 to more than 6 feet. The vegetation consists of thick strands of black needlerush. The upland areas that are ad-jacent to these areas of tidal marsh generally are dominated by poorly drained, sandy textured Haplaquods that are on flatwoods positions, and by somewhat poorly to excessively drained Quartzipsamments that are on the dune-like ridges and knolls.

Areas of the tidal marsh in Levy County that are between East Pass and Suwannee River are dominated by Terric and Typic

Sulfihemists (organic soils) that have sandy textured substrata. Limestone is deeper than 2 meters in this area. The vegetation generally consists of a mixture of black needlerush and sawgrass. The adjacent inland soils are Terric and Typic Medisaprists that are in cypress and bay swamps and sawgrass marshes on the floodplain of the Suwannee River.

Areas of the tidal marsh in Levy County that are between Jones Creek and the Withlacoochee River are dominated by thick Sulfaquents and by Psammaquents that are shallow to bedrock. Areas immediately adjacent to the Withlacoochie River are dominated by Terric and Typic Sulfihemists that have sandy textured substrata. The upland areas that are adjacent to these areas of tidal marsh are dominated by poorly drained, sandy textured Haplaquods that are on flatwoods positions. These soils have bedrock at depths ranging from less than 1 foot to 6 feet or more.

The strong correlations between the soils of the tidal marsh in Levy County and the adjacent upland soils suggest that the tidal soils may have once been on similar landscape positions, and supported similar vegetation as the present day upland soils have now, but were gradually drowned and covered by loamy textured marine sediments or organic deposits.

The largest area of intertidal marsh along Florida's Gulf coast occurs in Citrus County. Individual series were not separated in the Soil Survey Report because of the intricacy of their occurrence, but they were mapped as associations or complexes. The Weekiwachee–Durban complex consists of shallow to thick organic soils over sandy materials or rock. These organic soils contain high S.

Rock outcrop–Homosassa–Lacoochee complex is the largest map unit. Rock outcrop is limestone material at the surface, and this occurs at higher elevations in the marsh. The Homosassa series occurs in a lower position and is a medium-textured (sandy loam) soil with a dark-colored surface and a high S content. The Lacoochee series is in a relatively high position and consists of marl over sand over limestone. Extensive areas of this soil were noted near Ozello, and these were subsequently studied by Dr. Kelley Brooks and Sandy Nettles (graduate student) of the Department of Geology at the University of Florida. Nettles (1976) found this marly material to be composed of a variety of marine foraminifera. He concluded that "Paleontologically, the mud [marl] is characterized by a low diversity brackish water foraminiferal and salt marsh molluscan fauna and associated estuarine ostracodes. These data suggest a supratidal depositional environment influenced by a lower sea level, periodic tidal inundation, high evaporation rates, little dilution from rainfall and surface runoff resulting in hypersalinities and calcium carbonate deposition in supratidal pool and peat complexes."

Pasco County has soils similar to those in Hernando County. Homosassa series is the most widespread soil and Lacoochee the least.

Physical and Chemical Properties

Scientists from the Wetland Ecology Program at Florida A&M University have studied the productivity (plants and marine organisms) and the physical and chemical properties of intertidal marsh soils since 1967. Research has been conducted in Franklin, Wakulla, Taylor, Dixie, Citrus, Levy, and Hernando counties (Coultas, 1969, 1970, 1978, 1980; Hsieh and Weber, 1984; Subrahmanyam and Coultas, 1980; Coultas and Gross, 1975, 1977; Coultas and Calhoun, 1976). Six subgroups of soil have been studied: Typic Sulfihemists, Typic Sulfaquents, Typic Fluvaquents, Acric Haplaquods, Typic Psammaquents, and Lithic Ochraqualfs.

The relationship between soils, elevation, and vegetation is shown in Figure 3.2. This particular area is in Taylor County near Dallus Creek. Sulfaquents occur in the lowest position and Haplaquods or Psammaquents adjoin the uplands. *Spartina alterniflora* is found in the lowest position in the marsh but is much more extensive than usual in this area along travere B. This plant usually occurs in a band seldom more than 50 m wide adjoining the tidal streams and Gulf. Needlerush occupies the largest portion of the marsh.

Physical and chemical data on these soils are shown in Table 3.2. The Sulfaquents have a thick, very dark gray clay loam and clay surface horizon underlain by sand. The surface layers contain somewhat higher levels of organic C than is usual and they almost qualify as organic soils. The pH is slightly acidic under field conditions but becomes very acidic upon drying, probably because of the high S content, which ranges from 1.62 to 6.79%. Cation exchange capacity, field moisture content, and total N are related to the organic C and clay contents. Calcium is the predominant extractable cation and Na is more prevalent than K. All three soils have a high soluble salt content as indicated by the high electrical conductivity (EC) of the saturation extract.

The Psammaquents have a thin very dark gray surface (but with discrete spots of light gray material), but organic C, CEC, S, N, and clay levels are much less than in the Sulfaquents. The pH drops significantly upon drying, but less than with the Sulfaquents, probably because of their lower S content.

Haplaquods in the Dallus Creek area have a dark gray sandy surface horizon underlain by light gray sand. This is followed, abruptly, by a very dark

Figure 3.2A   Distribution of soils and location of leveling transects of Dallus Creek, Taylor County, Florida. (From Coultas and Gross, 1975.)

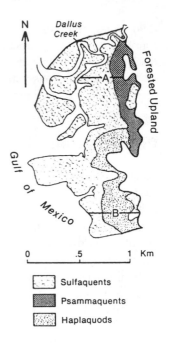

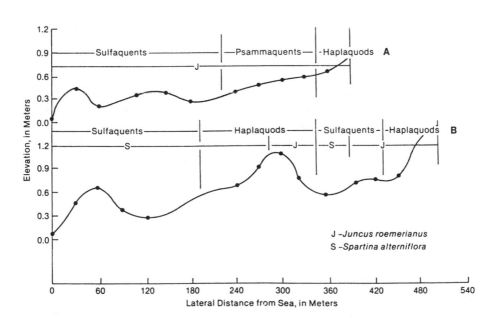

Figure 3.2B   Elevation of the tidal marsh soils and the associated vegetation along two traverses. Scale of figure spans average low and high tides.

Table 3.2 Some Physical and Chemical Properties of Soils in the Dallus Creek Marsh

Horizon	Depth (cm)	pH in H₂O Field moist	pH in H₂O Air dry	Electrical conductivity (mmhos/cm)	Extractable cations (meq/100 g) Ca	Mg	K	Na	Total N	% Organic C	Total S
Sulfaquent											
A11	0–18	6.3	4.0	68	41.5	16.4	1.8	1.2	1.17	10.55	1.62
A12	18–43	6.5	3.2	83	34.1	6.5	1.2	2.2	1.18	9.04	3.57
A13	43–64	6.0	3.0	96	31.4	7.1	0.7	2.2	0.93	6.80	4.49
A14	64–89	5.8	2.8	88	26.0	4.6	0.7	4.1	0.73	7.27	6.79
A15	89–114	5.6	2.5	66	8.5	7.4	0.2	3.5	0.49	4.94	4.76
AC1	114–140	6.1	3.7	31	0.5	0.1	0.3	0.8	0.04	1.09	0.77
AC2	140–152	6.1	3.1	14	0.5	0.1.	0.1	0.5	0.01	0.39	0.35
Psammaquent											
A11	0–18	7.0	4.5	66	5.4	2.7	0.3	0.8	0.28	1.21	0.48
A12	18–38	6.7	4.4	46	3.3	2.1	0.3	0.4	0.14	0.35	0.27
AC1	38–56	6.7	3.5	62	2.2	0.3	0.0	0.0	0.07	0.27	—
AC2	56–76	6.1	3.7	36	1.5	0.2	0.0	0.0	0.03	0.63	0.19
C1	76–94	6.5	3.9	29	1.3	0.3	0.0	0.0	0.03	0.50	—
C2	94–118	6.6	4.0	26	1.0	0.3	0.0	0.0	0.02	0.41	—
C3	118–137	7.0	4.0	16	0.9	0.2	0.0	0.1	0.01	0.34	—
Haplaquod											
A11	0–2	7.7	7.7	43	8.0	7.2	1.0	8.9	0.74	7.32	0.03
A12	2–22	7.8	5.6	29	1.5	2.9	0.1	1.3	0.21	1.59	0.14
A13	22–38	—	5.2	33	1.0	2.9	0.1	0.4	0.07	0.87	0.10
A2	38–53	7.3	3.6	32	0.5	0.8	0.2	0.2	0.07	0.52	0.09
B2h1	53–68	6.9	4.6	32	1.5	1.2	0.2	0.7	0.14	1.41	0.22
B2h2	68–93	6.8	4.6	34	2.0	2.0	0.2	2.2	0.07	1.12	0.16
B2h3	93–116	—	5.4	27	2.5	0.9	0.2	2.0	0.02	0.68	0.15
C	116–167	7.0	6.1	33	1.0	0.8	0.1	0.5	0.04	0.39	0.05

Adapted from Coultas and Gross, 1978.

gray to black spodic horizon (Bh) at 53 cm. The spodic layer contains much higher organic C, Fe, Al, and CEC than the superjacent horizon.

Most intertidal marsh soils have a low load-bearing capacity. One method of assessing this property is by calculating the "n" value: n = (A − 0.2R)/(L + 3H), where A = % moisture under field conditions, R = % silt + sand, L = % clay, H = % organic matter (organic C × 1.724) (Soil Survey Staff, 1975). For these Psammaquents, the "n" values are >1 in the upper 50 cm, indicating a low capacity to bear loads. Similar results were found for Psammaquents studied in Wakulla and Dixie counties. This raises a question concerning the Soil Conservation Service classification of similar soils as Mollisols (Haplaquolls, Bayvi series). Although these soils meet the criteria of dark color, thickness, and high base saturation, their "n" values are too high for such classification.

The mineralogy of the clay fraction in the soils of Dallus Creek are principally kaolinite, vermiculite, and vermiculite intergrades with minor amounts of quartz and gibbsite. This mineralogical suite suggests that these materials are of local origin, as these minerals, as well as kaolinite–metahalloysite, are typical of many of the adjoining upland soils. An exception to this is the estuaries, such as the Apalachicola River estuary, where montmorillonitic clays are common, indicating fresher, less-weathered conditions.

A shallow, light gray, loamy soil (marl) over sands was found in Hernando County (Coultas and Calhoun, 1976) (Table 3.3). Limestone was encountered at 38 cm. This soil is similar to those found in Citrus County, where the bulk of the carbonitic material was composed of testes of marine foraminifera. This soil has a very high pH, high electrical conductivity, high levels of extractable Ca, and low levels of organic C, N, and S. Calcite was the predominant clay-sized mineral at the surface. Density is relatively high and load-bearing capacity, as indicated by "n" values, is good.

Usually, this soil developed from marl occurs in a high position in the marsh. In Citrus County, however, it also occurs on a low-lying island in the Gulf. Similar soils (marl/sand/limestone) or erosional remnants thereof are found from Taylor to Pasco counties. In some areas, the marl has been eroded, leaving behind only calcareous sandy or loamy material. This occurs in Taylor County (an Ochraqualf) where secondary $CaCO_3$ concretions occur in the loamy substrate but the marl (presumed) has been eroded (Table 3.4). A gray and dark gray sandy clay argillic horizon occurs at 68 cm and limestone follows at 76 cm. The pH is high in this soil and Mg is the principal cation. Nitrogen and organic C are low except at the soil surface and S is very low. Clay content decreases from 24.0% at the surface to 19.7% at the 30- to 46-cm depth and then increases to 38.0% just above the limestone.

The Apalachicola River delta marsh soils illustrate the effect of fresh water on their development and chemistry (Coultas, 1980). Sulfihemists, sulfidic

Table 3.3 Some Chemical and Physical Properties of an Intertidal Marsh Soil in Hernando County, Florida

Horizon	Depth (cm)	pH in H₂O Field moist	pH in H₂O Air dry	Electrical conductivity (mmhos/cm)	Extractable cations (meq/100 g) Ca	Mg	K	Na	Σ	% Organic C	Total N	Total S
C1	0–12	8.2	8.4	48.3	15.5	4.8	1.7	0.2	22.2	0.9	0.06	0.36
C2	12–17	8.2	8.4	37.3	16.4	3.8	0.9	0.3	21.4	0.8	0.06	0.21
IIA1b	17–22	7.8	8.3	36.0	16.2	3.9	0.9	0.5	21.5	0.9	0.04	0.06
IIBb	22–38	7.8	8.4	32.1	10.7	2.3	0.3	0.4	13.7	0.2	*	0.00

Horizon	Depth (cm)	Water content (%) Saturation Wt.	Vol.	0.06 bar Wt.	Vol.	15 bar Wt.	Vol.	n value	Bulk density (g/cm)	Saturated hydraulic conductivity (cm/hr)	Sand (mm) 2–1	1–0.5	0.5–0.25	0.25–0.1	0.1–0.05	% Silt (50–2 μ)	Clay (2 μ)	Textural Class
C1	0–12	30.0	43.6	25.8	37.6	14.2	14.2	0.3	1.46	7.2	2.2	7.4	12.5	31.4	16.3	14.4	15.8	sl
C2	12–17	29.4	44.1	26.0	39.0	15.8	15.8	0.4	1.50	11.8	0.4	1.6	6.4	35.3	19.2	17.0	20.1	scl
IIA1b	17–22	20.8	36.0	19.5	33.7	6.6	6.6	0.2	1.73	0.2	0.2	0.9	4.9	54.0	21.2	4.7	14.1	sl
IIBb	22–38	18.5	32.7	17.8	31.5	6.6	6.6	0.1	1.77	0.4	0.5	1.0	5.6	59.0	21.9	2.5	9.5	lfs
IIIR	38+	Limestone																

Adapted from Coultas and Calhoun, 1976.

Table 3.4 Description and Some Physical and Chemical Properties of an Ochraqualf Near Hickory Mound in Taylor County, Florida

Vegetation—*Juncus roemerianus* and *Distichlis spicata* predominate.

Horizon	Depth (cm)	Description
A1	0–15	Very dark gray (5Y 3/1) sandy clay loam: abundant roots and rhizomes
A2	15–30	Olive gray (5Y 5/2) sandy loam: abundant limestone concentrations
B1	30–46	Light olive brown (2.5Y 5/6) sandy loam with light brownish-gray mottles (2.5Y 5/2)
B21 t	46–68	Grayish-brown (2.5Y 5/2) sandy clay loam with dark yellowish-brown (10YR 4/4) mottles
B22 t	68–76	Gray (5Y 5/1) and dark gray (5Y 4/1) sandy clay
IIR	76+	Limestone

Horizon	Depth (cm)	pH in H₂O Field moist	pH in H₂O Air dry	Electrical conductivity (mmhos/cm)	Extractable cations (meq/100 g) Ca	Mg	K	Na	Organic N	% Total C	Total S
A1	0–15	7.2	6.3	21	6.9	12.4	1.2	2.1	0.18	2.22	0.08
A2	15–30	7.2	6.8	24	10.9	6.8	0.7	0.9	0.07	0.64	—
B1	30–46	7.0	6.8	26	2.6	5.1	0.9	0.7	0.03	0.22	0.02
B21 t	46–68	7.0	6.6	25	3.1	8.2	1.2	1.8	0.03	0.36	0.06
B22 t	68–76	7.0	6.6	18	4.6	13.8	1.4	1.9	0.04	0.76	—
IIR	76+	Limestone									

Horizon	Depth (cm)	Very coarse	Coarse	Medium	Fine	Very fine	Silt	Clay	Texture	"n" value
A1	0–15	0.3	3.0	14.0	12.0	27.3	19.4	24.0	scl	0.9
A2	15–30	0.5	3.8	15.4	44.9	3.6	13.8	18.0	sl	0.6
B1	30–46	0.7	6.0	20.0	31.7	21.9	0.0	19.7	sl	0.6
B21 t	46–68	0.9	5.1	16.8	26.9	16.3	0.0	34.0	scl	0.7
B22 t	68–76	0.7	3.3	12.4	34.1	7.3	4.2	38.0	sc	0.7
IIR	76+	Limestone								

organic soils over clays, and sands at about 1 m occur near the mouth of the river. Vegetation is primarily sawgrass with abundant needlerush interspersed. The organic material is well decomposed, dark colored, and nonsaline or very slightly saline (a saline soil has an EC of >4 mmhos/cm). Farther from the main course of the river, the soils become lower in organic matter, more clayey, and more saline. Sulfaquents are soils representative of this area. In the eastern portion of East Bay, some distance from the influence of the river, a clayey saline Sulfaquent, vegetated primarily with *Spartina patens,* occurs. Two other soils in this area are covered with black needlerush and are located near the mouth of small local streams. One soil was not sulfidic and organic C did not decrease uniformly with depth, indicating stream deposition (Fluvaquent). In another soil, a Sulfaquent, organic C also decreased erratically but S content was high. Both soils are saline.

A few areas of Sulfihemists occur that are not associated with an estuary. One occurs about 3 mi northwest of Steinhatchee (SW1/4 sec 20, R9E, T9S, Taylor County). Sawgrass predominates but *Lythrum lancealatum* and *Kosteletskya virginica* occur. Sand occurs at 69 cm, and frequent pieces of wood are found 30 cm above and below this junction. Bulk densities (g/cm^3) range from 0.10 to 0.17 in the organic material. Cation exchange capacity (sum of cations) ranges from around 37 to about 48 meq/100 g. Nitrogen and organic C range from 1.34 to 3.35% and 15.2 to 23.5%, respectively. This area is in a middle to upper marsh position and receives fresh water from a localized stream and spring not connected to the Gulf.

Another estuarine area of interest is at the mouth of the Suwannee River (Table 3.5). This area has a very thick, dark-colored, loamy surface layer to 145 cm underlain by gray sand. The low bulk densities of the surface layers indicate the high content of plant material. The pH is near neutral under field conditions but drops drastically upon aeration. The soils are moderately saline and have a relatively high CEC, organic C, and N content for a mineral soil. Sulfur is highest (1.43%) at a depth of 30 to 64 cm. Clay content decreases with depth.

Buried wood (stems and branches) was found in many areas of the intertidal marsh, further demonstrating rising sea level and allowing an approximation of the rate of rise or at least the rate of accretion. Buried wood was found within 100 m of a tidal stream at 1 m depth in a middle marsh position in Levy County. This material was dated at 2949 ± 90 years BP by ^{14}C techniques, thus indicating an accretion rate of about 3.3 cm/100 years. Another dating of buried wood was made on a tree trunk found at a depth of 145 to 175 cm along the lower part of the St. Marks River (sec 14, T4S, R1E, Wakulla County). This was dated at 3075 ± 150 years, which indicates an accretion of 5.2 cm/100 years. This area could receive materials from the river and materials developed *in situ,* which could account for the greater accretion rate. Buried wood was also dated from the Sulfaquent in Taylor County (Dallus Creek). It was found at a depth of 114 to 140 cm and dated 3350 ± 95 years BP. This indicates accretion at a rate of 3.8 cm/100 years.

Productivity

This discussion will deal only with the productivity of needlerush. Observations and limited research (Kruczynski et al., 1978; Coultas and Weber, 1980) indicate that *Juncus roemerianus* grows best in low marsh positions. This suggests that the higher organic C, N, and CEC of the low marsh soils contribute to greater growth. Longer periods of tidal inundation may also be a contributing factor.

Kruczynski et al. (1978), using several methods to determine productivity near St. Marks Lighthouse in Wakulla County, found that aerial productivity decreased landward from 949 $g/m^2/year$ in the low marsh, to 595 $g/m^2/year$ in middle marsh, to 243 $g/m^2/year$ in high marsh. Belowground biomass (living and dead roots and rhizomes) ranged from 4063 g/m^2 in middle marsh to 5140 g/m^2 in high marsh. The low marsh soil was a Sulfaquent or Psammaquent, the middle marsh was an Ochraqualf, and the high marsh was a Haplaquod with a surface horizon thicker and darker than modal. Some surface (15 cm) properties of these soils are shown in Table 3.6. Cation exchange capacity, total N, and organic C decreased in a landward direction from the coast and may well explain differences in productivity. Salinity, as measured by conductivity, and pH varied only slightly.

Kruczynski et al. (1978) suggest that salinity may be one controlling factor in decreasing productivity in a landward direction. Certainly needlerush grows poorly near salt flats and not at all within them. Surface salinity measurements that I have made, however, do not substantiate this in major portions of the marsh. Perhaps extreme fluctuations in salinity may be more harmful to needlerush growth than absolute high levels. Salinity was relatively low in a sand flat near Horseshoe Beach (Dixie County) just after heavy rains had fallen. In the research cited above (Wakulla County), there was no significant increase in salinity in a landward direction.

Using standing crop as an estimate of productivity, Coultas and Weber (1980) found that several surface soil parameters correlated with productivity. Aboveground standing crop was negatively correlated with redox potential ($r = -0.45$ to -0.53) and positively correlated with time of year ($r = 0.40$ to 0.42). Belowground biomass (to a depth of 8 cm) was negatively correlated with redox potential ($r = -0.22$ to -0.29). In this study and in that by Kruczynski et al. (1978), there is a good possibility that the *Juncus* in the high marsh position

TABLE 3.5 Description and Some Physical and Chemical Properties of a Sulfaquent in Dixie County, Florida

Vegetation—needlerush (*Juncus roemerianus*) predominates; also 10% *Distichlis spicata.*

Location: SE ¼ of the SE ¼ sec. 23, R11E. T13S, Dixie County. This is an island at the mouth of the Suwannee River. Several islands in this area have similar soils.

Horizon	Depth, cm (in.)	Description
A11	0–2 (0–1)	Very dark gray (10YR 3/1) clay loam; few roots; sticky
A12	2–30 (1–12)	Very dark gray (5Y 3/1) very fine sandy loam; abundant roots and rhizomes
A13	30–64 (12–25)	Very dark gray (10YR 3/1) very fine sandy loam; frequent roots
A14	64–109 (25–43)	Black to dark olive gray (5Y 2.5/2) sandy loam; frequent dead roots
B15	109–145 (45–57)	Black to dark olive gray (5Y 2.5/2) sand; few roots
C1	145–175 (57–69)	Dark grayish-brown (10YR 4/2) sand with dark gray spots; rare roots
C2	175–203 (69–80)	Gray (10YR 5/1) sand; no roots

Horizon	Depth (cm)	Bulk density (g/cc)	pH in H₂O Field moist	Air dry	Electrical conductivity (mmhos/cm)	Extractable cations (meq/100 g)				Total N	% Organic C	Total S
						Ca	Mg	K	Na			
A11	0–3	—	6.1	5.8	19	12.0	21.6	1.8	8.9	1.17	8.72	0.64
A12	3–30	0.42	6.6	5.5	24	6.3	10.7	1.2	3.2	0.42	5.55	0.23
A13	30–64	0.53	6.5	3.9	28	6.3	4.9	0.5	0.3	0.33	5.17	1.43
A14	64–109	0.35	6.6	3.2	28	9.4	4.1	0.4	0.2	0.48	9.02	1.18
A15	109–145	1.03	6.7	3.4	19	5.6	0.7	0.1	0.2	0.09	2.44	—
C1	145–175	—	6.8	4.2	14	0.7	0.4	0.0	0.1	0.02	0.63	—
C2	175–190	—	7.1	6.1	11	0.4	0.6	0.1	0.1	0.01	0.26	—

%

Horizon	Depth (cm)	Very coarse	Coarse	Medium	Fine	Very fine	Silt	Clay	Texture	"n" value
A11	0–3	0.2	1.4	4.6	11.0	17.6	26.2	39.0	cl	3.6
A12	3–30	0.0	1.5	5.0	19.3	38.7	17.3	18.2	vfsl	3.1
A13	30–64	0.0	0.4	1.7	1.8	59.1	23.9	13.1	vfsl	4.3
A14	64–109	0.2	4.8	11.7	15.9	20.5	35.0	11.6	sl	18.5
A15	109–145	0.6	7.1	32.6	3.0	46.1	5.6	5.0	s	2.8
C1	145–175	0.7	10.5	39.2	12.0	32.8	3.1	1.7	s	1.5
C2	175–190	1.1	13.3	38.1	15.7	30.4	0.1	1.3	s	1.5

Table 3.6 Some Surface Soil Properties in an Intertidal Marsh Near St. Marks Lighthouse, Wakulla County, Florida

Site	pH	Electrical conductivity (mmhos/cm)	Cation exchange capacity (meq/100 g)	% Total N	% Organic C
Low marsh	6.3	24.5	17.0	0.37	9.42
Middle marsh	6.2	27.3	8.3	0.16	5.02
High marsh	6.3	24.4	3.4	0.06	1.58

Adapted from Coultas and Weber, 1980.

was genetically different from those plants in middle and low marsh (by far the major portion of the marsh). Plants in low and middle marsh were taller, thicker in diameter, and less dense than those in the high marsh. Transplanting plants from high marsh to low marsh and vice versa did not affect their density or morphology, but those plants moved from high to low marsh did not survive well, indicating that they were not adapted to the more reduced conditions of the low marsh. Luce and Coultas (1976) fertilized a needlerush stand with three levels of N and P (0, 75, and 150 lb/acre) and found very little growth response. This was in a middle marsh position in Wakulla County (probable Psammaquent). A year after fertilization, aboveground standing crop had not increased. Nitrogen concentration of the foliage had increased, however. Phosphorus uptake was not affected.

Total P in the marsh soils appears to be very low (Coultas, 1969, 1970), seldom over 100 ppm. The oxidation–reduction status of the soil affects many soil properties including P availability (Phung and Fiskell, 1973; Redman and Patrick, 1965). Available P would probably be highest in low and middle marsh positions, where reducing conditions are greatest.

Observations throughout the study area indicate that the largest standing crop of needlerush occurs in low marsh positions, where soils tend to be thicker and higher in many nutrients and organic matter. Clayey and loamy Sulfaquents at the mouth of the Fenholloway and Suwannee rivers support exceptionally productive stands. Probably the greatest standing crop of needlerush I have seen occurs just north of highway US 90 where it crosses a large area of marsh in the delta of the Escambia River. The *Juncus* was well over 2 m in height in the spring of 1989. The soils were clayey Sulfaquents (Bohicket series), but most vegetation was more typical of brackish water areas. The road embankment, no doubt, has affected the salinity of the soil and growth of these plants.

Why does needlerush predominate in the Gulf coast marshes of Florida but *Spartina alterniflora* in the broad marshes of Georgia? Tidal amplitude is sometimes suggested as one reason, and this is supported by the fact that *S. alterniflora* grows best at lowest elevations in the needlerush marshes of Florida, where the greatest tidal amplitude occurs. This is the zone where soils are also richest in many growth factors. Is *S. alterniflora* a more productive plant than needlerush? Kruczynski et al. (1978) summarized data from New England to Louisiana, and it appears that it is. Their own data in Florida do not support this, however. Some research indicates that *S. alterniflora* is higher in protein and breaks down more readily to provide detritus to more organisms.

Management of Soils

The obvious management problems in intertidal marshes are extreme wetness, salinity, the rising sea level, and occasional storm tides. The soils are shallow over limestone, saline, corrosive to metal and to materials such as concrete, and have a poor load-bearing capacity for animals or machinery. The soils of the low marsh contain high levels of S in a reduced form. Upon drainage and aeration, these materials become oxidized to form sulfuric acid. This may result in conditions that most higher forms of life cannot tolerate.

Intertidal marshes have been used for agriculture. Since these soils usually contain substantial quantities of organic matter, N, and other elements essential for plant production, they have some attractive properties. But the "richest" soils contain the highest S content and a pH drop from near neutral under flooded conditions to a pH of 3 or less upon drainage and aeration.

Intertidal marshes are being used for cattle grazing in Taylor County, Florida. Historically, marshes have been used for hay production in New England. Marshes have been filled and used for building sites and marinas, thus destroying the marshes, altering the soils, and eliminating the capacity of the marshes to support marine organisms.

In terms of C fixation, the marsh soils support a productive plant population in their unaltered state. The plants provide an energy source for a multitude of marine organisms. The marsh provides protection for the immature stages of these organisms and a buffer to the destructive force of storm tides. Serious consideration should be given to these facts before altering the soils and plants of the intertidal marshes.

References

Bureau of Coastal Zone Planning. 1978. *Statistical Inventory of Key Biophysical Elements in Florida's Coastal Zone.* Florida Department of Environmental Regulation, Tallahassee.

Coultas, C. L. 1969. Some saline marsh soils in North Florida. Part I. *Soil Crop Sci. Soc. Fla.* 29:111–123.

Coultas, C. L. 1970. Some saline marsh soils in North Florida. Part II. *Soil Crop Sci. Soc. Fla.* 30:275–282.

Coultas, C. L. 1978. Soils of the intertidal marshes of Dixie County, FL. *Fla. Sci.* 41:62–90.

Coultas, C. L. 1980. Soils of marshes in the Apalachicola, Florida estuary. *Soil Sci. Soc. Am. J.* 44(2):348–353.

Coultas, C. L. and F. G. Calhoun. 1976. Properties of some tidal marsh soils of Florida. *Soil Sci. Soc. Am. J.* 40(1):72–76.

Coultas, C. L. and E. R. Gross. 1975. Distribution and properties of some tidal marsh soils in Apalachee Bay, FL. *Soil Sci. Soc. Am. Proc.* 39:914–919.

Coultas, C. L. and E. R. Gross. 1977. Tidal marsh soils of Florida's middle Gulf coast. *Soil Crop Sci. Soc. Fla.* 37:121–125.

Coultas, C. L. and O. J. Weber. 1980. Soil characteristics and their relationship to growth of needlerush. *Soil Crop Sci. Soc. Fla.* 39:73–77.

Hsieh, Y. P. and O. J. Weber. 1984. Net aerial primary production and dynamics of soil organic matter formation in a tidal marsh ecosystem. *Soil Sci. Soc. Am. J.* 48(1):65–72.

Kruczynski, W. L., C. B. Subrahmanyam, and S. H. Drake. 1978. Studies on the plant community of a North Florida salt marsh. Part I. Primary production. *Bull. Mar. Sci.* 28(2):316–334.

Lucc, H. D. and C. L. Coultas. 1976. *Effect of N and P Fertilization on the Growth, N Content, and P Content of a Natural Stand of Juncus roemerianus Near St. Marks, FL.* Agron. Abstracts 149–150. American Society of Agronomy, Madison, Wisconsin.

Nettles, N. S. 1976. Intertidal calcitic muds along the west coast of Florida. M.S. thesis. University of Florida, Gainesville.

Phung, H. T. and J. G. A. Fiskell. 1973. A review of redox reactions in soils. *Soil Crop Sci. Soc. Fla. Proc.* 32:141–145.

Redman, F. H. and W. H. Patrick, Jr. 1965. *Effect of Submergence on Several Biological and Chemical Soil Properties.* Louisiana State University Bulletin 592. Baton Rouge.

Soil Conservation Service. 1960. *Soil Survey, Escambia County, FL.* U.S. Department of Agriculture, Washington, D.C.

Soil Conservation Service. 1977. *Soil Survey of Hernando County, FL.* U.S. Department of Agriculture, Washington, D.C.

Soil Conservation Service. 1980. *Soil Survey of Santa Rosa County, FL.* U.S. Department of Agriculture, Washington, D.C.

Soil Conservation Service. 1982. *Soil Survey of Pasco County, FL.* U.S. Department of Agriculture, Washington, D.C.

Soil Conservation Service. 1984. *Soil Survey of Bay County, FL.* U.S. Department of Agriculture, Washington, D.C.

Soil Conservation Service. 1988. *Soil Survey of Hernando County, FL.* U.S. Department of Agriculture, Washington, D.C.

Soil Conservation Service. 1989. *Soil Survey of Jefferson County, FL.* U.S. Department of Agriculture, Washington, D.C.

Soil Conservation Service. 1991. *Soil Survey of Wakulla County, FL.* U.S. Department of Agriculture, Washington, D.C.

Soil Survey Staff. 1975. *Soil Taxonomy: A Basic System of Classification for Making and Interpreting Soil Surveys.* Agric. Handbook No. 436. U.S. Government Printing Office, Washington, D.C.

Subrahmanyam, C. B. and C. L. Coultas. 1980. Studies on the animal communities in two North Florida salt marshes. Part III. Seasonal fluctuations of fish and macroinvertebrates. *Bull. Mar. Sci.* 30(4):790–818.

Vegetation*

4

Andre F. Clewell

Tidal marshes occupy terrain influenced by tidal amplitude and protected from erosive wave action. Tidal marshes may be conceptualized conveniently, although somewhat artificially, into three types according to salinity: salt marshes, brackish marshes, and freshwater tidal marshes. *Salt marshes* occur near the coast and range from mesohaline (5 to 18‰) to hyperhaline (>40‰). *Brackish marshes* occur primarily at the ends of river deltas and along the shores of bays, where freshwater delivery from rivers and other sources of runoff maintain generally oligohaline conditions (0.5 to 5‰). *Tidal freshwater marshes* occupy shores and low floodplains of rivers within the zone of tidal influence and may experience oligohaline conditions only briefly during tidal surges from tropical storms. These three marsh types may intergrade over prolonged transitions.

Hydrology and salinity largely determine composition and abundance of tidal marsh vegetation. The most important hydrological consideration is aeration, i.e., whether the substrate is waterlogged to the extent that plants must overcome anaerobic conditions to survive. Aeration is a function of the periodicity of tidal inundation, elevation relative to that of high tide, permeability of the substrates, impediments to the return of tidal waters such as dense vegetation and levees along tidal creeks, and the amount of runoff from upland sources. Salinity varies widely. At one extreme, there may be virtually no salinity along marshy shorelines of tidal rivers. At the other extreme are hyperhaline salt flats. The vegetation responds both to average soil salinity and to seasonal extremes of salinity. Salinity in the water column is less important ecologically and is generally more variable and less saline than is the soil water (Hackney and de la Cruz, 1978).

---

* Plant nomenclature follows Clewell (1985).

Few species of vascular plants tolerate mesohaline conditions. The flora at any given site is further restricted by the hydrological regime and by seasonal extremes of salinity. The result is a tendency for salt marsh vegetation to occur in monospecific zones rather than intermixed in plant communities. For that reason, salt marsh ecosystems lack plant communities. This generalization applies to a lesser degree to the entire tidal marsh ecosystem, because brackish and tidal freshwater marshes also tend to be vegetated in zones or patches of one or few species. Patchiness is attributable to storm events that disrupt the vegetation every few decades, followed by a burst of opportunistic regrowth. The resulting stochastically affiliated species patches are not well integrated into a plant community of closely interacting species populations.

Salt Marshes

Occurrence

Salt marshes generally occur along the coast behind protective beaches and on the leeward shorelines of barrier islands. They occupy vast tidal flats along the Gulf coast in Florida's Big Bend region. Much of this region contains "zero energy" shoreline (Tanner, 1960), so-named for the extensive shallows that effectively dampen wave action. Florida's salt marshes form a nearly continuous band of coastal vegetation, which extends 225 km from the Ochlockonee River in Wakulla County to Tarpon Springs in Pinellas County and which is centered on the zero energy shoreline. This band is commonly from 1 to 7 km broad and widens to 12 km in Citrus County (Soil Conservation Service, 1980). From the Ochlockonee River westward, salt marshes are common but not nearly as extensive. South of Tarpon springs, salt marshes are much less frequent, and their potential habitats are commonly preempted by mangroves.

Topography

Salt marsh topography basically consists of a gently inclined plane. Salt marsh vegetation begins at or just below mean sea level. Along the Gulf coast, the inner edge of the salt marsh may lie 1 to 3 m above mean sea level. The innermost marsh elevations are generally higher along the Atlantic coast, where tidal amplitude is greater. With proximity to the coast, soils are kept wet and saline by exposure to daily tidal inundation. At increasing elevations with distance from the coast, tidal inundations are less regular and the period of any given episode of inundation is briefer. Both soil moisture content and salinity fluctuate accordingly. The degree of fluctuation depends on the frequency and duration of tidal inundation, the seasonality of rainfall that dilutes saline water and maintains soil moisture, and the duration of rainless periods during which time evaporation raises soil salinity. Near adjacent uplands, freshwater runoff ameliorates the extremes in soil moisture and salinity.

This basic plan of a salt marsh is modified by frequent tidal creeks that carry incoming tides into marshes and return runoff at low tide. Low levees that flank these creeks are sufficient to channel incoming tides well into marshes before overbank flooding begins and to retard the return of runoff at low tide. Another modification to the basic plan is low berms, which develop from wave action during storms and separate all but a fringe of marsh from coastal waters. Such berms act as levees and are prevalent on all but the zero energy coastline.

Zonation

Salt marshes typically consist of four vegetational zones (Figure 4.1):

1. *Spartina alterniflora* (smooth cordgrass) zone (Clewell, 1985)

2. *Juncus roemerianus* (black needlerush) zone

3. Salt flats (including any associated salt barrens)

4. High marsh

These zones are listed in their relative order of distance from the coast and elevation above sea level. The *Spartina* zone lies at low elevations nearest the coast, and high marsh occupies the most elevated terrain bordering upland forest. *Spartina* and *Juncus* zones are commonly monotypic. Salt flats and high marsh each consist of several species; however, these species are frequently concentrated in bands along salinity gradients within these zones. The more gradual and uniform the gradient, the more likely that all zones will be present and well developed. The extent and, to some degree, the species composition of each zone depends largely upon tidal amplitudes. The low tidal amplitudes along the Gulf coast favor only a narrow fringe of *Spartina* zone along beaches and creek banks and a vast *Juncus* zone. The wider tidal amplitudes along the Atlantic coast favor an extensive *Spartina* zone with *Juncus* limited in a narrow zone near the upland (Kurz and Wagner, 1957).

The *Spartina* zone occupies sites that are typically flooded daily or nearly so by tidal inundations. On Florida's Gulf coast, this range encompasses about 4 to 10 dm; see Stout (1984) for elevation data at selected localities. On Florida's Atlantic coast, where tidal amplitudes are greater, the range is as much as 21 dm near Jacksonville (McKee and Patrick, 1988). At a marsh in Mississippi, Eleuterius and Eleuterius (1979) determined that *S. alterniflora* occupied an elevational range from 0.24 m below to 0.54 m above mean low water (MLW).

Figure 4.1 Dark lines of *Spartina alterniflora* occupy the banks of tidal creeks. *Juncus roemerianus* comprises most marsh vegetation with a uniform, turf-like cover. Salt flats consisting largely of white sandy barrens contrast with *Juncus*. A narrow band of high marsh separates the salt flats from coastal hammocks (forests). (Wakulla County, courtesy of C. S. Gidden.)

Tidal flooding in the *Spartina* zone occurred from 87 to 10% of the year (7645 to 855 h). *Spartina* zone soils tend to be medium or fine textured and rich in organic matter.

S. alterniflora (Figure 4.2) also grows at higher elevations within the elevational range typical for the *Juncus* zone. Such plants of *Spartina* rarely flower, and they are distinctly shorter than where tidal inundation is regular. These so-called "short forms" of *Spartina* grow at a distance from tidal creeks, where the soil is either highly saline or is waterlogged from poor drainage.

The *Juncus roemerianus* zone occupies elevations that are inundated irregularly by tides. In the aforementioned Mississippi marsh, Eleuterius and Eleuterius (1979) determined that *J. roemerianus* occupied elevations from 0.54 to 0.75 m above MLW. The *Juncus* zone was flooded from 5 to 0.8% of the year (475 to 75 h). Soils of *Juncus* marshes are quite variable and frequently consist of sandy loam or loamy sand (Coultas and Gross, 1975). Organic matter is frequently present and in places forms a fibrous peat soil derived from the slowly

Figure 4.2 *Spartina alterniflora* with a periwinkle (*Littorina irrorata*) that feeds on the soil surface and escapes predation by climbing on *Spartina*.

decomposing roots and rhizomes of needlerush (Eleuterius and Eleuterius, 1979). As with *Spartina, J. roemerianus* may also grow as a "short form" with distance from tidal creeks (Figure 4.3).

The *Spartina* and *Juncus* zones cannot be differentiated solely on the basis of elevation and concomitant regimes of salinity and hydrology. Eleuterius and Eleuterius (1979) noted that the line of demarcation between zones, although abrupt, did not correspond nicely with elevational contours. To the contrary, clumps of *Juncus* occurred within the *Spartina* zone and vice versa. Once *Juncus* colonized the upper edge of the *Spartina* zone, *Spartina* plants were excluded thereafter. Similar results and conclusions were reported earlier by

Figure 4.3   Broad expanse of *Juncus* marsh in Taylor County.

Kurz and Wagner (1957) in Florida. They noted that the upper elevational limit of *Spartina* was higher on finer textured soils, due to greater water retention of those soils. Eleuterius and Caldwell (1985a, 1985b) searched, without success, for soil differences that might help explain species zonation that occurred at the same elevation. Instead, they discovered highly variable values for pH, organic matter content, and nutrients (N, P, K, S, Zn, Ca, Mg, PO_4, NH_3) within and between species populations and concluded that these soil characteristics alone could not explain distributions of *Spartina, Juncus,* and two additional species (*Spartina patens, Scirpus [olneyi=] pungens*).

Salt flats are sandy, hypersaline environments that support typically sparse growths of low-growing vegetation, mainly succulents and grasses. Only a few species are common in salt flats: *Aster tenuifolius, Batis maritima* (saltwort, a small repent shrub with succulent leaves), *Borrichia frutescens* (sea ox-eye), *Distichlis spicata* (saltgrass), *Limonium carolinianum* (sea lavender), *Salicornia bigelovii* (annual glasswort), and *Salicornia virginica* (perennial glasswort). *Monanthochloe littoralis* (keygrass) is locally common near limestone outcrops that occur in salt flats along parts of the Gulf coast southward from southern Taylor County (see Figures 4.4 and 4.5).

Salt flats are tidally inundated infrequently, usually less than once a month.

In the interim, evaporation causes soil salinity to increase, often dramatically. Salt flats occasionally occupy slight depressions within *Juncus* zones, into which tidal waters become trapped and evaporate during dry seasons. Generally, salt flats are elevated above the adjacent *Juncus* zone. These elevated sites are tidally inundated less frequently than the *Juncus* zone and experience more prolonged intervals of evaporation and elevated salinity. High salinity causes ion toxicity, osmotic stress, and interferes with nutrient uptake, all of which inhibit plant growth. Therefore, vegetation tends to be scarce and too sparse for organic matter to accumulate. Scarcity of vegetation allows solar radiation and

Figure 4.4   Salt flat in Taylor County, bordered by the edge of the *Juncus* zone (bottom) and dominated by a dense short growth of *Distichlis*. A white sandy barren contrasts with vegetated marsh, and a coastal hammock appears in the background.

Figure 4.5 *Batis maritima*, a low-growing succulent shrub typical of salt flats.

wind to reach the soil surface, thereby promoting evaporation. The evaporative surface is frequently increased by an abundance of fiddler crab burrows (*Uca* spp.). Salt flats are favored by fiddler crabs, which prefer the open, loose, sandy substrates for burrowing. Accelerated evaporation causes hyperhaline conditions to develop quickly following tidal inundation. Kurz and Wagner (1957) noted that highly organic root mats beneath *Juncus* zones are relatively impermeable to water and that tidal sheet flow glides across these mats and soaks into the porous sands of adjacent salt flats. Soil salinity in salt flats is elevated accordingly.

The central and, usually, more highly elevated region within a salt flat may become too saline in the prolonged intervals between tidal inundations to support any vascular plants. Such a site is called a *salt barren* or sometimes a *salt pan* or *panne*. Salt barrens may be small or large and are always conspicuous because of the contrast of their bright, white sand with surrounding vegetated terrain. Salt barrens may form a nearly continuous band that parallels the coast and that is manifestly evident in aerial photographs. Such a band separates "low marsh," which occurs seaward of the barrens, from "high marsh," which occurs landward of them.

The high marsh zone is generally less frequently inundated by tides than are salt flats and barrens. The proximity of upland forest subjects the high marsh to freshwater surface runoff and groundwater base flow. This runoff increases soil moisture and reduces soil salinity that would otherwise concentrate from evaporation between infrequent tidal inundations. With proximity to the upland forest, salinity decreases until the only sources of salt are aerosols and infrequent tropical storm surges.

If the gradient is low, the seaward portion of the upper marsh is merely an extension of the *Juncus* zone, sometimes called "high *Juncus* marsh." Other species appear as elevation increases, and these species generally replace *Juncus* near the upland forest edge (Figure 4.6). If the gradient is relatively steep, a

Figure 4.6 High marsh in Taylor County, dominated by *Juncus roemerianus* in the foreground and sawgrass (*Cladium jamaicense*) nearer the trees.

defined *Juncus* zone may be absent. Common species of the high marsh zone include *Agalinus maritima, Andropogon glomeratus, Chloris [Eustachys] petraea, Eleocharis cellulosa, Elyonuris tripsacoides, Fimbristylis castanea, Muhlenbergia capillaris, Panicum virgatum, Paspalum distichum [vaginatum], Pluchea odorata, Sabatia stellaris, Scirpus robustus, Solidago sempervirens, Spartina bakeri* (sand cordgrass), *Spartina spartinae, Sporobolus virginicus,* and those species already listed for salt flats, among others. More salt barrens may be interspersed near the interior edge of this high marsh. At the edge of the upland forest, herbaceous marsh species become intermixed with shrubs, such as *Baccharis angustifolia, B. glomeruliflora, B. halimifolia, Ilex vomitoria, Lycium carolinianum,* and *Myrica cerifera.*

The upland forest adjacent to a salt marsh is commonly a pine-palmetto flatwoods (*Pinus elliottii, Serenoa repens*), a cabbage palm thicket (*Sabal palmetto*), or a hydric hammock dominated by live oak (*Quercus virginiana*), cabbage palm, and red cedar (*Juniperus silicicola*). This forest is limited to those species that tolerate salt in the form of aerosols and that withstand hurricane-force winds (Craighead and Gilbert, 1962). Saltwater inundation during

Figure 4.7 *Juncus* marsh and pine island inundated by a storm tide during a hurricane in 1972, Wakulla County. (Photo by R. J. Bridges, courtesy of St. Marks National Wildlife Refuge.)

hurricanes (Figure 4.7) is diluted by heavy rains and runoff, which protects coastal forests from root kill by residual salinity. Road fills or other structures may impede runoff after a tidal surge, and the residual salt kills forest trees, as happened after Hurricane Hugo in South Carolina (William H. Conner, personal communication).

Zonation in salt marshes was documented by Kurz and Wagner (1957) at selected locations throughout Florida. At St. Josephs Bay in Gulf County, they determined that all vegetational zones from *Spartina alterniflora* through the transitional shrub zone at the upland edge may occupy an elevational gradient of only 0.93 m. They noted that a rise in elevation from a *Juncus* zone through a salt flat and into a salt barren may measure only a few centimeters. This slight difference in elevation, however, has profound effects on the salinity and soil moisture regimes that determine plant distribution (Figure 4.8).

Along portions of the Gulf coast that experience low wave energy, a beach resembling a levee along an estuarine river may complicate marsh zonation. *S. alterniflora* is generally common just seaward of such a beach. A barren-like

Figure 4.8 A low, shrubby berm separates the white sandy beach along the Gulf from the marsh to the left. A tidal stream interrupts the berm (top left). A narrow sandy barren and salt flat border the inner edge of the berm before yielding to *Juncus* marsh (lower left).

Figure 4.9 Berm (Taylor County) similar to that shown in Figure 4.8. *Spartina patens* occupies the foreground. Shrubs on higher ground are mostly *Lycium carolinianum.*

strip of saline sand terminates the *Spartina* zone, in which sparse plant growth may develop with such species as *Atriplex pentandra, Cakile constricta, Chenopodium berlandieri, Sesuvium portulacastrum,* and *Suaeda linearis. Spartina patens* may occur just above the barren, reminiscent of high marsh. Shrubs, such as species of *Iva, Baccharis,* and *Lycium,* replace *S. patens* at a higher elevation (Figure 4.9). The crest of the beach, if it reaches a height exceeding 1 m or so above mean sea level, may contain a strip of coastal hammock consisting of red cedar, cabbage palm, and live oak. Proceeding down the landward side of the beach, shrubs reappear and then herbaceous zones, but not necessarily of the same species as on the seaward side. For example, *S. spartinae* and *Sporobolus virginicus* may dominate the landward side.

Although vascular plants are conspicuous, the soil surfaces of salt marshes, including salt barrens, are densely covered with algae, mainly diatoms and blue-green algae (see Stout [1984] and Montague and Wiegert [1990] for discussions on the algal flora). Sometimes algal density is sufficient to give a greenish cast to the soil surface. Rates of turnover in these algae are quite high, and much of the net primary production in salt marshes is attributed to the algae

(Pomeroy, 1959). The ecological role of marsh algae was partially elucidated in a dissertation by B.W. Ribelin and summarized by Clewell (1979; also see Ribelin and Collier, 1979). As *Juncus roemerianus* dies back from the leaf tip, pieces fall to the ground. These are anchored to the substrate within 5 to 7 days by gelatinous or mucilaginous secretions of diatoms and filamentous algae, which entirely cover the pieces in 2 or 3 weeks. The pieces decompose, presumably releasing nutrients that are absorbed by the algae. The detrital aggregates resulting from decomposition comprise more than 98% of the suspended particulates that are lifted from the marsh floor by tides, and they form an oily slick as they are transported to coastal waters.

Growth of Spartina alterniflora

Spartina growth is commonly quite variable from place to place within a salt marsh. Plant height ranges from 3 to 25 dm (Tanner and Dodd, 1985; Seneca et al., 1985). "Tall" plants frequently grow along coastal shorelines and tidal creeks (Figure 4.10), and "short" plants, which are less than 7 dm tall, generally grow further inland and remote from tidal creeks. At any given site within a

Figure 4.10 Tidal creek lined by *Spartina alterniflora* in Dixie County. In the background, *Juncus* marsh is seen towards the left and a salt flat and barren towards the right.

marsh, local populations generally consist of plants of uniform height. The distinctive height forms of *Spartina* have fascinated tidal marsh ecologists and have stimulated considerable effort in trying to elicit reasons for the differences in stature. Among the explanations proposed have been differences in genetic composition (ecotypes), plant age, latitude, salinity, available nitrogen, soil mineral content, toxic sulfide concentration, iron content, drainage of soil water, and the subsidy of energy from tidal flux (Stalter and Batson, 1969; DeLaune et al., 1979; King et al., 1982; Gallagher et al., 1988).

Ecotypic differentiation was suggested by Stalter (1973), based on reciprocal transplants that retained their initial heights after 23 months. Valiela et al. (1978) cited several other studies, however, which suggested that differences were ecophenic (without a genetic basis). Tanner and Dodd (1985) agreed with the ecophene hypothesis after measuring plants of both height forms that had been transplanted onto a mud flat 2 years previously. Seneca et al. (1985) obtained evidence for ecotypic differentiation after 17 months in a reciprocal transplant study, but after 5 years, the initial differences in height had disappeared, and the ecophene hypothesis was ultimately sustained. To the contrary, Gallagher et al. (1988) transplanted tall and short forms to a common garden and obtained evidence for ecotypic differentiation after 9 years. At this point, we can state that (1) *Spartina* plants are phenotypically plastic, (2) their plasticity is not immediately expressed upon transplanting, and (3) there is some evidence for localized ecotypic differentiation.

The possibility that differences in soil nutrients substantially affect *Spartina* growth has attracted considerable attention. DeLaune et al. (1979) discovered that *Spartina* responded positively in terms of growth to each of several soil nutrients (extractable P, K, Mg, Ca, Na, total N) when nutrient values were expressed in terms of soil volume. Nutrients are generally more abundant near tidal streams (tall *Spartina* habitat) than at interior or inland areas remote from streams (short *Spartina* habitat). The tidal flux presumably keeps streamside habitats supplied with nutrient-rich sediments. Correspondingly, plant tissue content is richer in nutrients for *Spartina* at streamside than it is inland (Buresh et al., 1980; Gallagher et al., 1980). Drainage of soils is more rapid at streamside (Wiegert et al., 1983); therefore, *Spartina* roots are able to grow more deeply and exploit a larger volume of soil for nutrients than would be possible inland (Buresh et al., 1980).

Of particular interest have been the results of nitrogen fertilization. Valiela et al. (1978) recorded an increase in height to short *Spartina* 3 to 4 years after the amendment of NPK fertilizer. Mendelssohn (1979) reported a similar positive response after short *Spartina* was fertilized with nitrate and more so with ammonium; however, tall *Spartina* did not respond to fertilization. Therefore, the short form is apparently a product of nitrogen deficiency. Buresh et al.

(1980) and Cavalieri (1983) also recorded increases in height following ammonium fertilization. In spite of these consistent results, soil nitrogen, especially ammonium, is sometimes abundant where short *Spartina* is prevalent (Buresh et al., 1980). This unexpected finding was explained by Hanson (1983), who discovered that most tidal marsh nitrogen was in the form of refractory (unavailable) organic nitrogen. Hanson's explanation gained credence by the fact that growth of *Spartina* negatively correlated with percent organic matter content in the soil (the source of refractory nitrogen) and also with an increasing C/N ratio (DeLaune et al., 1979). Hanson (1983) presented evidence that nitrogen fixation within the marsh was largely responsible for resupplying that critical nutrient.

Dacey and Howes (1984) noted that fertilization with nitrogen increased production, which, in turn, increased transpiration, lowering the water table and thereby improving aeration and increasing the volume of soil exploitable for nutrients by roots of *Spartina*. de la Cruz et al. (1989) presented experimental evidence for that sequence of events by clipping *Spartina* from plots that were remote from the nearest creek. They measured a higher water table beneath those plots than beneath unclipped control plots, where transpiration was occurring.

In contrast to nitrogen, fertilization with phosphorus elicited no response in growth (Buresh et al., 1980). DeLaune and Pezeshke (1988) determined, however, that phosphorus level in plant tissue was highly correlated with *Spartina* productivity. They suggested that phosphorus may be limiting at the start of the growing season and that other factors may control growth later. They measured nutrient levels (N, P, K, Na, Ca, Fe, Mn, Mg, Cu, Zn) in both the soil and in plant tissues. They concluded that nitrogen alone may not account for all variation in *Spartina* growth. Tissue concentrations of several nutrients were significantly associated with yield and height, and nutrient concentrations in the soil were commonly correlated with tissue concentrations. They warned that these correlations may not be directly related to growth.

Salinity has been investigated as a cause of differences in *Spartina* growth. Although *Spartina* occupies sites with a wide range of soil salinities (Stalter, 1973), researchers who made field measurements generally dismissed salinity as a factor contributing to differences in growth (Buresh et al., 1980; Mendelssohn and McKee, 1988). To the contrary, during the course of hydroponic studies, Cavalieri (1983) determined that shoot growth was reduced in response to increasing salinities. Cavalieri (1983) also obtained enhanced shoot growth by increasing the amount of nitrogen. With increasing salinity, he detected accumulations of free proline and glycinebetaine, which occur in response to salinity stress and function as cytoplasmic osmoticums. Both of these osmoregulators have a high (12%) nitrogen content. Their production may deplete available nitrogen, which is generally in short supply in salt marsh soils. The allocation of N for osmoregulation rather than for growth helps explain the occurrence of the short growth form of *Spartina*. This explanation was supported by the observation that proline accumulation began at relatively low salinities only if nitrogen was readily available.

In addition to nutrient availability and salinity, waterlogging of the soil causes two additional deterrents to growth in height and productivity: anaerobiosis in *Spartina* roots and the accumulation of free sulfides in the soil, which are toxic to those roots. Short forms of *Spartina* commonly occur on soils that experience prolonged waterlogging. The downward conveyance of oxygen from the shoot through aerenchymatous (spongy) tissue allows aerobic respiration in roots, even in moderately reduced soils. However, continuously flooded soils are highly reduced, and plants respire anaerobically with correspondingly depressed rates of growth. Mendelssohn et al. (1981) performed transect studies from tall *Spartina* marsh to short *Spartina* marsh. With proximity to short *Spartina* marsh, soil redox values (Eh) became increasingly positive, indicating the reduced conditions that favor anaerobiosis and sulfide concentration. They estimated anaerobiosis by measuring alcohol dehydrogenase (ADH) activity and concentrations of ethanol and malate in root systems. ADH is an enzyme in anaerobic ethanol production, and its activity is an index of root oxygen deficiency. Malate is a nontoxic end product of anaerobic respiration, and its accumulation is another index of oxygen deficiency. Both ADH and malate concentrations increased as Eh values increased. Ethanol diffused easily from *Spartina* roots and probably did not contribute to root toxicity.

The work of Mendelssohn et al. (1981) was performed in the extremely flat, expansive tidal marshes of Louisiana. Their transects began in marsh supporting tall *Spartina* near tidal creeks and extended inland away from the creeks into short *Spartina* marsh, where waterlogging (also organic matter content) increased. Florida's tidal marshes generally exhibit more topographic relief than those in Louisiana (Coultas and Weber, 1980), and therefore waterlogging is typically more prolonged in proximity to, rather than remote from, Florida's tidal creeks. Nonetheless, the ecophysiological principles elaborated from studies in Louisiana should apply universally, wherever specific site conditions are comparable.

DeLaune et al. (1979) reported that high sulfide concentrations of several hundred micrograms per gram of soil were not uncommon. Soluble species of sulfides include H_2S, $HS-$, and S_2- (Pezeshki et al., 1988). Sulfides inhibit active transport of N into roots, thereby impairing nitrogen metabolism in the plant. DeLaune et al. (1979) found that net photosynthetic rates declined significantly when H_2S concentrations reached a threshold of 34 mg/m^2. Some investigators, including King et al. (1982), feel that sulfide is the major limiting factor to *Spartina* growth. They noted that sulfide is produced at similar rates throughout a marsh, but that soluble sulfide is kept at tolerable levels by its combination with soluble iron to form pyrite. Water movement through the soil mixes sulfides with ionic iron, which combines with free sulfides to form nontoxic compounds, including pyrite. In waterlogged sites, however, water movement is retarded, and ionic iron is correspondingly scarce, thereby allowing the accumulation of sulfides to toxic levels. DeLaune et al. (1984) agreed that *Spartina* productivity was mainly controlled by the accumulation of free sulfide. They discounted, however, the effect of soil redox potential in the soil as a controlling factor, because increased ADH activity, which tracked the decreasing soil Eh, did not reduce growth. Furthermore, plant growth, as estimated from the uptake of CO_2 and N, was not affected by soil redox level or by anaerobic root respiration.

Mendelssohn and McKee (1988) also agreed that sulfide toxicity, in concert with extended periods of anaerobic metabolism in roots, appeared to be the major factor causing reduced growth. Their conclusion was based on reciprocal transplanting of streamside and inland plants of *Spartina*. Upon transplanting to an interior site, streamside plants caused a rapid decrease in Eh and increases in root ADH activity and in concentrations of sulfide and ammonium. After 1 year, standing crops of these plants were significantly reduced. In contrast, plants transplanted from the interior site to streamside displayed an increase in standing crop. Working in Louisiana, Pezeshki and DeLaune (1988) recorded a reduction in photosynthetic activity with distance from streamside, attributable to waterlogging and its effects of sulfides, nutrient uptake, and anaerobiosis.

The preceding discussion illustrates the complex ecophysiology determining plant height for one species—a species that grows monotypically under narrowly restricted site conditions, which are closely regulated by daily tidal flux. One can only wonder how much greater the complexity in multispecific plant communities may be, where site conditions are more variable.

Autecology of Juncus roemerianus

The aerial stem of *Juncus roemerianus* grows from an elongated rhizome and produces from one to four (rarely six) nearly basal, distichous leaves. The shoot continues growth as an erect, fertile scape. The terminal inflorescence is subtended by an erect, sharp-pointed bract that is scarcely distinguishable from the scape. This bract gives the plant its vernacular name, needlerush. Shoots of *Juncus* are produced continuously throughout the year, albeit at different rates according to the season. In contrast, in most other common salt marsh plants, including *Spartina alterniflora*, shoot growth largely occurs simultaneously in the spring, barring a mild winter, so that the shoots are of nearly uniform age.

Eleuterius and Caldwell (1981a) measured shoots in *Juncus*. They noted that plant height varied from place to place, as it did in *Spartina*. "Short" forms were as small as 3 dm. They produced the shortest leaves, the greatest density of shoots, and the fewest leaves per shoot. They exhibited the slowest rate of leaf growth and the shortest-lived leaves. Short forms were growing on sandy soils, where the salinity of the soil water was hyperhaline (60 to 300‰). "Medium" height plants were approximately 10 dm tall and grew in sandy clay soil, in which salinity of the soil water was 5 to 20‰. "Tall" plants averaged 14 dm tall (17 dm maximum) and grew directly on peat, in which the soil water salinity was 0 to 6‰.

Eleuterius and Caldwell (1981a) determined that a leaf grows for 4 to 9 months and that growth ceases upon the initiation of the next leaf on the same stem. Growth is faster in summer than in winter. Upon cessation of growth, the leaf slowly dies back from the apex. The average leaf lives 14 months and may live as long as 22 months. Dead leaves persist on the plant for at least several years, if they are not broken by wind action or swept away by strong tidal flows. A stand of *Juncus* is dark green if the plants contain mostly living leaves and becomes increasingly gray as more and more leaves die back and persist. In a population of "tall" plants, half of the aerial standing crop consisted of dead leaves. Persisting dead leaves cast shadows that reduce the amount of light reaching the soil surface, thereby making the habitat less hospitable for the establishment of competitive species.

Individual shoots may live 4 or more years. Shoot longevity is reduced to approximately 22 months in populations of "short" plants. Shoot turnover is much more rapid in populations of "tall" plants than in populations of "short" plants. Therefore, productivity in populations of tall plants may exceed productivity in populations of short plants, in spite of the greater plant density in populations of short plants.

Juncus exhibits very wide environmental tolerances relative to all other tidal marsh angiosperms (Eleuterius, 1984). Eleuterius noted that plants of *Juncus* grow on sand, silty clay, or peat. The soil may be unconsolidated and watery or densely compacted. Nutrient concentrations vary widely: P = 3 to 122 ppm, K = 33 to 245 ppm, and N = 0 to 3 ppm. pH ranges from 4.5 to 7 and did not correlate with growth. Eleuterius and Caldwell (1981b) grew *Juncus* in various nutrient solutions. They discovered that plants became stunted if there were mineral deficiencies of Mg, P, S, or K, all of which are generally abundant in highly organic salt marsh soils. There is little available Ca, N, and Fe in tidal marsh soils. If any of these were deficient, stunting was less distinct than for Mg, P, S, and K. Micronutrients (B, Mn, Zn, Cu) occur in very low concentration in tidal marsh soils. Nonetheless, if these nutrients were deficient, growth was impaired, more so than was typical for terrestrial plants. In contrast to Eleuterius and Caldwell (1981b), Coultas and Gross (1977) reported high levels of available calcium in Florida *Juncus* marsh soils.

Kruczynski et al. (1978) noted that plants had shorter leaves and higher density with increasing salinity. Eleuterius (1984) designated salinity as the single most important factor affecting *Juncus* occurrence and growth. Salinity of the soil water may reach 70‰ and may vary at different depths in the soil profile by as much as 20‰. *Juncus* plants cannot tolerate salinity that is in excess of 30‰ indefinitely, but they will survive brief exposures as high as 360‰. Fibrous (absorbing) roots are not produced near the surface in hyperhaline soils. In hyperhaline marshes, Eleuterius (1984) claimed that *Juncus* produces specialized roots that penetrate 6 to 8 ft deep, allowing plants to draw less saline water. This claim has not been reported by other workers and begs verification, particularly since Coultas and Hsieh (personal communication) have not observed living *Juncus* roots below a depth of 20 cm.

Salinity may not be the only important factor affecting *Juncus* growth. Coultas and Weber (1980) attempted reciprocal transplanting of *Juncus* between low and high marsh. After 21 months, they described differences in plant height, thickness, and growth form that were attributable to genetic control.

Juncus populations sometimes extend 15 to 25 km inland along river estuaries. Eleuterius (1984) found that *Juncus* plants growing at the inland limit of these low-salinity populations have poorly developed rhizomes. These rhizomes reveal damage from soil animals which remove protective scales and bore into rhizomes. Eleuterius speculated that these animals are relegated to the edge of the *Juncus* zone because of their intolerance of more saline soils. Where this fauna exists, *Juncus* plants are weakened and are subject to competition from less salt-tolerant plants, such as species of *Cladium, Nyssa*, and *Taxodium*. The greatest upstream penetration of salt water occurs in autumn and lasts only 3 to 6 weeks, but it is sufficient to kill or reduce the harmful fauna or competitive plants, thereby defining the landward extent of *Juncus* zonation. With increasing distance upstream, *Juncus* leaves lengthen and their shoot density decreases.

Eleuterius (1975, 1984) noted that extant populations of *Juncus* nearly always lack seedlings, even though *Juncus* populations produce approximately 200,000 seeds per hectare annually. Of these, an average of 60% germinate. Germination requires light and is inhibited wherever the salinity exceeds 15‰. Seedlings generally become established on bare sand. Upon reaching maturity, plants from these seedlings spread vigorously by rhizomes. They are persistent and highly competitive in relation to any of 50 species of tidal marsh plants with which they may grow. Tolerance to salinity is apparently controlled genetically, because seedlings from some sources exhibited higher tolerance to salinity than did those from others. Cavalieri and Huang (1979) observed that proline accumulation was of considerable importance to *Juncus* plants in regard to osmoregulation.

Organic matter accumulates beneath stands of *Juncus* because of its slow rate of decomposition of belowground tissue (16% annually) relative to that of other marsh plants, e.g., *Spartina cynosuroides* (27% annually) (Hackney and de la Cruz, 1980). Decomposition of aboveground parts is correspondingly slow. White et al. (1979) estimated standing crop loss to be 20% in 12 months for *Juncus*, compared to 100% in 7 months for *S. alterniflora*. Hackney and de la Cruz (1980) attributed the slow rate of decomposition to anaerobic conditions, the inability of *Juncus* plants to oxygenate their rhizospheres, and the lack of macerating soil fauna to shred dead roots and rhizomes. The greatest vertical zone of organic matter lay at a depth of 10 to 20 cm, where belowground production was high and decomposition low. Nonetheless, living *Juncus* belowground tissues were detected at a depth of 25 cm. de la Cruz et al. (1989) measured the Eh level beneath a stand of *Juncus* and determined that it was intermediate between that of a *Spartina* short and a *Spartina* tall marsh.

Other Salt Marsh Vegetation

Spartina patens occupies aerated soils with relatively low salinities. Burdick and Mendelssohn (1987) reported that *Spartina patens* tolerated short-term soil waterlogging by increasing its root porosity. This aerenchymatous growth, however, was inadequate to prevent anaerobiosis, as detected by ADH formation. Pezeshki et al. (1987) reported that *S. patens* generally occupied sites with a salinity from 1 to 10‰. Plants survived in a controlled environment, in which the salinity was raised to as much as 22‰, which caused a 43% reduction in the rate of photosynthesis. Salt excretion later occurred from both leaf surfaces.

Sporobolus virginicus grows in salt flats and high marsh. Gallagher (1979) demonstrated that *Sporobolus* tolerated interstitial soil salinities at a salt flat, which ranged from 30 to 185‰ over the course of 1 year. He recorded only a small correlation between biomass and soil salinity. This correlation was more pronounced for plants growing in sand and less so in loam soils. As salinity rose, decreases were recorded in culm length, leaf length, shoot density, internode length of the rhizome, and the root:shoot ratio (Gallagher, 1979; Donovan and Gallagher, 1985). Seedlings that were already 5 cm tall survived indefinitely after being placed in a solution with 80‰ salinity; however, growth was virtually nil. Best growth occurred at zero salinity. Salinity influenced nutrient concentrations. Potassium decreased in plants that were exposed to increasing salinity, perhaps due to competition with sodium ions during uptake or due to Na replacing K in metabolic activities. Manganese concentration increased with increasing salinity, and iron decreased. When ammonium nitrate was added to marsh soil with high salinity, nitrogen concentration in plant tissue increased, but biomass did not. *Sporobolus* adjusts to waterlogged conditions and anoxia by accelerating its metabolic rate to counteract the effects of inefficient anaerobiosis and by diversifying the end products of glycolysis. If the only end product were ethanol, not all of it could diffuse from the roots without causing toxicity. Therefore, malate, which is nontoxic, is also produced as an end product (Donovan and Gallagher, 1985).

Salicornia bigelovii is a succulent of hyperhaline salt flats. In response to highly saline soils, growth is slow and its seedlings retain their cotyledons as the primary photosynthetic organs for 25 to 30 days following germination. Its seedlings are inhibited when grown in fresh water but not in water with a salinity of 10 or 30‰. Glycinebetaine and soluble amino nitrogen comprised most of the tissue osmolality, and glycinebetaine level remained high, even when the plant was grown in fresh water (Stumpe et al., 1986).

Distichlis spicata is another common species of salt flat. Parrondo et al. (1978) demonstrated that *Distichlis* tolerated saline environments better than *Spartina alterniflora*. They noted, however, that *Spartina* dominated the most saline sites in a marsh, suggesting that salinity was not an important factor in determining the zonation of these two grasses. Kemp and Cunningham (1981) showed that growth of *Distichlis* was reduced significantly in low light intensities but not at high light intensities. Perhaps shade from the taller *Spartina* is key to zonation of these species.

Tidal Wrack

Tidal wrack consists of rafts of detrital materials within marshes, much of it dead shoots of *Juncus* and *Spartina*. These rafts develop in swirling eddies during storm tides and become stranded as water recedes. The raft may measure tens of meters in diameter and may be several decimeters thick. Vegetation trapped beneath a raft is killed. The wrack takes more than 1 year to decompose and may be lifted by a later storm and deposited elsewhere, smothering more vegetation. The trend is for wrack in the lower marsh to be transported to the upper marsh. Revegetation may not occur for 2 or more years, and some sites may become salt barrens, causing anomalies in vegetational zonation. Reidenbaugh and Banta (1980) reported an 11% loss of extant vegetation in one marsh by wrack.

Brackish Marshes

Brackish marshes are commonly associated with river deltas and other habitats where freshwater discharge regularly and substantially dilutes the saline waters of incoming tides. Salinity decreases upstream from the coast. Salinity may also decrease laterally on the floodplain with distance from the riverbank, depending upon the relative influences of tidal inundation from the coast, overbank flooding of the river, and other sources of runoff. Plant zonation generally reflects salinity. Relative to salt marshes, very little has been published on brackish marsh vegetation, its zonation and community structure, and the ecophysiology of its species. Most studies have taken place along the northern Gulf coast, west of Florida.

Some investigators (Chabreck, 1972; Eleuterius, 1972), have recognized two kinds of brackish marshes. One kind is more saline, occurs nearer the coast, and has few plant species. The other kind is less saline, occurs further inland, and has numerous plant species. O'Neil and Mettee (1982) called the more saline kind Brackish I Marsh and the less saline kind Brackish II Marsh. Brackish I marshes contained mostly *Juncus roemerianus*, some *Spartina cynosuroides* (big cordgrass) and *Scirpus americanus* (bulrush), and little else. *Spartina cynosuroides*, although conspicuous for its height, is not particularly common in Gulf coastal Florida. Brackish II marshes included a little *J. roemerianus* and a wealth of additional species, including *Cladium jamaicense* (sawgrass). Chabreck (1972) measured the salinity values from about 1 to 7‰ in stands of *Cladium*. Steward and Ornes (1975) demonstrated that *Cladium* is generally much more tolerant of low nutrient levels in the soil, including phosphorous, than are its associated species. Low nutrient requirements help explain the considerable abundance of *Cladium* on the sandy, infertile substrates common to most Florida tidal rivers.

Perhaps the most intensive study of brackish marshes in Florida was that of Thompson (1977), who described the vegetation of tidal marshes along the St. Marks River and its tributary, the Wakulla River (Figure 4.11), in Wakulla County. She sampled marsh vegetation quantitatively at river miles 5, 8, 10, and 12 (distance from coast) at six stations each, three on either side of the river. She also sampled riverbank vegetation at mile 15, which represented tidal freshwater marsh. The most widespread dominant species was *C. jamaicense*. *Typha domingensis* and *Phragmites australis* were abundant in patches. *Sagittaria lancifolia, Sagittaria subulata,* and *Spartina cynosuroides* were particularly common on riverbanks. Seaward of mile 5, the Brackish I marshes were dominated by *Juncus roemerianus*. River mile 5 marshes were probably equivalent to the interior limit of Brackish I Marsh of O'Neil and Mettee (1982). Common species in terms of plant density at mile 5 were *Cladium jamaicense, Crinum americanum, Distichlis spicata, Eleocharis cellulosa, Fimbristylis castanea, Juncus roemerianus, Lilaeopsis chinensis, Limonium carolinianum, Lythrum lineare, Mikania scandens, Sagittaria subulata, Spartina alterniflora, Spartina spartinae,* and *Sporobolus virginicus*. Common species in the probable equivalent of Brackish II Marsh were *Cladium jamaicense, Crinum americanum, Juncus roemerianus, Ludwigia repens, Lythrum lineare, Mikania scandens, Pontederia cordata, Sagittaria lancifolia, Sagittaria subulata, Saururus*

Figure 4.11 Brackish marsh along the Wakulla River at low tide. *Sagittaria subulata* covers the tidal flat in the foreground. The taller grassy vegetation is a mixture of *Cladium jamaicense* and *Juncus roemerianus.*

cernuus, and *Scirpus pungens* [*S. olneyi*]. The total flora consisted of 74 species, of which several were limited to tidal freshwater marshes.

Phragmites and *Typha* may be indicators of nutrient-rich substrates. Plants of both taxa have been shown to respond vigorously to fertilization (Ulrich and Burton, 1985; Grace, 1988). The reduced salinity at the mouths of larger tidal rivers favors seedling establishment in *Typha* (Beare and Zedler, 1987).

In southwestern Florida from Tampa Bay to Port Charlotte, *Juncus roemerianus* is the predominant brackish marsh species (Figure 4.12). There are several differences between marshes from this region and those further north. One is the fringe of mangroves on the banks of the more saline streams. Mangroves disappear a short distance from the coast, and the riverbanks are commonly lined with a bulrush, *Scirpus validus.* The southwestern brackish marshes are also characterized by occasional patches of leather fern (*Acrosticum danaeifolium*) and sometimes *A. aureum,* both tropical taxa. These marsh systems have recently attracted the attention of ecologists (Estevez et al., 1990).

Figure 4.12 Brackish marsh on a tidal island near the mouth of the Myakka River, Sarasota County. A tall band of *Scirpus validus* lines the shore in front of a live oak-cabbage palm hammock in the lower right portion of the photo. *Juncus* marsh covers most of the island.

Brackish marshes are known to burn naturally, ignited by lightning or by spontaneous combustion (Hackney and de la Cruz, 1981). Fires may also escape into marshes from adjacent pine flatwoods. *Juncus roemerianus* burns vigorously with a characteristic crackling sound. Hackney and de la Cruz (1981) measured an increase of about 50% in net primary production in the regrowth from rhizomes following fires in *J. roemerianus* and *Spartina cynosuroides*. Intentionally set, annual fires caused a decline in *Juncus* and a corresponding increase in other species, such as *Panicum virgatum*. McAtee et al. (1979) reported that *S. spartinae* doubled in biomass the year following a fire and increased its production of flowering culms.

Construction of fill roads across salt marshes has resulted in the reduction of tidal inundations and the replacement of salt marsh flora with brackish marsh flora (Clewell et al., 1976). Species replacement may begin within 4 years after construction, and complete replacement by *Cladium* and other taxa occurs within 40 years (Figure 4.13).

Figure 4.13 High *Juncus* marsh being colonized by tall clumps of *Spartina bakeri* in Taylor County. The marsh became isolated from tidal exchange by the construction of a fill road nearer the Gulf about a decade earlier. Figure 4.3 was taken in similar habitat nearby, where the fill road was adequately culverted to allow passage of tidal flows.

Tidal Freshwater Marshes

Tidal freshwater marshes are restricted to wetlands along tidal rivers. They are affected hydrologically by tidal damming but not by salinity, except rarely and briefly during tropical storms. Tides guarantee a perennially high water table, and river flow maintains a constant supply of fresh water. The permanently waterlogged soils preclude deep rooting by trees and make trees that do establish susceptible to toppling in hurricanes. Occasional fires in tidal marshes further discourage tree establishment, and herbaceous species and a few species of shrubs dominate these sites. Odum et al. (1984) attributed species zonation in tidal freshwater marshes to a combination of physical variables and ecological processes, mainly elevation relative to water level, interspecific competition, and herbivory. Substrate factors were considered less important. Flows in streams and from tides were thought to modify seed beds, exclude some species, and encourage others. Physical descriptions of Florida's tidal rivers were compiled by McPherson and Hammett (1990). The flora includes strictly freshwater species and usually some brackish marsh species. Thompson (1977) listed the following common species in terms of plant density in tidal freshwater marsh at mile 15 along the St. Marks River: *Juncus polycephalus*, *Physostegia purpurea*, *Polygonum punctatum*, *Pontederia cordata*, *Sagittaria subulata*, and *Zizania aquatica*.

Broad ecotones develop up river on floodplains between tidal freshwater marshes and forest (Clewell, 1990). *Cladium jamaicense* commonly dominates these sites, although numerous other marsh and forest herbs are interspersed. Trees are scattered about, commonly *Taxodium distichum* (bald cypress) or *Nyssa biflora* (swamp tupelo). Tree density, height, and tree species richness all increase at the expense of herbaceous species with distance upstream, until forested conditions are apparent. Clewell (1989) described the ecotone, termed a tidal woods, along the Choctawhatchee River. *C. jamaicense* dominated the ground cover, which also included *Dichromena colorata*, *Hydrocotyle verticillata*, *Hymenocallis rotata*, *Ipomoea sagittata*, *Mikania scandens*, *Osmunda regalis*, *Pluchea rosea*, *Sagittaria lancifolia*, and *Solidago sempervirens*. A few shrubs were scattered about (*Baccharis* spp., *Ilex vomitoria*, *Myrica cerifera*, *Sesbania punicea*). Trees were young, widely spaced, and mostly 3 to 11 m tall. They included *Acer rubrum*, *Fraxinus caroliniana*, *Juniperus silicicola*, *Magnolia virginiana*, *Nyssa biflora*, *Persea palustris*, *Pinus elliottii*, and *Taxodium ascendens*. The hardwoods were represented by coppice sprouts following the most recent fire. Frequent fires of perhaps every 5 to 10 years would be necessary to maintain this ecotone. Further upriver, the ultimate remnants of tidal freshwater marsh vegetation were beds of *Zizaniopsis miliacea* (cutgrass), sometimes flanked by *Nuphar luteum* (spatterdock), which grew in shallows near the forested riverbank.

The tidal freshwater marsh may extend almost or to the coast along the banks of those rivers with large discharges. The distinction is blurred between Brackish II and tidal freshwater marsh on the delta of the Apalachicola River (Clewell, unpublished). Much of this delta is covered by *Cladium jamaicense*. The tip of the delta contains *C. jamaicense*, *Phragmites australis*, *Pontederia cordata*, *Scirpus californicus*, *Sesbania punicia*, and *Typha latifolia*. The banks of the river channels within the delta are flanked by *Juncus roemerianus*, *Nuphar luteum*, *P. cordata*, and *T. latifolia*. A low levee behind these banks contains shrubs, especially *Cephalanthus occidentalis*, *Leitneria floridana*, and *Sesbania punicia*. The vast *Cladium* marshes that occupy the interior of the delta are in places monotypic, in other places interrupted by patches of *J. roemerianus*, and elsewhere mixed with other species, notably *Cicuta mexicana*, *Ipomoea sagittata*,

Rumex verticillatus, *Sagittaria lancifolia*, *Spartina patens*, and *Teucrium canadense*. With the exception of a brushy zone of *Salix nigra* (black willow), there is no transition between tidal freshwater marsh and forest. Additional descriptions of tidal marshes in Apalachicola Bay were presented by Clewell (1986) and Edmiston and Tuck (1987).

Paleoecology of Tidal Marsh Vegetation

Kurz and Wagner (1957) described two striking aspects of Gulf Florida's tidal marshes. One was the abundance of tree stumps in shallow estuaries just offshore. The other was the presence of pine islands (patches of upland slash pine and saw palmetto flatwoods) abruptly elevated above, and surrounded by, salt marshes. They ascribed these occurrences as evidence for a recent rise in sea level. Tanner et al. (1989) presented strong evidence for two decreases of approximately 1.0 m below and two increases of approximately 1.0 m above existing mean sea level during the past 5000 years. The latest change was a rise in sea level that occurred 800 years ago. William L. Tanner (personal communication) suggested that at least some of the rise in sea level along the Gulf coast during this century is better explained in terms of localized redistributions of coastal sediments. Redistribution could be caused at least in part by differential wave energies from place to place along the coast. Each rise or fall in sea level of 1 m during the past 5000 years must have influenced all vegetation in Florida tidal marshes, causing considerable shifting of zones. The boundaries of tidal marshes would also have changed. As sea level rose, some salt marsh became estuary and some coastal forest became tidal marsh. The reverse would have happened as sea level fell.

The possibility of sediment redistribution can be appreciated by viewing an aerial photograph showing the numerous, frequently large, and commonly meandering creeks that head within tidal marshes. These creeks flow bidirectionally, carrying large volumes of tidal flow, in spite of the nearly flat topography. Levees along these creeks are evidence of the sediment load in tidal flows. Although there is opportunity for considerable redistribution of sediments, especially during spring tides, the truly important times of sediment redistribution may occur during hurricanes. Only then would there be sufficient volume of water to overcome the resistance to sediment transport afforded by dense tidal marsh vegetation. If indeed substantial sediment redistribution is occurring between the cyclic fluctuations in eustatic sea level documented by Tanner et al. (1989), then tidal marsh ecosystems are more dynamic than generally realized.

Tidal Marsh Restoration

The technology for tidal marsh restoration has been developing during the past two decades in Florida and elsewhere. The impetus for such work has been to mitigate the loss of marsh from dredging and filling, to vegetate dredge spoils, and to stabilize eroding shorelines. *Spartina alterniflora* has been the preferred species, due to its ecological importance, the rapidity of its establishment, and the relative ease of growing nursery stock. Although many of the attempts at creating *Spartina* marshes have been less than satisfactory, restorationists have learned from their mistakes. Today, *Spartina* marshes can be successfully created routinely, but only with proper planning and design, careful implementation, and remediation as needed.

Crewz and Lewis (1991) inspected 18 Gulf coastal marshes created in peninsular Florida; 5 were considered successful and 6 were partially successful in terms of vegetational establishment. Crewz and Lewis (1991) presented reasons in instances when success was not attained. Factors they identified as critical to success were proper elevation, slope, drainage, substrates, adequate fertilization (generally little or none), adequate site design (e.g., protection from erosive waves), limitation of access (to avoid damage by trespassers), suitability of the nursery stock (e.g., out-planting of nursery stock grown in saline conditions), appropriate planting density, and frequency of monitoring (so as to correct problems promptly).

Crewz and Lewis (1991) stated that nearly all Florida Gulf coastal marsh creation projects have focused on establishing *S. alterniflora*; however, there have been a few attempts to plant other species, such as *Juncus roemerianus, Distichlis spicata*, and *S. patens*. Coultas et al. (1978) noted the establishment of *S. patens, Paspalum vaginatum*, and other marsh species on a spoil island in Wakulla County. Lewis (1982) reported success in transplanting *Juncus* plugs in Manatee County, wherever mechanical grading resulted in suitable elevations. Plugs averaged 12 shoots when planted and 300 shoots after 22 months. Coalescence of neighboring transplants had not yet occurred, due to wide spacing of the transplants and the slow rhizomatous development typical of *Juncus*. Eleuterius and Gill (1982) reported coalescence by *Juncus* in 5 years after planting on 1.2-m centers in Mississippi. Coultas (1980) discovered that *Juncus* can be transplanted in any season, but growth was best for material transplanted in February and March. Establishment was more rapid for those transplants that contained a bud on their rhizomes.

Marsh creation is an active discipline, in which technological development is advancing rapidly. For example, the potential for creating brackish and tidal freshwater marshes has been given a considerable boost by the recent development of methods for creating large-scale sawgrass marshes on mined and reclaimed lands in central Florida. Joel Butler (personal communication, 1990) has directed this effort for Seminole Fertilizer Corporation, and there is no compelling reason why his techniques could not be applied in tidal areas.

References

Beare, P. A. and J. B. Zedler. 1987. Cattail invasion and persistence in a coastal salt marsh: the role of salinity reduction. *Estuaries* 10(2):165–170.

Burdick, D. M. and I. A. Mendelssohn. 1987. Waterlogging responses in dune, swale, and marsh populations of *Spartina patens* under field conditions. *Oecologia* 74: 321–329.

Buresh, R. L., R. D. DeLaune, and W. H. Patrick, Jr. 1980. Nitrogen and phosphorus distribution and utilization by *Spartina alterniflora* in a Louisiana Gulf coast marsh. *Estuaries* 3(2):111–121.

Cavalieri, A. J. 1983. Proline and glycinebetaine accumulation by *Spartina alterniflora* Loisel. in response to NaCl and nitrogen in a controlled environment. *Oecologia* 57:20–24.

Cavalieri, A. J. and A. H. C. Huang. 1979. Evaluation of proline accumulation in the adaptation of diverse species of marsh halophytes to the saline environment. *Am. J. Bot.* 66(3):307–312.

Chabreck, R. H. 1972. *Vegetation, Water and Soil Characteristics of the Louisiana Coastal Region.* Bulletin No. 664. Louisiana State University, Agricultural Experiment Station, Baton Rouge.

Clewell, A. F. 1979. What's known or should be known about upper salt marsh ecology. Pages 33–37 in *Proceedings of the Florida Anti-Mosquito Association, 50th Annual Meeting.*

Clewell, A. F. 1985. *Guide to the Vascular Plants of the Florida Panhandle.* Florida State University Press, Tallahassee. 605 pp.

Clewell, A. F. 1986. *Natural Setting and Vegetation of the Florida Panhandle.* COESAM/ PDEI-86/001. U.S. Army Corps of Engineers, Mobil District; title page dated 1981. 773 pp.

Clewell, A. F. 1989. *Botanical Inventory of the Choctawhatchee River Valley, Florida.* Technical Report. Northwest Florida Water Management District, Havana, Florida. 65 pp.

Clewell, A. F. 1990. Florida rivers: the vegetational mosaic. Pages 47–63 in R. A. Livingston, Ed. *The Rivers of Florida.* Springer-Verlag, New York.

Clewell, A. F., L. F. Gainey, Jr., D. P. Harlos, and E. R. Tobi. 1976. *Biological Effects of Fill-Roads Across Salt Marshes.* Florida Department of Transportation, Tallahassee. 16 pp. plus illustrations.

Coultas, C. L. 1980. Transplanting needlerush (*Juncus roemerianus*). Pages 117–124 in D. P. Cole, Ed. *Proceedings of the Seventh Annual Conference on the Restoration and Creation of Wetlands.* Hillsborough Community College, Tampa, Florida.

Coultas, C. L. and E. R. Gross. 1975. Distribution and properties of some tidal marsh soils of Apalachee Bay, Florida. *Soil Sci. Soc. Am. Proc.* 39(5):914–919.

Coultas, C. L. and O. J. Weber. 1980. Soil characteristics and their relationship to growth of needlerush. *Proc. Soil Sci. Soc. Fla.* 39:73–77.

Coultas, C. L., G. A. Breitenbeck, W. L. Kruczynski, and C. B. Subrahmanyam. 1978. Vegetative stabilization of dredge spoil in North Florida. *J. Soil Water Conserv.* 33(4):183–185.

Craighead, F. C., Sr. and V. C. Gilbert. 1962. The effects of Hurricane Donna on the vegetation of southern Florida. *Quart. J. Fla. Acad. Sci.* 25:1–28.

Crewz, D. W. and R. R. Lewis, III. 1991. *An Evaluation of Historical Attempts to Establish Emergent Vegetation in Marine Wetlands in Florida.* Florida Sea Grant Technical Paper TP-60. Florida Sea Grant College, University of Florida, Gainesville.

Dacey, J. W. H. and B. L. Howes. 1984. Water uptake by roots controls water table movement and sediment oxidation in short *Spartina* marsh. *Science* 224:487–489.

de la Cruz, A., C. T. Hackney, and N. Bhardwaj. 1989. Temporal and spatial patterns of redox potential (Eh) in three tidal marsh communities. *Wetlands* 9(2):181–190.

DeLaune, R. D. and S. R. Pezeshki. 1988. Relationship of mineral nutrients to growth of *Spartina alterniflora* in Louisiana salt marshes. *Northeast Gulf Sci.* 10(1):55–60.

DeLaune, R. D., R. J. Buresh, and W. H. Patrick, Jr. 1979. Relationship of soil properties to standing crop biomass of *Spartina alterniflora* in a Louisiana marsh. *Estuarine Coastal Mar. Sci.* 8:477–487.

DeLaune, R. D., C. J. Smith, and M. D. Tolley. 1984. The effect of sediment redox potential on nitrogen uptake, anaerobic root respiration and growth of *Spartina alterniflora* Loisel. *Aquatic Bot.* 18:223–230.

Donovan, L. A. and J. L. Gallagher. 1985. Morphological responses of a marsh grass, *Sporobolus virginicus* (L.) Kunth, to saline and anaerobic stresses. *Wetlands* 5:1–13.

Edmiston, H. L. and H. Λ. Tuck. 1987. *Resource Inventory of the Apalachicola River and Bay Drainage Basin.* Office of Environmental Services, Florida Game and Fresh Water Fish Commission, Apalachicola.

Eleuterius, L. N. 1972. The marshes of Mississippi. *Castanea* 37:153–168.

Eleuterius, L. N. 1975. The life history of the salt marsh rush, *Juncus roemerianus.* *Bull. Torrey Bot. Club* 102(3):135–140.

Eleuterius, L. N. 1984. Autecology of the black needlerush *Juncus roemerianus. Gulf Res. Rep.* 7(4):339–350.

Eleuterius, L. N. and J. D. Caldwell. 1981a. Growth kinetics and longevity of the salt marsh rush *Juncus roemerianus. Gulf Res. Rep.* 7(1):27–34.

Eleuterius, L. N. and J. D. Caldwell. 1981b. Effect of mineral deficiency on the growth of the salt marsh rush *Juncus roemerianus. Gulf Res. Rep.* 7(1):35–39.

Eleuterius, L. N. and J. D. Caldwell. 1982. Colonizing patterns of tidal marsh plants and vegetational succession on dredge spoil in Mississippi. Pages 58–73 in R. H. Stovall, Ed. *Proceedings of the Eighth Annual Conference on Wetlands Restoration and Creation.* Hillsborough Community College, Tampa, Florida.

Eleuterius, L. N. and J. D. Caldwell. 1985a. Soil characteristics of four *Juncus roemerianus* populations in Mississippi. *Gulf Res. Rep.* 8(1):9–13.

Eleuterius, L. N. and J. D. Caldwell. 1985b. Soil characteristics of *Spartina alterniflora, Spartina patens, Juncus roemerianus, Scirpus olneyi,* and *Distichlis spicata* populations at one locality in Mississippi. *Gulf Res. Rep.* 8(1):27–33.

Eleuterius, L. N. and C. K. Eleuterius. 1979. Tide levels and salt marsh zonation. *Bull. Mar. Sci.* 29(3):394–400.

Eleuterius, L. N. and J. I. Gill, Jr. 1982. Long term observations on seagrass beds and salt marsh established from transplants. Pages 74–86 in R. H. Stovall, Ed. *Proceedings of the 8th Annual Conf. on Wetlands Restoration and Creation.* Hillsborough Community College, Tampa, Florida.

Estevez, E. D., L. K. Dixon, and M. S. Flannery. 1990. West-coastal rivers of peninsular Florida. Pages 187–221 in R. A. Livingston, Ed. *The Rivers of Florida.* Springer-Verlag, New York.

Gallagher, J. L. 1979. Growth and element compositional responses of *Sporobolus virginicus* (L.) Kunth to substrate salinity and nitrogen. *Am. Midl. Nat.* 102(1): 68–75.

Gallagher, J. L., R. J. Reimold, R. A. Linthurst, and W. J. Pfeiffer. 1980. Aerial production, mortality, and mineral accumulation—export dynamics in *Spartina alterniflora* and *Juncus roemerianus* plant stands in a Georgia salt marsh. *Ecology* 61(2):303–312.

Gallagher, J. L., G. F. Somers, D. M. Grant, and D. M. Seliskar. 1988. Persistent differences in two forms of *Spartina alterniflora*: a common garden experiment. *Ecology* 69(4):1005–1008.

Grace, J. B. 1988. The effects of nutrient additions on mixtures of *Typha latifolia* L. and *Typha domingensis* Pers. along a water-depth gradient. *Aquatic Bot.* 31:83–92.

Hackney, C. T. and A. A. de la Cruz. 1978. Changes in interstitial water salinity of a Mississippi tidal marsh. *Estuaries* 1(3):185–188.

Hackney, C. T. and A. A. de la Cruz. 1980. In situ decomposition of roots and rhizomes of two tidal marsh plants. *Ecology* 61(2):226–231.

Hackney, C. T. and A. A. de la Cruz. 1981. Effects of fire on brackish marsh communities: management implications. *Wetlands* 1:75–86.

Hanson, R. B. 1983. Nitrogen fixation activity (acetylene reduction) in the rhizosphere of salt marsh angiosperms. *Bot. Mar.* 26:49–59.

Kemp, P. R. and G. L. Cunningham. 1981. Light, temperature and salinity effects of growth, leaf anatomy and photosynthesis of *Distichlis spicata* (L.) Greene. *Am. J. Bot.* 68(4):507–516.

King, G. M., M. J. Klug, R. G. Wiegert, and A. G. Chalmers. 1982. Relation of soil water movement and sulfide concentration to *Spartina alterniflora* production in a Georgia salt marsh. *Science* 218:61–63.

Kruczynski, W. L., C. B. Subrahmanyan, and S. H. Drake. 1978. Studies on the plant community of a north Florida salt marsh. Part I. Primary production. *Bull. Mar. Sci.* 28:316–334.

Kurz, H. and D. Wagner. 1957. Tidal marshes of the Gulf and Atlantic coasts of North Florida and Charleston, South Carolina. *Fla. St. Univ. Stud.* 24:1–168.

Lewis, R. R., III. 1982. Restoration of a needlerush (*Juncus roemerianus* Scheele) marsh following interstate highway construction. II. Results after 22 months. Pages

69–83 in *Proceedings of the Ninth Annual Conference on Wetlands Restoration and Creation.* Hillsborough Community College, Tampa, Florida.

McAtee, J. W., C. J. Scifres, and D. L. Drawe. 1979. Improvement of gulf cordgrass range with burning or shredding. *J. Range Manage.* 32(4):372–375.

McKee, K. L. and W. H. Patrick, Jr. 1988. The relationship of smooth cordgrass (*Spartina alterniflora*) to tidal datums: a review. *Estuaries* 11(3):143–151.

McPherson, B. F. and K. M. Hammett. 1990. Tidal rivers of Florida. Pages 31–46 in R. A. Livingston, Ed. *The Rivers of Florida.* Springer-Verlag, New York.

Mendelssohn, I. A. 1979. The influence of nitrogen level, form, and application method on the growth response of *Spartina alterniflora* in North Carolina. *Estuaries* 2(2):106–112.

Mendelssohn, I. A. and K. L. McKee. 1988. *Spartina alterniflora* dieback in Louisiana: time-course investigation of soil waterlogging effects. *J. Ecol.* 76:509–521.

Mendelssohn, I. A., K. L. McKee, and W. H. Patrick, Jr. 1981. Oxygen deficiency in *Spartina alterniflora* roots: metabolic adaptation to anoxia. *Science* 214:439–441.

Montague, C. L. and R. G. Wiegert. 1990. Salt marshes. Pages 481–516 in R. L. Myers and J. J. Ewel, Eds. *Ecosystems of Florida.* University of Central Florida Press, Orlando.

Odum, W. E., T. J. Smith, III, J. K. Hoover, and C. C. McIvor. 1984. *The Ecology of Tidal Freshwater Marshes of the United States East Coast: A Community Profile.* FSW/OBS-83/17. Office of Biological Services, U.S. Fish and Wildlife Service, Washington, D.C.

O'Neil, P. E. and M. F. Mettee. 1982. *Alabama Coastal Region Ecological Characterization.* Vol. 2. A Synthesis of Environmental Data. FWS/OBS-82/42. Office of Biological Services, U.S. Fish and Wildlife Service, Washington, D.C.

Parrondo, R. T., J. G. Gosselink, and C. S. Hopkinson. 1978. Effects of salinity and drainage on the growth of three salt marsh grasses. *Bot. Gaz.* 139(1):102–107.

Pezeshki, S. R. and R. D. DeLaune. 1988. Carbon assimilation in contrasting stream side and inland *Spartina alterniflora* salt marsh. *Vegetatio* 76:55–61.

Pezeshki, S. R., R. D. DeLaune, and W. H. Patrick, Jr. 1987. Response of *Spartina patens* to increasing levels of salinity in rapidly subsiding marshes of the Mississippi River deltaic plain. *Estuarine Coastal Shelf Sci.* 24:389–399.

Pezeshki, S. R., S. Z. Pan, R. D. DeLaune, and W. H. Patrick, Jr. 1988. Sulfide-induced toxicity: inhibition of carbon assimilation in *Spartina alterniflora. Photosynthetica* 22(3):437–442.

Pomeroy, L. R. 1959. Algal productivity in salt marshes of Georgia. *Limnol. Oceanogr.* 4:386–398.

Reidenbaugh, T. G. and W. C. Banta. 1980. Origin and effects of *Spartina* wrack in a Virginia salt marsh. *Gulf Res. Rep.* 6(4):393–401.

Ribelin, B. W. and A. W. Collier. 1979. Ecological considerations of detrital aggregates in the salt marsh. Pages 47–58 in R. J. Livingston, Ed. *Ecological Processes in Coastal and Marine Systems.* Plenum Press, New York.

Seneca, E. D., S. W. Broome, and W. W. Woodhouse. 1985. Comparison of *Spartina alterniflora* Loisel. transplants from different locations in a man-initiated marsh in North Carolina. *Wetlands* 5:181–190.

Soil Conservation Service. 1980. *General Map of Ecological Communities, State of Florida.* U.S. Department of Agriculture, Washington, D.C.

Stalter, R. 1973. Transplantation of salt marsh vegetation. II. Georgetown, South Carolina. *Castanea* 38:132–139.

Stalter, R. and W. T. Batson. 1969. Transplantation of salt marsh vegetation, Georgetown, South Carolina. *Ecology* 50:1087–1089.

Steward, K. K. and W. H. Ornes. 1975. The autecology of sawgrass in the Florida Everglades. *Ecology* 56:162–171.

Stout, J. P. 1984. *The Ecology of Irregularly Flooded Salt Marshes of the Northeastern Gulf of Mexico: A Community Profile.* Biological Report 85(7.1). U.S. Fish and Wildlife Service, Washington, D.C.

Stumpe, D. K., J. Tarquinio P., J. R. Weeks, V. A. Lindley, and J. W. O'Leary. 1986. Salinity and *Salicornia bigelovii* Torr. seedling establishment. Water relations. *J. Exp. Bot.* 37:160–169.

Tanner, G. W. and J. D. Dodd. 1985. Effects of phenological stage of *Spartina alterniflora* transplant culms on stand development. *Wetlands* 4:57–74.

Tanner, W. F. 1960. Florida coastal classification. *Gulf Coast Assoc. Geol. Soc. Trans.* 10:259–266.

Tanner, W. F., S. Demirpolat, F. W. Stapor, and L. Alvarez. 1989. The "Gulf of Mexico" late Holocene sea level curve. *Gulf Coast Assoc. Geol. Soc. Trans.* 39: 553–562.

Thompson, S. M. 1977. Vascular plant communities and environmental parameters under tidal influence on the Wakulla and St. Marks rivers, Florida. Master's thesis. Florida State University, Tallahassee.

Ulrich, K. E. and T. M. Burton. 1985. The effects of nitrate, phosphate and potassium fertilization on growth and nutrient uptake patterns of *Phragmites australis* (Cav.) Trin. ex Steudel. *Aquatic Bot.* 21:53–62.

Valiela, I., J. M. Teal, and W. G. Deuser. 1978. The nature of growth forms in salt marsh grass *Spartina alterniflora. Am. Midl. Nat.* 112:461–470.

White, D. A., T. E. Weiss, J. M. Trapani, and L. B. Thien. 1979. Productivity and decomposition of the dominant salt marsh plants in Louisiana. *Ecology* 59(4): 751–759.

Wiegert, R. G., A. G. Chalmers, and P. F. Randerson. 1983. Productivity gradients in salt marshes: the response of *Spartina alterniflora* to experimentally manipulated soil water movement. *Oikos* 41:1–6.

Aboveground Net Primary Productivity of Vascular Plants

5

Yuch-Ping Hsieh

Aboveground net primary productivity (ANPP) is a measure of energy and nutrient flows in a marsh ecosystem. Estimation of ANPP in a marsh has proven difficult because it can only be assessed indirectly from the storage and flows of biomass. A variety of methods, most of them developed for other ecosystems, have been employed to assess the ANPP of marshes. The values of ANPP resulting from different methods often differ greatly for the same site. For example, Linthurst and Reimold (1978) examined primary production in salt marshes of the U.S. Atlantic coast by five methods and found that the values of assessments vary as much as ten times. Kruczynski et al. (1978) also used five methods to estimate the aboveground primary production of coastal marshes in North Florida and reported fivefold differences in results. Direct comparison of ANPP values obtained by different methods in marshes could be misleading and confusing (e.g., Linthurst and Reimold, 1978; Stout, 1984), yet ANPP values are constantly cited in literature without mentioning the uncertainty.

   Different ANPP methods are based on varying principles and assumptions. After evaluating five commonly used methods, Linthurst and Reimold (1978) concluded that all the methods except the Wiegert–Evans method (1964) underestimated ANPP in salt marshes. The Wiegert–Evans method, on the contrary, may have overestimated ANPP. Williams and Murdoch (1972) have developed a nondestructive growth monitoring method for the ANPP of *Juncus roemerianus* marshes. This method monitors the growth and dieback of *J. roemerianus* leaves during most of their life histories and uses the information to calculate the most probable ANPP. The Williams–Murdoch method is probably the most direct and accurate ANPP method for marshes, although it has two main disadvantages: (1) the procedure is very tedious and time consuming and (2) its application is limited to *J. roemerianus*. Perhaps due to those disadvantages, the Williams–Murdoch method has not been widely adopted in ANPP studies.

Current methods are either inadequate or impractical for a reliable ANPP assessment. The reason they are still being used is because no better methods are available. A better method needs to be developed for ANPP in marshes. In this chapter, the most commonly used ANPP methods for marshes are critically reviewed and current knowledge on ANPP is presented, with emphasis on the Gulf coastal marshes. A new ANPP procedure for marshes is recommended, with a case study to demonstrate its usefulness. *J. roemerianus* is chosen as a primary example in this chapter because it represents a major species in the Gulf coastal marshes.

Common Methods

Peak Living Standing Crop Method

The average of the peak living standing crop over 2 or more consecutive years is used to represent the ANPP. The advantage of this method is its simplicity. It assumes, however, no carryover of living standing crop from the previous year, no significant mortality during the growing season, no significant growth after the peak of living standing crop, and no significant grazing. The assumption of no carryover of living standing crop is probably valid for the aboveground part of plants in the northern states, where all living standing crop dies in winter. This assumption is not valid for most of the Gulf coastal states, where growth of living standing crop is almost continuous throughout a year. In order to correct this shortcoming of the method, some workers suggested that the lowest values of living standing crop be subtracted from the peak living standing crop to account for the carryover (Ovington et al., 1963; Singh and Yadava, 1972). The correction measure still does not account for the mortality, disappearance of living standing crop, and growth after peak standing crop is reached. This method, therefore, tends to underestimate ANPP.

Maximum–Minimum Method

This method sums up the positive changes per unit time of the living standing crop for 1 year and uses this value for ANPP (Milner and Hughes, 1968). This method ignores the dieback of living standing crop, which is a significant part of the ANPP, and therefore underestimates ANPP.

Smalley Method

This method was originally presented by Smalley (1959) and is probably the most widely used method for estimating ANPP in salt marshes because of its simplicity. The procedure involves periodically harvesting plant materials in field plots and separating them into living and dead components. The ANPP is the sum of the net production in each harvest calculated by the following rules:

1. If there is an increase in both the living and dead standing crops, net production is the sum of these increases.

2. If both living and dead standing crops decrease, the net production is zero.

3. If living standing crop increases and dead standing crop decreases, net production is equal to the increase in the living standing crop.

4. If living standing crop decreases and dead standing crop increases, the net production is the amount of increase in the dead standing crop minus the amount of decrease in the living standing crop. If the result is negative, the net production is zero.

The Smalley method attempts to account for the mortality of living crop. It does not account, however, for the disappearance of dead material and tends to underestimate ANPP.

Loss of Dead Material Method

This method assumes that the standing crop in field plots varies little year to year (in a steady state); therefore, the sum of the losses of dead matter over the course of a growing season ought to equal net annual production (Valiela et al., 1975). It was necessary to calculate losses, rather than merely accumulate increments of living biomass, because growth, death, and disappearance of dead material took place between sampling intervals. The loss of dead material during a sampling interval is calculated as follows:

1. If there is an increase in living standing crop and a decrease in dead standing crop, the loss of the dead material equals the decrease in the dead standing crop.

2. If there is a decrease in living standing crop, then the loss of dead material is the net loss in living and dead standing crops. If the increase

in dead standing crop is greater than the decrease in living standing crop, the loss of dead material is zero.

This procedure underestimates ANPP since it only estimates the net loss rather than the absolute loss of dead material. For example, if mortality of living standing crop equals loss of dead standing crop between sampling periods, no productivity would be observed with this method.

Wiegert-Evans Method

Wiegert and Evans (1964) developed this method originally for the ANPP of a grassland exclusive of that consumed by herbivores. The method entails the measurement of the disappearance rate of dead standing crop at intervals over a year. ANPP is calculated by the sum of the product of the disappearance rate, mean dead standing crop, and time interval plus changes in living and dead standing crops at the beginning and end of 1 year. In order to minimize the error of estimating the disappearance rate of dead standing crop, Wiegert and Evans used a paired-plot approach. A pair of plots sharing a common border was chosen in the field so that the conditions and composition of the plots are generally similar. The living standing crop of the paired lots is removed by clipping. The dead standing crop of one plot is harvested first; then, after a period of time (usually 1 month), the other plot is harvested also. The disappearance rate of the dead standing crop (r_i) in that interval is calculated as:

$$r_i = \ln \frac{(w_1/w_2)}{t_1} \tag{5.1}$$

where w_1 and w_2 are the weights of dead standing crop in a pair of plots harvested at the beginning and end of the time interval t_i, respectively. The amount of dead material (x_i) disappearing during an interval t_i is calculated:

$$x_i = \frac{(r_i)(t_i)(w_1 + w_2)}{2} \tag{5.2}$$

ANPP is then calculated:

$$\text{ANPP} = b + a + \text{sum}(x_i) \tag{5.3}$$

where b and a are the changes in living and dead standing crops between the beginning and end of a 1-year period, respectively. If the system is in a steady state, a and b in the above equation approach zero and the ANPP equals the last term of the equation.

The Wiegert–Evans method is based on several assumptions. Two of them may be questionable in a salt marsh environment. First, it assumes the pair of plots are homogenous, and no dead plant material is added to the second plot after the first plot of the pair is harvested. Tidal action may redistribute dead plant materials and deposit some on the second plot and contributes to an underestimation of the disappearance rate. Second, it assumes that the removal of living standing crop in plots will not influence the rate of disappearance of the dead standing crop. This assumption is not guaranteed in a salt marsh since tidal activity could increase the disappearance rate of the dead standing crop once the living standing crop is removed. These two opposite factors in the influence of the disappearance rate of dead standing crop are difficult to evaluate and depend on the settings of each individual salt marsh. Linthurst (1977) concluded in his study that the Wiegert–Evans method overestimated ANPP due to increased removal of dead standing crop in the paired plots after living crop was clipped. Some marsh plants have an insignificant monthly disappearance rate of dead standing crop (e.g., disappearance rates of *Juncus roemerianus* are usually less than 5% per month) which cannot be measured with certainty by the Wiegert–Evans method.

Even with these disadvantages, the method is considered one of only a few available that incorporate the components necessary for a better estimate of ANPP in a salt marsh environment and has been widely applied in studies of salt marshes.

Radiocarbon Tracer Method

This method estimates ANPP using ^{14}C as a tracer to estimate the net photosynthesis rates of plants (Giurgevich and Dunn, 1982; UNESCO, 1973). ^{14}C-labeled carbon dioxide is commonly released in an enclosed transparent chamber with living plants under field conditions. A pyrheliometer is used to measure solar radiation energy. After a brief period of time, the plant is harvested, soaked in formalin to stop metabolism, and processed to measure the ^{14}C activity in the plant. Net preliminary production is then calculated from the rate of ^{14}C incorporated in the plant, taking into consideration solar radiation energy, temperature, living standing crop, and density over the whole year.

The procedure has many shortcomings. It requires complicated equipment to handle and measure ^{14}C activity. Radiocarbon is hazardous to public health and the environment. It requires a good estimation of maintenance energy of a plant population in order to get a good net photosynthesis for ANPP. Many factors that influence rates of photosynthesis, such as nutrient status, age, species, carbon dioxide concentration, solar radiation, temperature, and moisture, are difficult to represent throughout a year. This method usually overestimates

ANPP and may include part of the productivity later transferable to the belowground productivity.

Nondestructive Monitoring Method

This method was specifically developed for *Juncus roemerianus* by Williams and Murdoch (1972). It is based on height–weight relationship of *Juncus roemerianus* (weight of a leaf can be estimated by its height). The procedure involves continuously monitoring the growth (height) of a number of individual plants over their life histories and uses the observations to calculate the population mean weight ($\overline{B}$), mean maximum weight ($\overline{B}_{max}$), and mean residence time (MRT) of the living standing crop (LSC). The aboveground ANPP of the plants can be calculated as:

$$\text{ANPP} = (\text{LSC})(\overline{B}_{max}/\overline{B})(365/\text{MRT}) \tag{5.4}$$

This method is close to a direct measurement of plant population growth under field conditions and therefore is the most ideal method, in principle, for ANPP. Unfortunately, it is also the most tedious method and was developed specifically for *J. roemerianus*.

Current Status of ANPP Measurements in Marshes

Comparing ANPP of marshes directly from the literature is difficult because of the drawbacks of current ANPP techniques. It is possible, however, to obtain some relative magnitude of ANPP by examining a vast number of documented studies. For example, Turner (1976) compared ANPP estimates of *Spartina alterniflora* from different methods using peak living standing crop as a master variable (Figure 5.1). His results showed that the maximum–minimum method was consistently lower than other methods, as expected. The Smalley method gave higher estimates than the maximum–minimum method because it attempts to account for the turnover of living and dead standing crops. Other methods such as the Wiegert–Evans method give even higher ANPP estimates because they also attempt to correct for losses that occur between sampling periods.

A latitudinal gradient in ANPP values in *S. alterniflora* marshes is found regardless of the method used (Figure 5.2). This latitudinal effect, however, is considered relatively small and often less than the difference between streamside and higher locations within a marsh (Turner, 1976). The difference in ANPP between streamside and higher locations has been noted by many workers (Linthurst and Reimold, 1978; Kruczynski et al., 1978; Turner, 1976) (Figure 5.3). The differences in tidal action, nutrient status of sediments, and plant

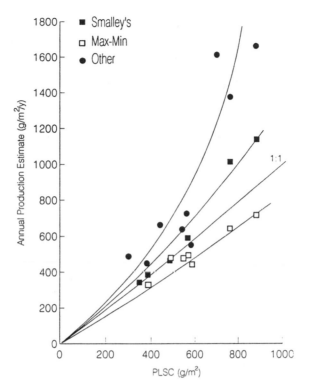

Figure 5.1 A comparison of marsh average end-of-the-season aboveground live (EOSL) estimates with other methods used to estimate annual production of *Spartina alterniflora*. (After Turner, 1976.)

genetics have been used to explain the effects, although no clear consensus has been reached by researchers. Application of nitrogen fertilizer increased biomass production of the short form but not the tall form of *S. alterniflora* or *Juncus roemerianus* in Georgia marsh (Gallagher, 1975). Application of phosphorus and nitrogen fertilizers to a *J. roemerianus* marsh in Florida did not increase its biomass production (Luce and Coultas, 1976). Transplanting *J. roemerianus* from a higher location to streamside and vice versa did not affect original differences in height and diameter, but the high marsh plants survived poorly streamside, indicating their poor adaption to more reduced environment (Coultas and Weber, 1980).

Table 5.1 lists some documented values of ANPP in marshes of the northern Gulf of Mexico reported by various authors using different methods. In general, the mean ANPP of the Gulf coastal marshes is higher than that of northern counterparts. For example, the mean ANPP estimates of *J. roemerianus*

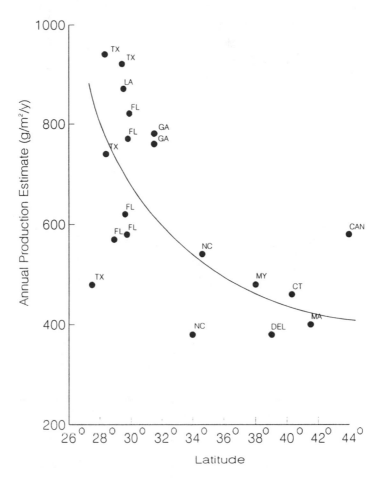

Figure 5.2   The relationship between latitude and annual production estimates of *Spartina alterniflora*. (After Turner, 1976.)

in the Gulf coastal marshes (1508 g/m^2/year, from Table 5.1) is almost twice the value for North Carolina reported by Turner (1976).

Searching for a Reliable ANPP Method

Current methods for assessing ANPP are either inadequate or impractical. A practical and reliable method for ANPP assessment in marshes needs to be identified. In principle, the method suggested by Williams and Murdoch (1972)

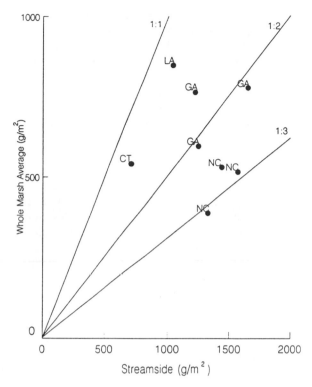

Figure 5.3 The relationship between average marsh biomass and streamside biomass in late summer (end-of-the-season live values). (After Turner, 1976.)

Table 5.1 **ANPP Estimations of the Gulf Coast Marshes (g/m²/year)**

Location	PLSC	PTSC	SMALLEY	WE	Reference
Juncus roemerianus					
Mississippi		2000			Eleuterius, 1972
Mississippi	1300		1697		de la Cruz, 1974
Alabama			3078		Stout, 1978
Florida	728	1237	949	1140	Kruczynski et al., 1978
	584	1063	595	954	Kruczynski et al., 1978
	256	601	245	243	Kruczynski et al., 1978
Florida	232	849			Kruczynski et al., 1978

Table 5.1 (continued) ANPP Estimations of the Gulf Coast Marshes (g/m²/year)

Location	PLSC	PTSC	SMALLEY	WE	Reference
Spartina alterniflora					
Louisiana	1018	1960	1409		Kirby, 1972
	782	1544	1005		Kirby, 1972
	845	1629	1138		Kirby, 1972
Louisiana	1056	1944			Allen, 1974
	742	1490			Allen, 1974
	452	906			Allen, 1974
	639	1961			Allen, 1974
Louisiana	532	1201			Gosselink et al., 1975
	394	1018			Gosselink et al., 1975
	798	1778			Gosselink et al., 1975
	650	1441			Gosselink et al., 1975
	594	1352			Gosselink et al., 1975
Louisiana				2645	Kirby & Gosselink, 1976
		754		2658	Hopkinson et al., 1980
		1070	1527		Hopkinson et al., 1980
				2895	Hopkinson et al., 1980
Alabama			2029		Stout, 1978
Mississippi	1473		1964		de la Cruz, 1974
			1084		de la Cruz, 1974
Texas	428	735			Turner & Gosselink, 1975
	938	1551			Turner & Gosselink, 1975
	735	1143			Turner & Gosselink, 1975
	923	1846			Turner & Gosselink, 1975
Florida	772	1201			Turner & Gosselink, 1975
	824	1201			Turner & Gosselink, 1975
Florida			700		Kruczynski et al., 1978
			335		Kruczynski et al., 1978
			130		Kruczynski et al., 1978
Spartina patens					
Louisiana	895	1685	2128		Payonk, 1975
Louisiana		1350	1342	1420	White et al., 1978
		1356	2000	5812	Hopkinson et al., 1980
Mississippi	1242		1922		de la Cruz, 1974
Spartina cynosuroides					
Mississippi	1862		2190		de la Cruz, 1974
Cladium jamaicense					
Florida	750	2315		2643	Hsieh and Weber, 1984

Note: PLSC = peak living standing crop; PTSC = peak total standing crop; SMALLEY = Smalley's method (1959); WE = Wiggert–Evans method (1964).

should give a reasonable assessment of ANPP in *J. roemerianus* marsh. The problems with the Williams–Murdoch method are tediousness and species limitation. The ANPP method presented here is based on the principles of the Williams–Murdoch method without the disadvantages of the original method (Hsieh, 1996). The usefulness of this new procedure is demonstrated in a north Florida *J. roemerianus* marsh.

The Study Area

The study area is located on the Gulf coast in the St. Marks National Wildlife Refuge and has been described by Kruczynski et al. (1978) as consisting of low, upper, and high marsh zones. The low marsh (LM) zone was within 120 m of the riverbank and 1 m or less above the mean low water level (MLW). The upper marsh (UM) is between 120 and 240 m away from the riverbank and between 1 and 1.4 m above the MLW. The high marsh (HM) is located between 400 and 600 m (the flatwoods inland border) from the riverbank and between 1.4 and 1.6 m above the MLW. A salt barren zone at the border between UM and HM probably represents the mean high water level of the marsh. This study was carried out in the UM and HM zones, which both have a lush growth of *J. roemerianus* with occasional *Spartina alterniflora*. Tidal water inundates the UM twice daily and the HM less frequently. The sediment has a loamy sandy texture with high organic matter content overlaying limestone at a depth of 90 to 120 cm.

Experiments

The nondestructive growth monitoring method of Williams and Murdoch (1972) was used to evaluate the ANPP of *J. roemerianus* in the HM and UM of the St. Marks Wildlife Refuge. The ANPP of this site had been studied by Kruczynski et al. (1978) using five other methods. The growth and longevity of leaves of *J. roemerianus* were monitored by monthly measuring heights of tagged leaves from their emergence to complete dieback. In each season, 50 emerging leaves were tagged and monitored, and a total of 200 leaves were tagged and monitored in a year. The height–weight relationship of the plants was constructed by regressing the height and weight of 150 leaves collected from the experimental sites, i.e., for HM, $wt(g) = 0.002 (ht, cm)^{1.310}$, and for UM, $wt (g) = 0.006 (ht, cm)^{1.226}$. The mean weight of an individual leaf (B) and of the population ($\overline{B}$) in the growing standing crop (GSC) and the mean residence time of the GSC (MRT) were calculated using the definition provided by Williams and Murdoch (1972), i.e.,

$$B = \frac{\Sigma(b \times \Delta t)}{\Sigma \Delta t} \tag{5.5}$$

$$\overline{B} = \frac{\Sigma B}{n} \tag{5.6}$$

where b is the mean weight of a leaf between two successive measurements in the GSC, Δt is the interval between the two measurements, and n is the number of leaves measured. The mean residence time of GSC (MRT) is defined as:

$$MRT = \frac{\Sigma(B \times \Sigma \Delta t)}{\Sigma B} \tag{5.7}$$

The annual ANPP is calculated as the product of mean annual GSC, ratio of mean maximum weight to mean growing leaf weight, and mean annual frequency of GSC turnover, i.e.,

$$ANPP = (GSC)\left(\frac{\Sigma \overline{B}_{max}}{\Sigma \overline{B}}\right)\left(\frac{365}{MRT}\right) \tag{5.8}$$

A 0.25-m^2 aboveground standing crop was harvested monthly over a period of 1 year. The leaves were sorted into three categories: GSC (green leaves with no significant dieback portion), dying standing crop (DYSC, green leaves with significant dieback portion), and dead standing crop (DSC, brown leaves that were completely dead). Leaves in each category were oven-dried, counted, and weighed.

Results and Discussion

The population mean weight per leaf in GSC ($\overline{B}$), mean maximum weight per leaf ($\overline{B}_{max}$), mean residence time of GSC (MRT), mean annual GSC, and the calculated annual ANPP for the HM and UM of the marshes are listed in Table 5.2. Kruczynski et al. (1978) reported ANPP of the HM and the UM to be in the range of 100 to 763 and 350 to 954 g/m^2/year, respectively (Table 5.3). The wide ranges of ANPP values they reported were caused by the difference in methods rather than true ANPP variation. The ANPP values estimated for the HM and UM using the Williams–Murdoch method were 612 and 993 g/m^2/year, respectively, which were significantly higher than the values of 243 and 595 g/m^2/year reported by Kruczynski et al. (1978) using the Smalley method. The Smalley method underestimates ANPP because it does not include mortality of DSC. Linthurst and Reimold (1978) have favorably considered the Wiegert–Evans method (Wiegert and Evans, 1964). The Wiegert–Evans method is probably not suitable for a *J. roemerianus* marsh because its monthly disappearance rate of DSC (a major variable measured in the Wiegert–Evans method) is usually too low (approximately 2% per month [Christian et al., 1990]) to be detected.

Table 5.2 The Seasonal and Annual Means of the Population Parameters of Mean Weight per Leaf ($\overline{B}$), Mean Maximum Weight per Leaf ($\overline{B}_{max}$), Mean Residence Time of Growing Standing Crop (MRT), Mean Annual Growing Standing Crop (GSC), and the Calculated ANPP of a North Florida *Juncus roemerianus* Marsh

High Marsh

	Spring	Summer	Fall	Winter	Overall
$\overline{B}$	0.493 ± 0.121[a] g	0.282 ± 0.123 g	0.266 ± 0.110 g	0.200 ± 0.102 g	0.310 ± 0.116 g
$\overline{B}_{max}$	0.660 ± 0.177 g	0.429 ± 0.145 g	0.367 ± 0.187 g	0.293 ± 0.169 g	0.437 ± 0.169 g
MRT	190 days	192 days	176 days	140 days	175 days
GSC					208 g/m²
ANPP					612 g/m²/year

Upper Marsh

	Spring	Summer	Fall	Winter	Overall
$\overline{B}$	1.164 ± 0.259 g	1.240 ± 0.310 g	0.577 ± 0.187 g	0.606 ± 0.179 g	0.901 ± 0.253 g
$\overline{B}_{max}$	1.811 ± 0.300 g	1.870 ± 0.430 g	1.207 ± 0.405 g	1.280 ± 0.400 g	1.542 ± 0.378 g
MRT	268 days	234 days	266 days	172 days	253 days
GSC					402 g/m²
ANPP					993 g/m²/year

[a] Mean ± standard deviation.

Table 5.3 **Comparison of ANPP (g/m²/year) in a North Florida** *Juncus roemerianus* **Marsh by Various Methods**[a]

Marsh zone	Method							
	WE	Smalley	WEQ	RC	CD	WM1	NEW1	NEW2
Low marsh	1140	949	372	585	2582		1255	
Upper marsh	954	595	350	528	808	993	952	1052
High marsh	245	243	100	358	763	612	686	706

Note: WE = Wiegert and Evans (1964); Smalley (1959); WEQ = Wiegert–Evans quick method (Wiegert and Evans, 1964), RC = recut plot method (Milner and Hughes, 1968); CD = carbon dioxide fixation (UNESCO, 1973); WM1 = Williams–Murdoch method by this study; NEW1 = Williams–Murdoch method using the data provided by Kruczynski et al. (1978), and NEW2 = the new procedure of this study.

[a] The first five methods were done by Kruczynski et al. (1978) and the last three methods were done in this study.

This study and that of Kruczynski et al. (1978) took place approximately 10 years apart. I compared the status of the living standing crop (LSC) and DSC in those two studies (Figure 5.4). The peak LSC in those two studies occurred at different months of the year, but the mean annual LSC and DSC were about the same. I also estimated the Williams–Murdoch ANPP values of the marsh by using the data presented by Kruczynski et al. (1978). This was done using the following steps. First, I estimated the population parameters of mean weight per leaf in the LSC ($\overline{B}_1$) by the ratio of LSC and its density of leaves, i.e.,

$$\overline{B}_1 = \frac{LSC}{\text{density of LSC}} \tag{5.9}$$

The values of $\overline{B}_{max}$ can be obtained from the maximum height and the height–weight relationship. The values of MRT_1 can be obtained from the longevity data of the LSC. With these parameters in hand, the ratio of maximum growth to the mean LSC (R_1) and the frequency of LSC turnover (F_1) can be calculated, followed by the values of ANPP, using Equation 5.8. Notice that the pair B_1 and MRT_1 was used since $\overline{B}$ and MRT were not available in the report of Kuczynski et al. (1978). The parameters $\overline{B}_1$ and MRT_1 are different from B and MRT by definition, as depicted in Figure 5.5. By definition, B_1 normally is smaller than B and MRT_1 is greater than MRT. Using either pair, the results of ANPP estimation should not differ significantly. The ANPP values estimated by this

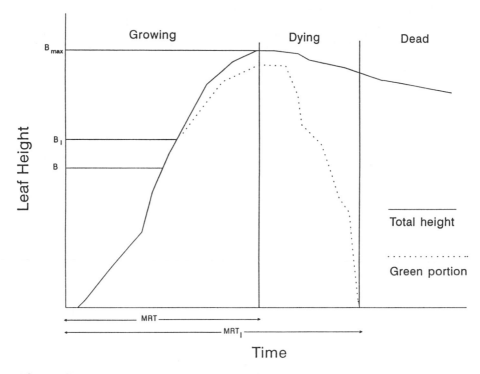

Figure 5.4 The monthly variation in standing crops measured by Kruczynski et al. (1978) and by this study at the St. Marks *Juncus roemerianus* marsh of North Florida.

method were 686 g/m²/year for HM and 952 g/m²/year for UM (Table 5.3). The original Williams–Murdoch procedure found 612 g/m²/year for HM and 993 g/m²/year for UM. The results of the above analysis and the status of LSC and DSC indicated that the ANPP has not changed significantly in the last 10 years. The ANPP values obtained by the Smalley method were lower, especially in the HM, where turnover time of the GSC was short. The above analysis illustrates that the population parameters used by Williams and Murdoch (1972) can be estimated without actually tagging and monitoring individual plants.

Recommended ANPP Procedure for Marshes

Williams and Murdoch (1972) tried to assess the ANPP of a *Juncus roemerianus* marsh by the population parameters $\overline{B}$, $\overline{B}_{max}$, MRT, and GSC. These population parameters can be estimated using a much simpler procedure, i.e., by harvesting monthly a 0.25-m² aboveground standing crop over a period of 1

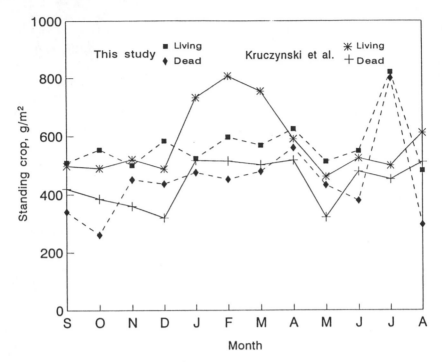

Figure 5.5 A conceptual diagram that depicts the growth and dieback kinetics and longevity of a *Juncus roemerianus* leaf. The solid line indicates the total height and the broken line indicates the green portion of a leaf.

year (Hsieh, 1996). The monthly harvest is divided into GSC and DYSC categories, counted for their respective densities, oven-dried, and weighed. The GSC is the plants that are still growing, e.g., the leaves of *J. roemerianus* that have an all green or little dieback (brown) tip. The DYSC is the plants that have reached maximum height and starts to die back significantly, e.g., the leaves of *J. roemerianus* that have a significant dieback (brown) tip.

The parameter $\bar{B}$ can be estimated by the ratio of GSC and its density, i.e.,

$$\bar{B} = \frac{GSC}{\text{density of GSC}} \qquad (5.10)$$

The parameter $\bar{B}_{max}$ can be estimated by the plants that have attained maximum height or weight (i.e., plants in transition between the GSC and DYSC categories). Different species may show different signs of maturity when they reach maximum growth. For example, the sign of a mature leaf of *J. roemerianus* is usually a significant dieback portion of the tip (Eleuterius and Caldwell,

TABLE 5.4 Selected Mean Resident Times (days) of Growth Standing Crop (MRT) and Live Standing Crop (MRT$_l$) for Marsh Species in the Southeastern United States

Species[a]	Location	MRT	MRT$_l$	Reference
Juncus	Louisiana	240	420	Eleuterius & Caldwell (1981)
roemerianus (T)	Florida	278	517	Kruczynski et al. (1978)
	Florida	253		This study
Juncus	Mississippi	270	480	Eleuterius & Caldwell (1981)
roemerianus (M)	N. Carolina	259		Christian et al. (1990)
	Florida	315	456	Kruczynski et al. (1978)
	N. Carolina	237		Williams & Murdoch (1972)
	N. Carolina	238		Hopkinson et al. (1980)
Juncus	Mississippi	135	225	Eleuterius & Caldwell (1981)
roemerianus (S)	Florida	210	332	Kruczynski et al. (1978)
	Florida	175		This study
Spartina alterniflora	Louisiana	222		Hopkinson et al. (1980)
Spartina patens	Louisiana	231		Hopkinson et al. (1980)
Spartina cynosuroides	Louisiana	210		Hopkinson et al. (1980)
Distichlis spicata	Louisiana	247		Hopkinson et al. (1980)
Sagittaria falcata	Louisiana	85		Hopkinson et al. (1980)

[a] Letters in parenthesis: T = tall form; M = medium form; S = short form.

1981). The parameter MRT is a function of species and the environment. It needs to be estimated indirectly (e.g., from data documented in the literature). Table 5.4 lists some MRT and MRT$_l$ values from the literature. The MRT of *J. roemerianus* seems to relate to height. The short form of *J. roemerianus* has significantly shorter MRT than the medium and tall forms. The medium and tall forms of *J. roemerianus* seem to have similar MRTs. In the Gulf coastal marshes, the average MRT for the medium and tall forms of *J. roemerianus* (mean maximum height = 80 to 150 cm) was 261 days. The average MRT for the short form *J. roemerianus* (mean maximum height = 50 to 60 cm) was 193 days. For the very short form of *J. roemerianus* such as the example cited by Eleuterius and Caldwell (1981) in Mississippi (mean maximum height = 37 cm), the MRT was 137 days. Using the new procedure and the data collected from 0.25-m^2 monthly harvests, I calculated the ANPP of HM and UM as 706 and 1052 g/m^2/year, respectively (see Table 5.3).

The new ANPP procedure described has several advantages over the original Williams–Murdoch method. First, it eliminates the need to monitor the growth kinetics of hundreds of living plants in the field, which greatly reduces the labor involved in and the cost of an ANPP study. Second, it improves the accuracy of estimating weights of plants by direct weighing rather than indirect assessment using the height–weight relationship. This advantage enables the technology to be applied to species other than *J. roemerianus* that do not have a good height–weight relationship. The precision and accuracy of the new procedure can be improved by increasing the sampling size of the experiment, which is difficult with the Williams–Murdoch method because of the labor involved. One disadvantage of the new procedure is that it does not measure MRT directly. Future ANPP studies should make an effort to determine the MRT of different species in different marshes. For certain species (e.g., *Spartina alterniflora*), the $\overline{B}_{max}$ may tend to be underestimated due to the loss of leaves before a plant attains its maximum growth. In this case, an estimated correction factor should be made to account for the loss of fallen leaves during the growth of a plant.

Conclusion

The study of ANPP has been impaired by lack of an appropriate ANPP technique. Current ANPP values cited in the literature rely on a methodology whose accuracy and precision can hardly be assessed. Generally, the maximum–minimum method gives the lowest ANPP estimates, followed by the Smalley method and the Wiegert–Evans method. There is a latitudinal effect on the ANPP of marshes. As the latitude increases, the ANPP value decreases. There is also a streamside effect on the ANPP of marshes. The ANPP of streamside marsh is usually higher than that of the off-streamside marsh.

The nondestructive growth kinetics monitoring method used by Williams and Murdoch (1972) provided the most ideal and reasonable assessment of ANPP in marshes. Unfortunately, the Williams–Murdoch method was extremely tedious and was restricted only to *J. roemerianus* marshes. I have recommended a new ANPP procedure that retains the advantages of the Williams–Murdoch method and eliminates its tediousness and species limitation. The new ANPP procedure requires only monthly harvests of aboveground plants over a period of 1 year.

References

Allen, R. 1974. Aquatic primary production in various marsh environments in Louisiana. M.S. thesis. Louisiana State University, Baton Rouge.

Christian, R. R., W. L. Bryant, and M. M. Brinson. 1990. *Juncus roemerianus* production and decomposition along gradients of salinity and hydroperiod. *Mar. Ecol. Prog. Ser.* 68:137–145.

Coultas, C. L. and O. J. Weber. 1980. Soil characteristics and their relationship to growth of needle rush. *Proc. Soil Crop Sci. Fla.* 39:73–77.

de la Cruz, A. A. 1974. Primary productivity of coastal marshes in Mississippi. *Gulf Res. Rep.* 4:351–356.

Eleuterius, L. N. 1972. The marshes of Mississippi. *Castanea* 37:153–168.

Eleuterius, L. N. and J. D. Caldwell. 1981. Growth kinetics and longevity of the salt marsh rush *Juncus roemerianus*. *Gulf Res. Rep.* 7:27–34.

Gallagher, J. L. 1975. Effect of an ammonium nitrate pulse on the growth and elemental composition of natural stands of *Spartina alterniflora* and *Juncus roemerianus*. *Am. J. Bot.* 62:644–648.

Giurgevich, J. R. and E. L. Dunn. 1982. Seasonal patterns of daily net photosynthesis, transpiration and net primary productivity of *Juncus roemerianus* and *Spartina alterniflora* in a Georgia salt marsh. *Oecologia* 52:404–410.

Gosselink, J. G., C. Hopkinson, and R. T. Parrondo. 1975. *Productivity and Stress Physiology of Marsh Plants in Louisiana*. Annual Report, Corps of Engineers, Vicksburg, Mississippi.

Hopkinson, C. S., J. G. Gosselink, and R. T. Parrondo. 1980. Production of coastal Louisiana marsh plants calculated from phenometric techniques. *Ecology* 61: 1091–1098.

Hsieh, Y. P. 1996. Assessing net primary production of vascular plants in marshes. *Estuaries.* In press.

Hsieh, Y. P. and O. J. Weber. 1984. Net aerial primary production and dynamics of soil organic matter formation in a tidal marsh ecosystem. *Soil Sci. Soc. Am. J.* 48: 65–72.

Kirby, C. J. 1972. The annual net primary production and decomposition of the salt marsh grass *Spartina alterniflora* Loisel in Barataria Bay estuary of Louisiana. Ph.D. dissertation. Louisiana State University, Baton Rouge.

Kirby, C. J. and J. G. Gosselink. 1976. Primary production in a Louisiana Gulf coast *Spartina alterniflora* marsh. *Ecology* 57:1052–1059.

Kruczynski, W. L., C. B. Subrahmanyam, and S. H. Drake. 1978. Studies on the plant community of a north Florida salt marsh. Part I. Primary production. *Bull. Mar. Sci.* 28:316–334.

Linthurst, R. A. 1977. An evaluation of biomass, stem density, net aerial primary production (NAPP), and NAPP estimation methodology for selected estuarine angiosperms in Maine, Delaware, and Georgia. M.S. thesis. North Carolina State University, Raleigh.

Linthurst, R. A. and R. J. Reimold. 1978. An evaluation of methods for estimating the net aerial primary productivity of estuarine angiosperms. *J. Appl. Ecol.* 15: 919–931.

Luce, H. D. and C. L. Coultas. 1976. *Effect on N and P Fertilization on the Growth, N Content, and P Content of a Natural Stand of Juncus roemerianus Near St. Marks, Florida.* Agronomy Abstracts. American Society of Agronomy, Madison, Wisconsin. pp. 149–150.

Milner, C. and R. E. Hughes. 1968. *Methods for the Measurement of the Primary Production of Grasslands.* IBP Handbook No. 6. Blackwell Scientific Publications, Oxford.

Ovington, J. D., J. Heitkamp, and D. B. Lawrence. 1963. Plant biomass and productivity of prairie, savanna, oakwood, and maizefield ecosystems in central Minnesota. *Ecology* 44:52–63.

Payonk, P. 1975. The response of three species of marsh macrophytes to organic enrichment. M.S. thesis. Louisiana State University, Baton Rouge.

Singh, J. S. and P. S. Yadava. 1972. Biomass structure and net primary productivity in the grassland ecosystem at Kurukshetra. Pages 59–74 in P. M. Golley and F. B. Golley, Eds. *Papers from a Symposium on Tropical Ecology with Emphasis on Organic Productivity.* University of Georgia Press, Athens.

Smalley, A. E. 1959. The role of two invertebrate populations, *Littorina irrorata* and *Orchelimum fidicunum,* in the energy flow of a salt marsh ecosystem. Ph.D. thesis. University of Georgia, Athens; University Microfilm, Ann Arbor, Michigan.

Stout, J. P. 1978. An analysis of annual growth and productivity of *Juncus roemerianus* Scheele and *Spartina alterniflora* Loisel in coastal Alabama. Ph.D. dissertation. University of Alabama, Tuscaloosa. 95 pp.

Stout, J. P. 1984. *The Ecology of Irregularly Flooded Salt Marshes of the Northeastern Gulf of Mexico: A Community Profile.* Biological Report 85 (7.1). U.S. Fish and Wildlife Service, Washington, D.C.

Turner, R. E. 1976. Geographic variations in salt marsh macrophyte production: a review. *Contr. Mar. Sci. Univ. Texas* 20:47–68.

Turner, R. E. and J. G. Gosselink. 1975. A note on standing crop of *Spartina alterniflora* in Florida and Texas. *Contr. Mar. Sci. Univ. Texas* 19:113–118.

UNESCO. 1973. *A Guide to the Measurement of Marine Primary Productivity Under Some Special Conditions.* Monographs on Oceanography Methodology, No. 3. UNESCO, Paris

Valiela, I., J. M. Teal, and W. J. Sass. 1975. Production and dynamics of salt marsh vegetation and the effect of experimental treatment with sewage sludge. *J. Appl. Ecol.* 12:973–981.

White, D. A., T. E. Weiss, J. M. Trapani, and L. B. Thein. 1978. Productivity and decomposition of the dominant salt marsh plants in Louisiana. *Ecology* 59: 751–759.

Wiegert, R. G. and F. C. Evans. 1964. Primary production and the disappearance of dead vegetation on an old field in southeastern Michigan. *Ecology* 45:49–63.

Williams, R. B. and M. B. Murdoch. 1972. Compartmental analysis of the production of *Juncus roemerianus* in a North Carolina salt marsh. *Chesapeake Sci.* 13:69–79.

Fishes and Invertebrates

6

William L. Kruczynski and Barbara F. Ruth

Intertidal marshes are among the most productive ecosystems on earth. In many localities, the primary productivity of marsh habitats approaches or exceeds that exhibited in cultivated fields (Odum, 1971; Teal, 1962). Tidal marshes support a diverse and abundant fauna (Kraeuter and Wolf, 1974; Daiber, 1982) because of the abundance of available energy, as well as sheltered habitats and reduced predation (Thayer and Ustach, 1981; Dardeau et al., 1992; de la Cruz, 1980; Odum and Smith, 1981). Marsh fauna include animals that visit the marsh at high tide to feed and animals that permanently live on the marsh (Nichol, 1936; Nixon and Oviatt, 1973).

This chapter summarizes species composition and community dynamics of fishes and invertebrates, exclusive of insects and arachnids (see Chapter 7 and Appendix B), in tidal *Juncus*-dominated marshes of the Gulf coast of Florida. Thorough reviews of composition and dynamics of salt marsh fauna can be found in Ranwell (1972), Cooper (1974), Thayer et al. (1979), Pomeroy and Wiegert (1981), Heard (1982), Daiber (1982), Stout (1984), Durako et al. (1985), Mitsch and Gosslink (1986), Chabreck (1988), Adam (1990), and Wiegert and Freeman (1990).

The energy cycling and species composition of Gulf coast marshes differ from marshes of the south Atlantic coast. Secondary productivity is higher on the Gulf coast than the Atlantic coast: Penaeid shrimp yield was 33.9 kg/ha for the Gulf coast and 13.9 kg/ha for the Atlantic coast; total fish yield was 485.4 kg/ha for the Gulf coast and 320.2 kg/ha for the Atlantic coast (de la Cruz, 1981). The faunal characteristics described in this chapter for *Juncus* marshes of the Gulf coast of Florida will generally be applicable to similar habitats in

Alabama and Mississippi. Although studies summarized in this chapter are predominantly for Florida *Juncus* marshes, some studies from the Atlantic coast and other Gulf coast states are included for comparison and completeness.

Daiber (1982) summarized the literature on the biology and natural history of salt marsh fauna, from protozoa through mammals. He observed that in order to manage this important natural resource for future generations, it is imperative to understand the structure and function of tidal marsh ecosystems. To accomplish that goal, it is necessary to know what species are present, how they carry on their life functions, and how they interact with other members of the community and with physical, chemical, and climatological variables.

Tidal marshes generally exhibit marked plant and animal zonation in response to physical and chemical gradients within the tidal limits of this habitat (Macnae, 1957a, 1957b; Teal, 1962; Healy, 1975; Nixon and Oviatt, 1973). Dardeau et al. (1992) described plant zonation patterns in tidal wetlands and Kurz and Wagner (1957), Eleuterius (1972, 1973, 1976), and Stout (1984) summarized plant zonation in Gulf coast *Juncus* marshes. Differences in hydrology, soils, and plant communities within tidal marshes provide important microhabitats for the diverse marsh fauna (Teal, 1962; Kruczynski et al., 1978a, 1978b). However, there are few definitive studies which have defined the interaction of these variables in controlling either plant or animal zonation. Animals associated with tidal marshes must be able to withstand rapid changes in environmental conditions which can exist in marshes (Subrahmanyam et al., 1976).

Day et al. (1973) studied the community structure and carbon budget of a Louisiana salt marsh. Only about 10% of vascular plant material produced in tidal marshes is directly consumed by herbivores (Heard, 1982). Marsh snails and insects may be the only marsh fauna to feed directly upon marsh plants (Haines and Montague, 1979). Although most marsh organisms have historically been categorized as either detritivores or predators, Sullivan and Moncrieff (1990) recently demonstrated that benthic and planktonic microalgae form the basis of the food chain for the fish and invertebrate fauna of a *Spartina* marsh. Food habits of many species may change with age, and many species utilize tidal marshes as nursery areas; the young of many estuarine and marine species are associated with marsh habitats, while adults are found elsewhere.

Tidal marshes have a diverse fauna that changes spatially within the marsh and with tide and season. In addition, *Juncus*-dominated tidal marshes of the Gulf coast span a wide geographical area. Between Tarpon Springs and Cedar Key, Florida, there is a faunal break; more tropical animals are found south of this break. Animal associations are also dependent on the physical/chemical conditions and geographic location of the study site. Our summary of the literature on the animal community must be viewed with the understanding that each study characterizes the dynamics of the animal community of the particular marsh investigated.

This summary of the fauna of *Juncus* marshes relies on several important studies. Heard (1982) described the common tidal marsh macroinvertebrates of the northeastern Gulf of Mexico. Subrahmanyam et al. (1976) characterized the macroinvertebrate fauna of two north Florida salt marshes. There have been two comprehensive studies of fishes of *Juncus* marshes (Kilby, 1955; Subrahmanyam and Drake, 1975). Kilby (1955) sampled the fishes at Cedar Key and Bayport, Florida and found a more tropical association than the marsh investigated by Subrahmanyam and Drake (1975) in Wakulla County. Two summaries of the literature on Gulf coast salt marshes have been prepared. Stout (1984) summarized knowledge of the irregularly flooded (*Juncus*) marshes of the northwestern Gulf of Mexico and Durako et al. (1985) prepared a similar summary for all salt marsh habitats of Florida.

Meiofauna

Meiofauna was first used by Mare (1942) to describe small benthic organisms that pass through a 0.5-mm mesh sieve and includes nematodes, harpacticoid copepods, amphipods, polychaetes, oligochaetes, tubellarians, ostracods, foraminiferans, and minor phyla, such as kinorhynchs and gastrotrichs. Coull and Bell (1979) reported that the majority of meiofaunal species are found in the upper centimeter of substrate. Wieser and Kanwisher (1961) observed that most nematodes were found in the upper 4 cm of substrate.

Meiofauna is an important and often overlooked component of North American estuarine communities. Much of our knowledge of Atlantic coast meiofauna has been obtained by Bruce Coull and his students. Coull and Bell (1979) prepared an excellent summary of the ecology of this diverse assemblage of organisms, and Coull and Fleeger (1977), Bell et al. (1978), and Bell (1979) studied seasonal and spatial distribution of meiofauna in South Carolina tidal marshes and estuaries. Meiofauna process organic detritus and are a food source for higher trophic levels.

Coull and Bell (1979) reported that the mean density of meiofauna was approximately $10^6/m^2$, an average of between 1 and 2 g/m^2 in biomass. Many authors, however, have observed that meiofauna exhibit patchy distributions (Wieser and Kanwisher, 1961; Teal and Wieser, 1966; Bell et al., 1978; Coull and Bell, 1979). In a Sapelo Island, Georgia marsh, nematodes were more abundant at stations closest to the water than at more inland stations (Teal and Wieser, 1966). Wieser and Kanwisher (1961) and Coull and Bell (1979) ob-

served a negative correlation of total meiofauna abundance to root mass: more
organisms were found in core samples taken between marsh plants than were
found adjacent to marsh plants. Teal and Wieser (1966), however, found greater
nematode density with increased root material. Different environmental factors
may be controlling the distribution and abundance of different groups of
meiofauna. Bell et al. (1978) found higher densities of nematodes and lower
densities of copepods near *Uca* burrows and concluded that biogenic structure
significantly affects meiofauna distribution.

A study on the distribution and abundance of harpacticoid copepods in a
South Carolina tidal marsh–mudflat–tidal creek system revealed marked horizontal patterns in species preferences (Coull et al., 1979) (Figure 6.1). Copepods were most abundant in the low marsh and least abundant in the high
marsh. Some species, such as *Microarthidion littorale*, exhibited a very wide
distribution, while others, such as *Nitocera lacustris*, were restricted to specific
marsh zones.

Ivester (1978) studied the species composition of macrofauna and meiofauna
of three natural marshes in coastal Alabama (*Spartina alterniflora*, *Juncus
roemerianus*, and *Distichlis spicata*). There were 11 meiofauna taxa collected,
and nematodes accounted for more than 50% of the total fauna in each marsh;
harpacticoid copepods and oligochaetes were second and third in abundance.

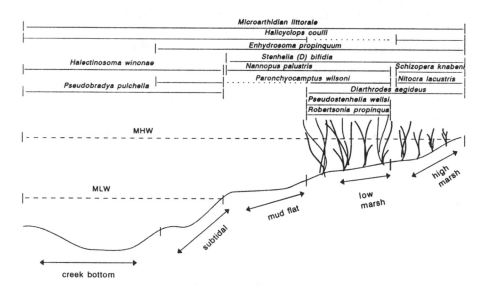

Figure 6.1 Zonation of meiobenthic copepods in a southeastern salt marsh. (Redrawn
from Coull et al., 1979; Daiber, 1982.)

Marked seasonal fluctuation of total meiofaunal abundance was observed for the three marsh types. During 1977, the *D. spicata* marsh had the greatest fluctuation in abundance, and in 1978, the *S. alterniflora* marsh exhibited the most fluctuation in abundance. The *Juncus* marsh exhibited peaks in abundance in spring and fall and low abundance in summer months. Total abundance of meiofauna was generally lower in the *Juncus* marsh than in the other two marsh types. Annual mean density for all taxa in the *Juncus* marsh was 516 cm^{-2} for 1977 and 602 cm^{-2} for 1978. Mean density for the same time periods was 104 and 2258 in the *Spartina* marsh and 1256 and 639 in the *Distichlis* marsh. Diversity (H') remained relatively constant during the 2-year study; average diversity in the *Juncus* marsh (0.55) was slightly higher than that observed in the *Spartina* (0.31) and *Distichlis* (0.50) marshes.

The *Juncus* marsh was numerically dominated by nematodes (55%), harpacticoid copepods (26%), oligochaetes (10%), and polychaetes (3%). Ten taxa were represented. Nematodes, copepods, and oligochaetes exhibited similar trends of spring and fall peaks and summer lows in densities.

Because of the difficulty in classifying meiofauna, information concerning seasonal and spatial distribution by species is not often available. Thus, in many studies, e.g., Ivester (1978), data are presented and ecological indices are calculated based upon the total numbers of the lowest recognizable taxon, often phylum or class. While generalities can be gleaned from such calculations, much information concerning seasonal and spatial distributions and abundance of individual meiofaunal species cannot.

Recent advances in the systematics of free-living marine nematodes have been made by Heip et al. (1982, 1985), Tarjan (1980), Keppner and Tarjan (1989), Keppner and Keppner (1988), and Keppner (1986, 1987a, 1987b, 1987c, 1987d, 1988a, 1988b, 1989). These authors describe and provide taxonomic keys for many species of free-living nematodes collected in estuaries of northwest Florida.

The systematics of estuarine oligochaetes has also received attention. M.R. Milligan (personal communication), of Mote Marine Laboratory, Sarasota, Florida, reports that approximately 350 species of marine and estuarine oligochaetes have been described from eastern North America. They are common in tidal and subtidal habitats and often dominate both meiofauna and macrofauna collections. Unfortunately, most comprehensive surveys have historically treated this diverse group as a single taxon. Erseus (1986, 1990) and Coates and Erseus (1986) have described and listed many species of oligochaetes that may be found in *Juncus* marshes on the west coast of Florida. Healy (1994) has recently described three new species of oligochaetes from salt marshes of coastal Georgia.

Macroinvertebrates

Macroinvertebrates of the *Juncus roemerianus* marshes of north Florida are exposed to frequently changing and often extreme environmental conditions. Slight variations of elevation in marshes (Eleuterius and Eleuterius, 1979) and irregular flooding due to periodic and aperiodic events (Provost, 1976) result in constantly changing habitat conditions. The relative shallowness of intertidal marshes causes more rapid heating and cooling than in nearby aquatic habitats, and thus water temperature, as well as dissolved oxygen and pH, can vary daily as well as seasonally. Rains and river discharge can rapidly reduce salinities, while droughts cause evaporation that can raise salinities above those of deeper waters nearby. Marshes have a semifluid substrate with high organic carbon content which, along with changing environmental conditions, limits permanent residents to poikilothermic, euryhaline, and metabolic regulatory species that are mainly burrowing infauna and mobile epifauna (Heard, 1982). Intertidal marshes do provide protection and habitat for a diverse invertebrate fauna, many of which are dependent upon wetland habitats for all or part of their life cycles (Peters et al., 1979; Bell, 1980; Boesch and Turner, 1984; Thayer et al., 1979; Zimmerman et al., 1990; Thomas et al., 1990).

The high vascular plant productivity of *Juncus* marshes (de la Cruz, 1974; Kruczynski et al., 1978a, 1978b) is largely accumulated and broken down in marsh and estuarine sediments (Haines, 1979). Sullivan and Moncreiff (1990) demonstrated that edaphic algae are the ultimate food source for marsh fish and invertebrates; vascular plants appear to be a minor direct source of carbon. In general, marsh fauna can be grouped into detritivores, which obtain all or part of their nutrition from ingesting sediment and detritus (selective and nonselective deposit feeders), and omnivores, which feed on a mixture of detritus and associated bacteria, protozoa, and other fauna (Cammen, 1979; Subrahmanyam and Coultas, 1980). Detritivores link primary production with higher trophic levels (Figure 6.2).

Faunal communities are structured by substrate and vegetation into recognizable groups; thus, characteristic species assemblages can be seen in intertidal emergent wetlands (Dardaeu et al., 1992). Literature on faunal components of *Juncus* marshes is relatively sparse, but many of the same species are found in both *Juncus*- and *Spartina*-dominated marshes on the Atlantic coast, with differences in densities rather than presence or absence (Weigert and Freeman, 1990). Studies of community structure and/or specific invertebrate species found in Gulf of Mexico *Juncus* marshes have been conducted by many authors, including Subrahmanyam et al. (1976), Harlos (1976), Kruczynski and Subrahmanyam (1978), Ivester (1978), Wilder (1979), Humphrey (1979), Subrahmanyam and Coultas (1980), and Zimmerman et al. (1990a, 1990b).

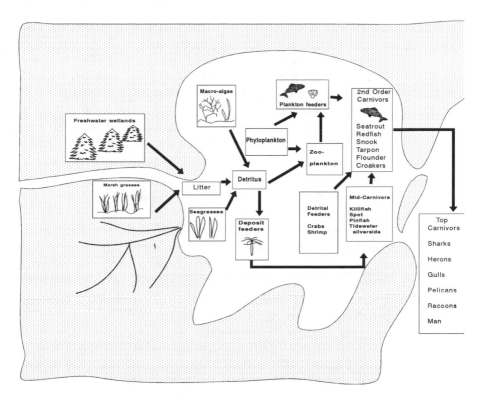

Figure 6.2 Conceptual diagram of a marsh–estuarine food chain. (Redrawn from Durako et al., 1985.)

Salt marsh macrofaunal communities are composed of both estuarine and resident species (Teal, 1962) that vary spatially as well as temporally (Gilmore and Trent, 1974; Cammen, 1979; Subrahmanyam et al., 1976; Subrahmanyam and Coultas, 1980). The majority of invertebrates found in north Florida *Juncus* marshes belong to three major groups: annelids (mainly polychaetes), crustaceans, and mollusks (Subrahmanyam et al., 1976). Distribution and abundance of these macrofauna are influenced by density of vegetation (Hamilton, 1978a; Wilder, 1979), light (Bingham, 1972; Hamilton, 1977, 1978b), salinity (Harlos, 1976; Subrahmanyam and Coultas, 1980), temperature (Young, 1973), and substrate characteristics (Chester et al., 1983). Biological factors such as predation, competition, density-dependent processes (Subrahmanyam and Kruczynski, 1979; Cammen, 1979; Subrahmanyam and Coultas, 1980), and reproductive cycles (Kruczynski and Subrahmanyam, 1978) can also influence community composition. Some invertebrates, such as shrimp and crabs, are transient residents whose presence varies with life history stage and tidal cycle (Zimmerman et al., 1982, 1990a, 1990b; Thomas et al., 1990; Sheridan, 1992).

A comprehensive study of the macroinvertebrates of north Florida *Juncus* marshes was conducted by Subrahmanyam et al. (1976). Two marshes (St. Marks and Wakulla, Wakulla County) were divided by elevation and soil characteristics into three zones: low, upper (middle), and high marsh. A fourth habitat, a marsh tidal creek, was also sampled. *Juncus* occurred in all three marsh zones. Trap and quadrat transect sampling methods were used to collect epifauna and infauna within the marsh zones. Distribution of major taxa by abundance in the marshes was: Crustacea, 44%; Mollusca, 31%; Annelida, 24%; and insect larvae, 1%. A total of 52 taxa were found in lower and upper marsh samples (Table 6.1). Species diversity (H') was statistically significantly different among seasons, with peaks in spring and fall and an overall mean of 2.49. Species richness (D) and equitability (E) showed similar seasonal trends.

Mean density for all organisms in the two *Juncus* marshes was 475 m^{-2}, with seasonal peaks in February–April and September. Mean density was lower than densities reported for salt marshes in most other areas, but densities of some individual species were comparable to other areas (Table 6.2). Mean overall biomass was 123 g/m^2, with summer and fall peaks. Densities of macroinvertebrates in a *Juncus* marsh in Alabama ranged from 378 to 4989 m^{-2} over 3 years (Ivester, 1978), had high diversity, and exhibited seasonal abundance patterns. Humphrey (1979) studied a low-salinity Mississippi *Juncus* marsh that yielded a mean density of 26 m^{-2} with low diversity. That marsh was dominated by mollusks and exhibited biomass values equal to those reported for north Florida marshes (Humphrey, 1979). *Eleocharis pavula* beds on the creekside of a Mississippi *Juncus* marsh yielded densities ranging from 2281 to 6668 m^{-2} and was dominated by crustaceans (LaSalle and Rozas, 1991). Cammen (1979) reported densities of 7600 m^{-2} in North Carolina *Spartina alterniflora* marshes, where seasonal patterns varied by species; total abundance was greatest in winter and spring. A Texas *Spartina* marsh had densities averaging 3000 m^{-2}, with some species exhibiting pronounced seasonal variation (Gilmore and Trent, 1974). Longer and more frequent tidal flooding in *Spartina* marshes may contribute to higher densities in these marsh systems compared to less frequently flooded *Juncus* marshes (Eleuterius and Eleuterius, 1979).

Macrofaunal components of north Florida *Juncus* marshes were characterized by Subrahmanyam et al. (1976) into two distinct invertebrate species assemblages: low/upper marsh and high marsh communities. The low/upper marsh community was dominated by the mollusk *Littorina irrorata*, the isopod *Cyathura polita*, tanaidaceans (*Hargeria* [=*Leptochelia*] *rapax* and *Halmyrapseudes bahamensis* [=*Apseudes* sp.]), and several abundant polychaetes (*Scoloplos*

Table 6.1 List of Invertebrate Species and Their Abundance in Low Marsh and Upper Marsh Samples at St. Marks and Wakulla Marshes

Species	Habitat[a]	N[b]	A[c]	Species	Habitat[a]	N[b]	A[c]
Bivalvia				**Isopoda**			
Cyrenoidea floridana	i, e	618	C	Cyathura polita	i	2986	C
Modiolus demissus	e	429	C	Sphaeroma quadridentatum	e	229	S
Polymesoda caroliniana	e	74	S	**Tanaidacea**			
Tagelus plebius	i	33	R	Apseudes sp.	i	2657	C
Lyonsia hyalina	i	32	R	Leptochelia sp.	i	1888	C
Amygdalum papyria	i	16	R	Littrodinops palustris	i	48	S
Pseudocyrena floridana	i	1	R	Epitonium rupicolum	i	32	S
Gastropoda				Melongena corona	e	16	R
Littorina irrorata	i, e	3680	C	Haminoea succinea	e	2	R
Heleobobs sp.	i	1614	C	**Oligochaeta**	i	176	C
Melampus bidentatus	i, e	196	C	**Polychaeta**			
Neritina reclivata	e	153	S	Scoloplos fragilis	i	2599	C
Ceritidea scalariformis	e	115	S	Laeonereis culveri	i	944	C
Ceritidea costata	i	65	S	Neanthes succinea	i	886	C
Amphipoda				Amphicteis gunneri	i	496	C
Grandidierella bonieroides	i	353	C	Namalycastis abiuma	i, e	288	S
Gammarus mucronatus	i, e	226	S	Notomastis sp.	i, e	16	R
Orchestia grillus	i, e	148	S	Capitella sp.	i	2	R
Melita nitida	i	144	S	Marphysa sanguinea[d]	i	4	R
Talitrid	i, e	139	S	Eteone heteropoda[d]	i, e	4	R
Corophium louisianum	i	128	R	Lycastopsis pontica[d]	i	55	R

Table 6.1 (continued)   List of Invertebrate Species and Their Abundance in Low Marsh and Upper Marsh Samples at St. Marks and Wakulla Marshes

Species	Habitat[a]	N[b]	A[c]	Species	Habitat[a]	N[b]	A[c]
Insect larvae	i	259	S	**Decapoda (continued)**			
				Uca speciosa	i, e	203	C
Mysidacea				*Sesarma reticulatum*	i, e	160	C
Taphromysis bowmani	p, e	100	R	*Uca longisignalis*	i, e	136	S
Mysidopsis almyra	p, e	9	R	*Eurytium limnosum*	e	47	S
				Panopeus herbstii	e	37	R
Cirripedia				*Paleomonetes pugio*	p	36	R
Chthalamus fragilis	e	24	R	*Uca pugilator*	i, e	22	R
				Sesarma cinereum	i, e	17	R
Decapoda				*Callinectes sapidus*	p	2	R
Uca spp.	i, e	333	C				

[a] e = epifauna, i = infauna, p = pelagic.
[b] Total number.
[c] Abundance: C = common, S = scarce, R = rare.
[d] Found only in transect samples.

From Subrahmanyam et al., 1976.

Table 6.2 Comparison of Densities of Some Common Marsh Macroinvertebrates from St. Marks and Wakulla, Florida, Juncus Marshes with Densities Reported in the Literature

Species	Density/m²	Locality	Authority		Density/m² (this study)	
					Apsuedes	Leptochelia
Littorina irrorata	5–90, x̄ = 38	Panama City, Florida	Bingham, 1972	LMª	88	
	6	Louisiana	Day et al., 1973	UM	65	
	40	Georgia	Frankenburg and Burbanck, 1963	HM	0	
Cyantha polita	32	Georgia	Frankenburg and Burbanck, 1963	LM	57	
	68	Massachusetts	Frankenburg and Burbanck, 1963	UM	6	
	100–200	Atlantic coast	Burbanck, 1959	HM	4	
	3364	North Carolina	Cammen et al., 1974			
Tanidacea	6	Rhode Island	Nixon and Oviatt, 1973	LM	80	47
	8	Georgia	Kuenzler, 1963	UM	30	31
	3	Louisiana	Day et al., 1973	HM	0	0
Modiolus demissus	8	Georgia	Kuenzler, 1963	LM	1	
	6	Rhode Island	Nixon and Oviatt, 1973	UM	0.6	
		Louisiana	Day et al., 1973	HM	0	
Melampus bidentata	3–300, x̄ = 6	Rhode Island	Nixon and Oviatt, 1973	LM	2	
	7	Virginia	Kerwin, 1972	UM	7	
	2	Louisiana	Day et al., 1972	HM	5.2	
	2	North Carolina	Cammen et al., 1974			
Neanthes succinea	52	Georgia	Frankenburg and Burbanck, 1963	LM	27	
	17–98	North Carolina	Cammen et al., 1974	UM	11	
Laeonereis culveri	87	Georgia	Frankenburg and Burbanck, 1963	LM	14	
	22–35	North Carolina	Cammen et al., 1974	UM	26	
Scoloplos fragilis	34	Georgia	Frankenburg and Burbanck, 1963	LM	47	
	630	Massachusetts	Frankenburg and Burbanck, 1963	UM	62	

ª LM = low marsh, UM = upper marsh, HM = high marsh.

From Subrahmanyam et al., 1976.

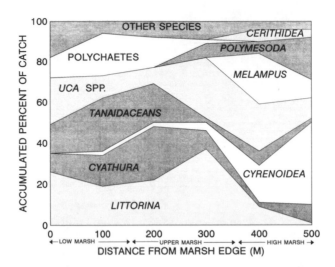

Figure 6.3   Horizontal distribution of macroinvertebrates in St. Marks and Wakulla tidal marshes based on transect sampling. (Redrawn from Stout, 1984; Subrahmanyam et al., 1976.)

[=*Leitoscoloplos*] *fragilis*, *Neanthes succinea*, and *Laeonereis culveri*). The high marsh community was characterized by fiddler crabs (*Uca* spp.), the bivalve *Cyrenoidea floridana*, and the gastropods *Melampus bidentatus* and *Cerithidea scalariformis* (Figure 6.3 and Table 6.3). Species were considered common if they occurred in at least three seasons at densities greater than 3 m^{-2}, scarce if they occurred in at least two seasons at densities greater that 1 m^{-2}, and rare if they occurred only in one or two seasons at densities less that 1 m^{-2}. Mollusks and arthropods also dominated a Mississippi *Juncus* marsh, but the bivalves *Polymesoda caroliniana* and *Geukensia demissa* and the gastropod *Neritina usnea* (=*reclivita*) (Humphrey, 1979) replaced *Littorina irrorata* in importance, possibly because of lower salinities. Oligochaetes dominated the macrofaunal community of an Alabama *Juncus* marsh, and along with *Nereis* (=*Neanthes*) *succinea*, an unidentified capitellid, *Melampus bidentatus*, *Polymesoda caroliniana,* and *Uca* spp. comprised 95% of the community (Ivester, 1978).

Macrofaunal zonation patterns were observed in north Florida *Juncus* marshes: biomass and diversity were greater in the low marsh than in the upper marsh (Subrahmanyam et al., 1976). Dominant species showed a differential gradient of abundance with respect to elevation (Figure 6.3 and Table 6.3). The low marsh exhibited densities of 540 m^{-2} compared to 381 m^{-2} for the upper marsh. Diversity values (H′) of the low and upper marsh zones were significantly different. Higher H′ values for the lower zones indicated a richer and

Table 6.3 Horizontal Distribution and Relative Abundance of Macroinvertebrates of Two *Juncus* Marshes

Abundance	LM^a (0–100 m)	LM and UM (0–300 m)	UM (200–300 m)	UM and HM (200–500 m)	HM (400–500 m)	Ubiquitous (0–500 m)
Abundant		L. irrorata			M. bidentatus	C. floridana
		C. polita				
		Apseudes sp.				
		Leptochelia sp.				
		S. fragilis				
		L. culveri				
		N. succinea				
		A. gunneri				
		M. demissus				
		S. reticulatum				
		G. bonneroides				
		Uca spp.				
		N. abiuma				

Table 6.3 (continued) Horizontal Distribution and Relative Abundance of Macroinvertebrates of Two *Juncus* Marshes

Abundance	LM[a] (0–100 m)	LM and UM (0–300 m)	UM (200–300 m)	UM and HM (200–500 m)	HM (400–500 m)	Ubiquitous (0–500 m)
Scarce	S. quadridentatum E. linmosum L. palustris	G. mucronatus U. longisignalis U. speciosa O. gillus Talidtrid	M. nitida	N. reclivata P. caroliniana U. pugilator	C. scalariformis	Insect larvae Oligochaetes L. pontica[b]
Rare	T. bowmani M. almyra P. pugio C. fragilis M. corona H. succinea A. papyria T. plebius E. heteropoda[b] M. sanguinea[b] C. capitata Notomastis sp.	C. louisianum P. herbstii E. rupicolum	S. cinereum L. hyalina	P. floridana		

[a] LM = lower marsh, UM = upper marsh, HM = high marsh.
[b] Denotes species found only in transect samples.

From Subrahmanyam et al., 1976.

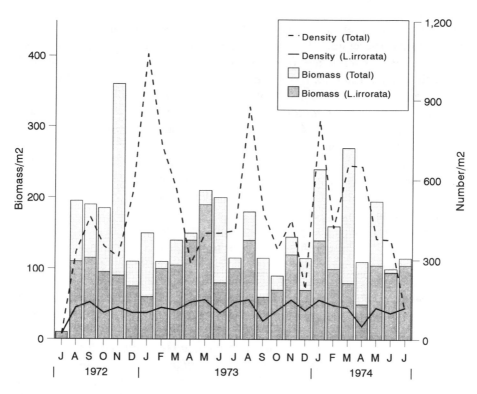

Figure 6.4 Seasonal fluctuations in biomass (bars, left axis) and density (lines, right axis) of *Littorina irrorata* and total macroinvertebrates sampled in Florida marsh zones. (Redrawn from Stout, 1984; Subrahmanyam et al., 1980.)

more diverse community in the low marsh. Similar zonation patterns were found in *Juncus* marshes in Alabama and Mississippi (Ivester, 1978; Humphrey, 1979). Horlick and Subrahmanyam (1983) found a 26% species similarity between the marsh benthic infauna of a north Florida *Juncus* marsh and that of a nearby tidal creek. Factors influencing macrofaunal zonation patterns may include frequency of flooding and proximity to subtidal waters and estuarine organisms, as well as greater availability of organic detritus (Coultas, 1969) in the lower marsh.

Bimodal seasonal peaks in abundance reported for north Florida *Juncus* marshes were the result of increased diversity and greater seasonal densities of some species (Subrahmanyam and Coultas, 1980) (Figure 6.4). Peaks in H′ nearly coincided with a high late-winter density of 575 m^{-2} and corresponded with species breeding patterns. Juvenile recruitment of *Littorina irrorata*, *Cyathura polita*, *Sesarma reticulatum*, *Melampus bidentatus*, *Geukensia*

(=*Modiolus*) *demissa*, and *Cyrenoidea floridana* was common during late winter. Seasonal succession of individual species by relative abundance was observed: *Cyathura polita* and *Neanthes succinea* were most abundant in the fall, while the tanaidacean *Halmyrapseudes bahamensis* and *Scoloplos fragilis* peaked in the spring. Biomass was highest in summer, when density and species diversity were low. *L. irrorata* contributed 81% of the total biomass in summer (Subrahmanyam et al., 1976). Similar seasonal abundance patterns were also observed in *Spartina* marshes (Gilmore and Trent, 1974; Cammen, 1979) and Alabama and Mississippi *Juncus* marshes (Ivester, 1978; Humphrey, 1979). Low summer densities may be attributed to fish predation (Cammen, 1979; Subrahmanyam and Coultas, 1980), high temperature (Young, 1973), or low oxygen and moisture (Stout, 1984).

The biology of a few of the more common macroinvertebrates found in north Florida *Juncus* marshes elucidates temporal and spatial variations observed in marsh faunal communities. *L. irrorata*, the marsh periwinkle, is a dominant animal in intertidal salt marshes in southeastern North America (Heard, 1982; Stout, 1984). *Littorina* feeds on the marsh floor during low tides and then climbs plant stems to escape rising tides, which is thought to be a behavioral response to avoid predation by blue crabs (Hamilton, 1976). The upslope movement pattern of *Littorina* is apparently guided by visual perception toward vegetation refuge (Hamilton 1977, 1978a). *Littorina* females shed fertilized eggs into the water at high tides. After hatching, pelagic larvae develop, gradually changing into small juveniles which settle onto the marsh floor (Heard, 1982). Subrahmanyam et al. (1976) found young snails (<11 mm) more abundant in the lower marsh and smaller animals (<5 mm) always occurred in infaunal samples, while larger juveniles were scarce in infauna. Individuals 1 to 4 mm in size were most abundant in fall and winter (Subrahmanyam et al., 1976).

Melampus bidentatus is a pulmonate snail found only within the tidal marsh (Harlos, 1976) that breaths air through a pneumostome and lungs (Russell-Hunter et al., 1972). The snail lacks an operculum to protect it from desiccation and is most active at night and on cloudy days (Hausman, 1936). Reproduction occurs in phase with spring tides, when these hermaphroditic snails aggregate, mate, and lay large numbers of eggs that develop into planktonic veliger larvae (Wilder, 1979). Wilder (1979) found a significant relationship between plant and snail densities; greatest densities were found in mixed vegetation dominated by *Juncus*. Subrahmanyam et al. (1976) reported higher densities and larger sizes in the upper marsh; length varied from 1 to 8 mm. Peak density occurred in the fall and consisted of mostly small snails.

The isopod *Cyathura polita* is widely distributed in many salinity, temperature, and sediment regimes from Maine to Louisiana (Burbanck, 1959;

Frankenberg and Burbanck, 1963). This infaunal animal occurs in north Florida *Juncus* marshes throughout the year. Larger individuals were more abundant in spring and smaller ones in fall (Subrahmanyam et al., 1976). Reproduction occurred from April through June. Males made up 100% of the sexually mature population in winter; their numbers dropped in spring as females with oostigites appeared (Kruczynski and Subrahmanyam, 1978). Kruczynski and Subrahmanyam (1978) found no significant variation in densities between low and upper marsh zones in the same marsh system.

The ribbed mussel *Geukensia* (=*Modiolus*) *demissa* often occurs in large epifaunal colonies on the tidal marsh floor and infaunal groups among marsh grass roots (Heard, 1982). Reproduction occurs in the spring, and shell growth is most rapid during warm months. Young mussels crawl freely until locating others of their kind and attach in clusters with byssal threads. Heaviest mortality occurs during summer and fall (Fotheringham, 1980). In north Florida *Juncus* marshes, this bivalve was found in the epifauna. *G. demissa* occurred more commonly and in smaller sizes in lower marsh zones (Subrahmanyam et al., 1976). Smaller mussels (5 to 35 mm) occurred during all seasons, with peak densities in spring; large ones (36 to 56 mm) occurred only in winter.

Cyrenoidea floridana, commonly called the Florida marsh clam, ranges from New Jersey to Louisiana (Leatham et al., 1976; Heard, 1982). *C. floridana* can be very abundant in brackish marshes, but its small size causes it to be often overlooked, and very little is known about its biology (Heard, 1982). Subrahmanyam et al. (1976) found this bivalve throughout the marsh, but it was most abundant in infaunal samples from upper and high marsh zones. Large individuals (4 to 5 mm) were only collected in winter, with peak densities occurring in fall and winter.

Fiddler crabs (*Uca* spp.) are common in, and adjacent to, tidal marshes. Fiddler crabs live in burrows along edges of the marsh. During high tides, the crabs plug burrow entrances and emerge at low tide to forage (Heard, 1982). Fiddlers form large colonies, and young individuals are usually found closer to the water's edge than older ones (Fotheringham, 1980). *U. pugilator* ranked first in abundance of macroinvertebrates in high marsh zones in north Florida *Juncus* marshes based on density, but no size ranges were reported (Subrahmanyam et al., 1976). Densities given for *U. pugilator* by Subrahmanyam et al. (1976) may have included several species of *Uca* (e.g., *U. panacea*).

Some animals commonly found elsewhere in estuarine systems can also be found in intertidal marshes. The temporal and areal extent of marsh available to estuarine organisms is determined by tidal range and marsh topography (Stout, 1984). Flooded marsh habitat is used by mobile invertebrates (Bell, 1980) including juveniles of transient species (Thayer et al., 1979) to feed (Bell, 1980) or take refuge (Zimmerman and Minello, 1984). Penaeid shrimp

(*Penaeus* spp.) and blue crabs (*Callinectes sapidus*) are commercially and recreationally important species that depend on marsh systems for portions of their life cycles (Rufison, 1981; Boesch and Turner, 1984; Thomas et al., 1990). Macroinvertebrates reported to occur in both the *Juncus* marsh and adjacent creeks and ponds in north Florida include mysids (*Taphormysis bowmani* and *Mysidopsis almyra*), shrimps (*Penaeus duorarum* and *Paleomonetes pugio*), and crabs (*Callinectes sapidus*), particularly in lower *Juncus* marsh zones (Subrahmanyam et al., 1976).

   Penaeus duorarum (pink shrimp) is the dominant commercial shrimp species found in north Florida *Juncus* marshes (Subrahmanyam et al., 1976). Pink shrimp spawn offshore and postlarvae use tidal currents and salinity gradients to guide them to marsh and tidal creek nursery areas that provide protection and food (Figure 6.5). Young adults emigrate from marshes and return to offshore breeding grounds as water temperatures decrease in the fall (Durako et al., 1985). Subrahmanyam et al. (1976) found *P. duorarum* most abundant in marsh

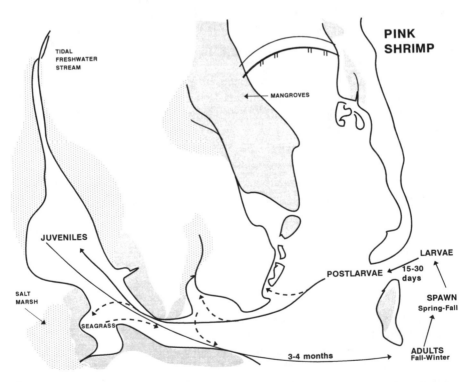

Figure 6.5   Life cycle of the pink shrimp, *Penaeus duorarum*. (Redrawn from Durako et al., 1985.)

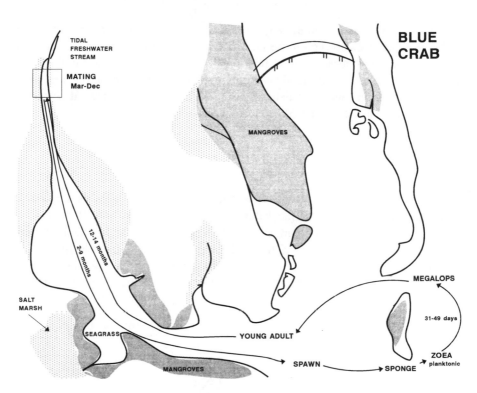

Figure 6.6   Life cycle of the blue crab, *Callinectes sapidus*. (Redrawn from Durako et al., 1985.)

zones during late spring and summer. Turner (1977) statistically linked commercial yields of penaeid shrimp to area of intertidal vegetation. He concluded that inshore yields are directly related to area of estuarine vegetation, and not to volume, area, or depth of estuarine water. Zimmerman and Minello (1984) found densities of *P. aztecus* (brown shrimp) to be nine times greater in a Galveston Bay (Texas) salt marsh than in adjacent unvegetated areas. Juvenile brown shrimp were always more abundant on flooded marsh surfaces than adjacent bare bottoms (Zimmerman et al., 1990a).

Blue crabs (*C. sapidus*) comprise the fourth largest commercial fishery in Florida, with the highest landings in north Florida (Durako et al., 1985). Marshes play an important role in crab reproduction (Figure 6.6). Blue crabs mate in the marsh, and megalops enter marshes, molt into first crab stage, and begin juvenile growth (Tagatz, 1968; Shirley et al., 1990). Subrahmanyam et al. (1976) reported *C. sapidus* in all marsh zones as well as nearby tidal creeks, with highest abundances in spring and summer. No distinction between juveniles and adults was made. Subrahmanyam and Coultas (1980) found blue crabs in isolated ponds located within the high marsh that could only be accessed via flooded marsh surfaces. A direct relationship between blue crab landings and wetland area (Lynne et al., 1981) is thought to exist, but other factors, including natural fluctuations and cohort success, also affect blue crab survival (Peters et al., 1979; Durako et al., 1985).

Fishes

An estimated 75% of Florida's recreational and commercial fishes are dependent upon marsh and estuarine habitats for at least a portion of their lives (Durako et al., 1985). Nakamura et al. (1980) found that marsh–estuarine habitats of the Gulf of Mexico had the highest proportion of juvenile fish of the habitats that they sampled. Life histories of three estuarine/marsh-dependent fishes are shown in Figures 6.7, 6.8, and 6.9. Various cyprinodontiform species

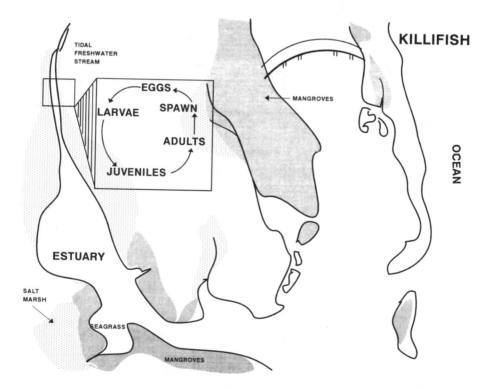

Figure 6.7   Life cycle of cyprinodontiform fish species indigenous to the salt marsh. (Redrawn from Durako et al., 1985.)

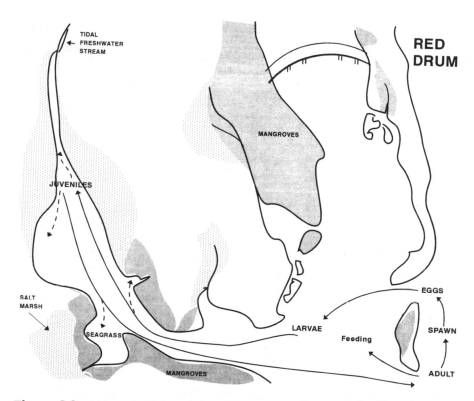

Figure 6.8   Life cycle of the red drum, *Sciaenops ocellatus*, which utilizes the marsh–estuarine ecosystem as a nursery ground. (Redrawn from Durako et al., 1985.)

are considered marsh residents and complete their life histories within the salt marsh. The red drum spawns offshore and the spotted seatrout spawns in the estuary. Both of these game fish utilize the marsh and estuary as nursery areas (Durako et al., 1985).

The relative shallowness of intertidal marshes and the dense stands of vegetation make adequate sampling of mobile fauna within the marsh community difficult. Thus, relatively little information is available on the fishes found directly within the marsh; rather, the majority of information deals with the structure of the wetlands and estuarine fishes in general (Thayer et al., 1979).

Recent studies performed at the National Marine Fisheries Service Laboratory at Galveston, Texas, and elsewhere have utilized drop nets within marsh habitats to sample fishes, crabs, and shrimp (Zimmerman et al., 1990a, 1990b). Other studies have used block nets or flumes constructed within the marsh to direct and funnel marsh fauna with falling tide (McIvor and Odum, 1986; Hettler, 1989). However, no studies are known that utilized these types of

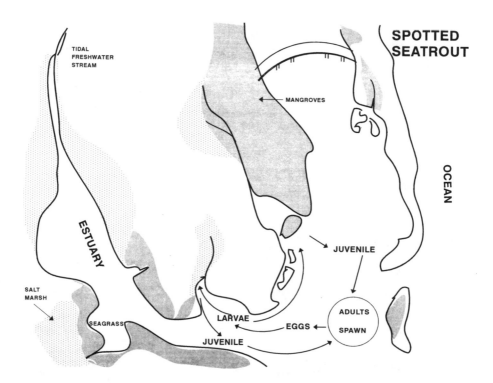

Figure 6.9 Life cycle of the spotted seatrout, *Cynoscion nebulosus*, which spawns in estuaries and utilizes marsh and tidal creek habitats as a nursery ground. (Redrawn from Durako et al., 1985.)

collecting gear for *Juncus* marshes on the west coast of Florida. Studies of the fishes of this portion of the coast are limited to collections by trawl or seine in aquatic habitats within and adjacent to salt marshes (i.e., tidal creeks, tidal ponds, and nearshore waters).

Fishes of tidal marshes have been categorized as permanent residents, juveniles of nonresident species, adult migrants, and rare sporadic visitors (Stout, 1984). The species composition of the community is influenced by seasons and tides, breeding behavior, feeding behavior, habitat diversity and available space, and proximity to estuarine waters, as well as other factors, such as pollution and water quality (Stout, 1984). Caution should be observed in making comparisons of fish data from different studies since sample gear and the geographical area of the marsh will affect the recorded occurrence and abundance of fish species.

A total of 90 species of fishes have been reported from creeks, ponds, and open waters of *Juncus* marshes (Table 6.4). The early literature on fishes of the

Table 6.4 Composite List of Fish Species Collected from *Juncus* Marshes of the Northeastern Gulf of Mexico

Species	Reference[a]	Species	Reference[a]
Achirus lineatus	1, 2, 3, 4, 5	Dorosoma petenense	4
Adinia xenica	1, 3, 4, 5, 6	Elassoma evergladei	1
Anchoa mitchilli	3, 4, 5	Eleotris pisonis	5
Anchoa hepsetus	3, 4	Elops saurus	6
Anguilla rostrata	5, 6	Eucinostomus argenteus	1, 2, 4
Archosargus probatocephalus	1, 5	Eucinostomus gula	1, 2, 3, 4
Arius felis (=Galeichthys felis)	5, 6	Evorthodus lyricus	5
Bagre marinus	6	Floridichthys carpio	1, 2
Bairdiella chrysura	1, 3, 4, 6	Fundulus chrysotus	1
Brevoortia smithi	6	Fundulus similis	1, 2, 3, 4, 6
Brevoortia patronis	3, 4, 5	Fundulus grandis	1, 2, 3, 4, 5, 6
Caranx hippos	4	Fundulus confluentus	1, 3, 4, 5
Centopristes melanus	6	Gambusia affinis	1, 6
Centrarchus macropterus	6	Gobionellus boleosoma	3
Chaenobryttus coronarius (=Lepomis gulosus)	1	Gobiosoma robustum	1, 2, 4
Chaetodipteris faber	4	Gobiosoma bosci	3, 4, 5
Chasmodes saburrae	1	Harengula jaguana	3
Chriopoeps goodei (=Lucania goodei)	1	Harengula pensacolae	4
Citharichthys spilopterus	4	Heterandria formosa	1
Cynoscion nebulosa	1, 4, 5, 6	Hyporhamphus unifasciatus	1, 3, 4
Cynoscion arenarius	3, 4, 5	Ictalurus punctasus	5
Cyprinodon variegatus	1, 2, 3, 4, 5, 6	Jordanella floridae	1
Dasyatis sabina	1, 4	Lagodon rhomboides	1, 2, 3, 4, 6
		Leiostomus xanthurus	1, 2, 3, 4, 5, 6
		Lepisosteus platyrhincus	1

Table 6.4 (continued) Composite List of Fish Species Collected from *Juncus* Marshes of the Northeastern Gulf of Mexico

Species	Reference[a]	Species	Reference[a]
Lepisosteus oculatus	5	*Paralichthys lethostigma*	3, **4**, 6
Lepomis punctatus	1	*Paralichthys albigutta*	2
Lepomis macrochirus	5	*Poecilia latipinna*	1, 3, 4, 5, 6
Lepomis microlophus	1, 5	(=*Mollienesia latipinna*)	
Lucania parva	1, 2, 3, 4, 5, 6	*Polydactulus octonemus*	2, 3, 4
Lutjanus griseus	1	*Prionotus scitulus*	**4**
Membras martinica	5	*Prionotus tribulus*	**3, 4**
Menidia beryllina	1, 2, 3, 4, 6	*Sciaenops ocellata*	1, **3**, 6
Menticirrhus americanus	**4**	*Selene vomer*	**4**
Micrognathus crinigerus	2	*Seriola zonata*	**4**
Microgobius gulosus	1, 2, 4	*Sphoeroides nephelus*	2, 4
Microgobius thalassinus	**5**	*Sphyraena barracuda*	1
Micropogon undulatus	**6**	*Strongylura notata*	1, 2, 3, 4
Micropterus salmoides	1, 5	*Strongylura marina*	3, 4, 5
Mugil cephalus	1, 3, 4, 5, 6	*Strongylura timicu*	1, 2
Mugil curema	**1**, 4	*Symphurus plagiusa*	2, **3**, 4, **6**
Myrophis punctatus	**4**	*Syngnathus scovelli*	1, 2, **3**, 4
Notemigonus crysoleucas	1	*Syngnathus louisianae*	5
Notropis petersoni	1	*Synodus foetens*	3, 4, **5**
Oligoplites saurus	2, **3**, 4, **5**	*Trachinotus falcatus*	2
Opsanus beta	1, 2	*Trinectes maculatus*	**6**
Orthopristis chrysopterus	**1, 3, 4, 6**		

[a] Uncommon occurrences in references appear in bold type. 1 = Kilby (1955), 2 = Cherr (1974), 3 = Subrahmanyam and Drake (1975), 4 = Subrahmanyam and Coultas (1980), 5 = Hackney and de la Cruz (1981), 6 = Zilberberg (1966).

Modified from Stout, 1984.

west coast of Florida consists largely of descriptive summaries of the life histories of individual species and annotated lists (Jordon, 1884; Breder, 1940; Kilby, 1948; Moody, 1950; Caldwell, 1954, 1955; Herald and Strickland, 1949; Townsend, 1956; Hansen, 1969). A comprehensive fish survey within this geographic area was published by Reid (1954), who sampled the shallow waters (flats) and tidal channels near Cedar Key (Levy County), Florida; 17 estuarine stations were sampled twice a month for a year (1950 to 1951), mostly by trawl. Reid collected 122 fish species and summarized their life histories and feeding information in an annotated list. Cyprinodontids were absent or collected in low numbers in habitats sampled in this study. Reid (1954) observed that cyprinodonts were generally most common in "pools and watercourses within the extensive *Juncus* marsh which borders the coastal shoreline in this region of the state." Absent from Reid's list are *Fundulus confluentus* and *Adinia xenica*, two known residents of *Juncus* marshes.

Kilby (1955) sampled the fish fauna of two coastal marshes on the west coast of Florida, Cedar Key and Bayport (Hernando County), monthly for 1 year (1948) using a seine. The Cedar Key marsh was vegetated predominantly by *Spartina alterniflora* with patches of *J. roemerianus* and thickets of *Avicennia nitida* (black mangrove). The Bayport marsh was vegetated predominantly by *J. roemerianus*. Marsh habitats sampled included inner pools, outer pools, water courses, and open waters. The Bayport marsh had a lower salinity range (0 to 24.7 ppt) than the Cedar Key marsh (1.2 to 35.6 ppt), and inner pools of the Bayport marsh were vegetated by predominantly freshwater species such as *Typha*, *Nais*, *Eleocharis*, and *Sagittaria*.

Kilby collected 55 species of fishes at Cedar Key and 49 species at Bayport (Table 6.5). At *both* Cedar Key and Bayport, 29 species were found. Fishes captured in pools were predominantly small individuals, whereas fishes taken in the water courses were almost always adults. Over 97% of the total catches at both study marshes (Table 6.6) was made up of 18 species. *Mollienesia* (=*Poecilia*) *latipinna* was the most common species collected at both sites, representing 22.9% of the catch at Bayport and 32.4% at Cedar Key. Both Reid (1954) and Kilby (1955) suggested that Cedar Key may be the dividing line between temperate and tropical faunas on the Florida Gulf coast.

Zilberberg (1966) studied the fishes of a tidal *Juncus* marsh near the mouth of the Econfina River (Taylor County); canal, pond, and tidal creek habitats were sampled monthly with a seine for 1 year (1964) (Table 6.7). *Mugil cephalus* and *Leiostomus xanthurus* were the dominant fish species in all habitats sampled. However, dominance varied seasonally and by habitat type.

Margins of roadside canals were vegetated predominantly by *J. roemerianus* and *S. alterniflora*. In general, upper portions of canals were dominated by *Lucania parva* and *Gambusia affinis* (=*G. holbrooki*) and lower portions by

Table 6.5 **Population Composition of Most Abundant Fish Species Collected at Bayport and Cedar Key Habitats**

	Percent of total catch	
Species	*Bayport*	*Cedar Key*
Poecilia latipinna	23	32
Lucania parva	20	1
Eucinostomus argenteus[a]	10	1
Gambusia holbrooki	10	3
Menidia beryllina[b]	9	6
Cyprinodon variegatus	7	14
Floridichthys carpio	3	1
Jordanella floridae	3	0
Fundulus confluentus	3	3
Fundulus grandis	2	4
Heterandria formosa	2	0
Adinia xenica	2	15
Fundulus similis	1	5
Mugil cephalus	1	6
Lagodon rhomboides	1	0
Lepomis punctatus	1	0
Leiostomus xanthurus	0	3
Anchoa mitchelli	0	3
Total	98%	97%

[a] May include *Eucinostomus harengulus* and/or *Eucinostomus jonesi*.
[b] May include *Menidia peninsulae*.

From Kilby, 1955.

Fundulus spp. and *Poecilia latipinna*. *M. cephalus* (1-year age group) and *Cyprinodon variegatus* were locally abundant within the canals. During spring, either *Mugil*, *Leiostomus*, or *Lagodon* replaced all or some of the other species in the canals. Zilberberg observed a fish kill in a tidal canal when water temperature reached 41°C in June 1964.

Ponds consisted of two types, those connected to the Gulf by shallow tidal creeks and those connected to tidal waters only during extreme high tides or periods of flooding. Dominant species varied from pond to pond, with *P. latipinna*, *Menidia beryllina*, *F. similis*, *Lucania parva*, and *G. holbrooki* being common. (It is probable that data for *M. beryllina* reported by Zilberberg and other authors reported herein included *M. peninsulae* as well.) During spring, *Mugil*, *Leiostomus,* and to some extent *Lagodon* became dominant.

Table 6.6 Composition of the Ten Dominant Fish Species in Apalachicola Bay, Salt Marsh Tidal Creeks, and Salt Marsh Pond Habitats in Northwest Florida

Apalachicola Bay	Marsh tidal creek	Isolated ponds
Anchoa mitchilli	Anchoa mitchilli[a]	Leiostomus xanthurus
Micropogon undulatus	Leiostomus xanthurus	Poecilia latipinna[b,c]
Cynosion arenarius[b]	Fundulus similis[b]	Cyprinodon variegatus[b]
Leiostomus xanthurus	Fundulus grandis	Adinia xenica
Chloroscombrus chrysurus	Cyprinodon variegatus[b]	Mugil cephalus[c]
Menticirrhus americanus[c]	Adinia xenica	Fundulus confluentus[b,c]
Symphurus plaguisa	Eucinostomus argenteus	Fundulus grandis[b]
Polydactylus octonemus	Menidia beryllina[b]	Lucania parva[c]
Arius felis	Lagodon rhomboides[b]	Fundulus similis
Prionotus tribulus[c]	Fundulus confluentus	Microglobius gulosus[b,c]

[a] Rare in isolated ponds.
[b] Rare in estuary.
[c] Rare in tidal creek.

From Subrahmanyam and Coultas, 1980.

The tidal creek habitat was stratified into upstream, midstream, and downstream segments. The biota of the upstream portion was sampled most frequently, and species abundance varied widely in accordance with tides. *Elops saurus, Sciaenops ocellatus,* and *Cynoscion nebulosus* were all seasonally abundant during high tide. The populations of the five most common species collected at this station were related to both salinity and temperature (Figure 6.10). In general, salinity and numbers of fish were lowest during winter.

During late summer to early fall (July to September), numbers and diversity of adult fishes peaked. Young of three species (*Mugil cephalus, Leiostomus xanthurus,* and *Lagodon rhomboides*) became common from January to June in all marsh areas studied and then departed the marsh. Zilberberg concluded that temperature was probably one of the factors responsible for the exodus of these species from the marsh and that seasonal use of the tidal marsh habitat was consistent with findings from other locales (Gunter, 1945; Kilby, 1948, 1955; Reid, 1954; Townsend, 1956).

Subrahmanyam and Drake (1975) sampled tidal creeks of two north Florida *Juncus* marshes (Wakulla County) by seine, monthly (1972 to 1974) at high and low tides on the same day; 55 species of fish were collected. The Biological Index (BI) (Sanders, 1960) was calculated to rank the 13 most important species based on numbers and biomass at both locales at high- and low-tide

Table 6.7 Fishes and *Callinectes sapidus* Collected at Various Tidal Marsh Habitats Near the Mouth of the Econfina

Species	Zones			Species total
	Canals	Ponds	Creeks	
Callinectes sapidus	52	27	186	265
Elops saurus	—	—	109	109
Brevoortia smithi	—	—	15	15
Bagre marinus	—	—	2	2
Galeichthys felis	—	—	155	155
Anguilla rostrata	—	6	—	6
Adinia xenica	34	42	11	87
Cyprinodon variegatus	88	265	64	317
Fundulus grandis	23	115	45	183
Fundulus similis	27	62	55	144
Lucania parva	154	180	106	440
Gambusia affinis	230	102	46	378
Mollienesia latipinna	211	361	119	691
Centropristes melanus	—	—	12	12
Centrarchus macropterus	—	—	27	27
Orthopristis chrysopterus	—	—	18	18
Bairdiella chrysura	—	—	32	32
Cynoscion nebulosus	—	—	128	128
Leiostomus xanthurus	830	972	1415	3217
Micropogon undulatus	—	—	5	5
Sciaenops ocellata	—	—	87	87
Lagodon rhomboides	327	250	924	1501
Mugil cephalus	1460	1131	2270	5861
Menidia beryllina	80	283	80	443
Paralichthys lethostigma	1	3	—	4
Trinectes maculatus	1	—	—	1
Symphurus plagiusa	1	—	—	1

From Zilberberg, 1966.

stages. The 11 most important species were 7 species of cyprinodontiforms (*Fundulus similis*, *F. grandis*, *Cyprinodon variegatus*, *Poecilia latipinna*, *Adinia xenica*, *Lucania parva*, and *F. confluentus*) and *Menidia beryllina*, *Leiostomus xanthurus*, *Eucinostomus argenteus*, and *Lagodon rhomboides*. There are three species of *Eucinostomus* now known to exist in coastal areas. *E. argenteus* reported by Subrahmanyam and Drake and others probably included *E. jonsesi* and *E. harengulus*.) The seven species of cyprinodonts dominated the catches

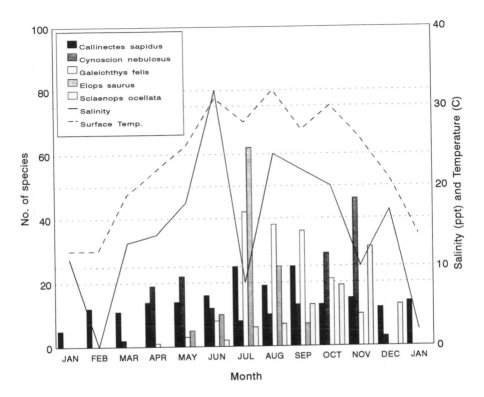

Figure 6.10   Monthly abundance of the five most common fish species and *Callinectes sapidus* and salinity and temperature in a tidal creek near the mouth of the Econfina River. (Redrawn from Zilberberg, 1966.)

at low tides at both localities, while the other four species, with some additional species, were dominant at high tides. When BI was recalculated excluding the seven cyprinodontiform species (the permanent marsh dwellers), the four species mentioned above were still dominant. However, *Mugil cephalus, Anchoa mitchilli, Polydactylus octonemus,* and *Oligoplites saurus* were also important in the catches (Table 6.8).

Mean total number of fish (catch per unit effort) at St. Marks marsh was 212 at high tide and 991 at low tide. At Wakulla marsh, the average catch was 394 at high tide and 1421 at low tide. Higher numbers of fish were captured in cooler months at high tide and in warmer months at low tide. Average biomass per unit effort was 1069 g at high tide and 1158 g at low tide at St. Marks and 886 g at high tide and 1127 g at low tide at Wakulla. Significantly higher biomass was found in warm months (May to November), which can be attributed to greater numbers of species and juveniles present and to larger

Table 6.8 Frequency of Occurrence (F/12), Biological Index (BI), and Ranking (R) of the 13 Most Important Fish Species, Excluding Cypinodontiform Species, Based on Weights in Sample at High and Low Tides at St. Marks and Wakulla Marshes

Species	St. Marks Low tide			St. Marks High tide			Wakulla Low tide			Wakulla High tide		
	F	BI	R	F	BI	R	F	BI	R	F	BI	R
Menidia beryllina	9	60	1	11	78	1	8	44	3	9	65	2
Leiostomus xanthurus	5	47	2	10	77	2	7	62	1	9	79	1
Eucinostomus argenteus	5	30	3	4	18	6	5	24	5	5	29	5
Mugil cephalus	3	19	4	2	15	9	3	8	7	3	26	8
Lagodon rhomboides	3	15	5	8	62	3	6	45	2	6	42	4
Anchoa mitchilli	2	14	6	3	16	7	4	24	4	7	46	3
Paralichthys albigutta	2	12	7									
Paralichthys lethostigma	3	10	8				1	6	11			
Polydactylus octonemus	2	10	9	4	35	4	4	24	6	3	28	7
Dasyatis sabina	1	10	10									
Dorosoma petenense	1	8	11				2	8	8			
Microgobius gulosus	2	7	12									
Oligoplites saurus	3	7	13	4	11	11	3	6	10	4	14	10
Syngnathus scovelli	2	7	14									
Anchoa hepsetus				4	16	8						
Strongylura marina				4	28	5				5	29	6
Cynoscion arenarius				2	15	10						
Harengula jaguana				1	9	12				2	13	11
Hyporamphus unifasciatus				1	9	13				1	9	12
Orthopristis chrysoptera				1	7	14				1	9	13
Gobiosoma bosci							1	7	9			
Prionotus tribulus							1	5	12			
Strongylura notata							2	5	13	3	16	9
Synodus foetens							2	4	14			
Eucinostomus gula										1	7	14

From Subrahmanyam and Drake, 1975.

individuals of resident fishes. Permanent residents were caught as adults and juveniles at high and low tide and included cyprinodontiform fishes; *F. similis, F. grandis,* and *C. variegatus* were the most common permanent residents. Nursery fish were caught at both stages of the tide and included *Menidia beryllina, Leiostomus xanthurus, E. argenteus, Mugil cephalus,* and *Anchoa mitchilli.* Foraging species were captured only at high tides and included adults of *Hyporhamphus unifasciatus, Bairdiella chrysoura,* and *Cynoscion arenarius.* Sporadic marsh visitors included *Polydactylus octonemus.*

Subrahmanyam and Drake (1975) analyzed food preferences of common fishes and grouped them into detritivores, detritivore/carnivores, and carnivores. Detritivores included *Cyprinodon variegatus, L. xanthurus,* and *M. cephalus.* Food of detritivores/carnivores consisted mostly of infaunal crustaceans and detritus and included *E.* spp., *F. grandis, F. similis, H. unifasciatus, Lagodon rhomboides, Menidia beryllina,* and *Strongylura* spp. Carnivores included *Cynoscion arenarius* and *Oligoplites saurus.* They concluded that feeding strategies may change with age and observed that *F. grandis, F. similis, H. unifasciatus, L. rhomboides,* and *Strongylura* spp. became more selective as they grew older; larger sizes of these species were found to feed mostly on crustaceans.

Other studies have investigated the feeding habits of marsh fish. Differences in stomach content of juveniles spot-collected on incoming and outgoing tides indicated that the marsh–creek system is much more valuable as a feeding area than adjacent deeper waters (Miller and Dunn, 1980). Carnivorous species, such as seatrout (*Cynoscion* spp.) and red drum (*Sciaenops ocellatus*), were found to move into marshes at high tides to feed on anchovies, silversides, and juveniles of many fish species (Subrahmanyam and Drake, 1975; Reid, 1954; Zilberberg, 1966). Estuarine populations of *L. rhomboides* were found to be omnivorous in all size classes and seasons (Hansen, 1969). Vegetation contributed 40 to 60% of total volume of food and was the dominant food in summer and fall. Crustaceans, polychaetes, and chordates were dominant in pinfish stomachs in winter and spring.

Rozas and LaSalle (1990) investigated the feeding habits of *F. grandis* and observed that this species consumed a greater volume of food when individuals had access to the marsh surface than when confined to subtidal areas. *F. grandis* is omnivorous, and the most important components of its diet were common marsh macroinvertebrates including *Uca longisignalis, Corophium louisianum, Hargeria rapax,* and *Littrodinops palustris.* Harrington and Harrington (1972) demonstrated that seven species of killifish invaded a flooded high marsh on Florida's east coast to feed on mosquito larvae.

Subrahmanyam and Coultas (1980) reanalyzed historic data on fish (Subrahmanyam and Drake, 1975) and macroinvertebrates (Subrahmanyam et al., 1976) collected from St. Marks and Wakulla tidal marshes and compared findings with the fauna of ponds located within the marsh zones. They found marked seasonal trends in abundance and species composition in both tidal creeks and ponds. In tidal creeks, peaks in abundance occurred in March and July, with a general increase in numbers in warmer months. *F. grandis* and *F. similis* dominated the catch in summer and fall and *C. variegatus* dominated in late summer. In ponds, fish were most abundant in February and July. Of the 19 species observed in the ponds, *C. variegatus*, *Adenia xenica*, and *Leiostomus xanthurus* were most abundant in February, and *Poecilia latipinna* and *Mugil cephalus* dominated the catch in July.

Relative importance of species was calculated for both tidal creeks and pond samples using BI value and numerical and biomass percentage ratios (Subrahmanyam and Coultas, 1980; Durako et al., 1985). In tidal creeks, the killifishes (*F. grandis*, *F. similis*, and *C. variegatus*) and the common estuarine fishes (*Leiostomus xanthurus* and *Lagodon rhomboides*) were found to be the most important species based upon BI values and numbers. The biomass percentage ratio sequence varied more than the other two methods due to the mixture of juveniles and adults in the samples and less abundant (rare) larger species.

In marsh ponds, *Leiostomus xanthurus*, *C. variegatus*, *M. cephalus*, *P. latipinna*, and *F. grandis* were among the top species based upon BI and numerical and biomass ratios. It is significant that freshwater species (*Lepomis microlophus*, *L. macrochirus*, and *Gambusia affinis* [=*G. holbrooki*]) occurred rarely in ponds sampled in north Florida. This may be because Subrahmanyam and Coultas (1980) grouped ponds with different flooding regimes into one habitat type and data from all ponds were pooled. Ponds located within the low and high marsh zones had different species composition (Kilby, 1955), and analyzing these data separately would have made a useful comparison of marsh ponds located in different geographic areas. Hackney and de la Cruz (1981) sampled an oligohaline (0.5 to 5.0 ppt) *Juncus* marsh in Mississippi and observed that *Micropterus salmoides* and *L. macrochirus* were the dominant predator species.

Temporal associations of fishes in tidal creeks were examined through cluster analysis (Subrahmanyam and Coultas, 1980; Durako et al., 1985). Two major and five subgroups were observed using this technique (Figure 6.11). Cyprinodontiform species occurred as two groups (1 and 2) and were found in most samples. The *Strongylura* group (3) represented species caught at high tide, and the *Menidia* (4) and *Leiostomus* (5) groups included all the migratory species. Cherr (1974) sampled tidal creeks in an intertidal marsh at Live Oak Island (Wakulla County, Florida) and determined that five species were year-round residents of that marsh, including *M. beryllina*, *Floridichthys carpio*,

Fundulus grandis, *F. similis*, and *C. variegatus*. Winter–spring residents included *Lucania parva*, *Paralichthys albigutta*, and *Syngnathus scovelli*. Summer–fall residents included *Strongylura* spp., *Eucinostomus* spp., and *Oligoplites saurus*.

Subrahmanyam and Coultas (1980) concluded that seasonal changes in community composition were profoundly influenced by breeding patterns, recruitment, and seasonal succession of dominants. However, they observed some correlations between seasonal abundance of some species with physical/chemical environmental parameters. For example, abundance of *F. grandis* correlated positively with temperature and *Lagodon rhomboides* with salinity. The abun-

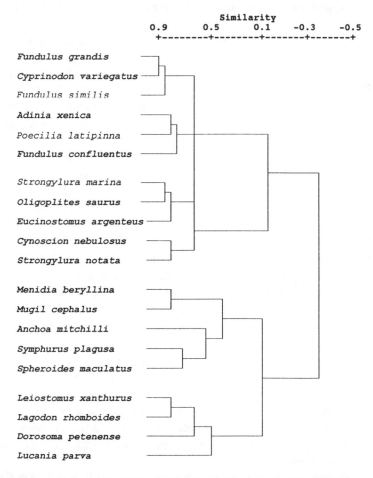

Figure 6.11 Faunal similarity of fishes collected at north Florida salt marshes. (Redrawn from Subrahmanyam and Coultas, 1980.)

dance of *Leiostomus xanthurus* correlated negatively with temperature and salinity. The abundance of *O. saurus* correlated positively with dissolved oxygen, while the abundances of *E. argenteus* and *Lagodon rhomboides* correlatedly negatively with dissolved oxygen.

There is no question that physical/chemical parameters play a major role in fish community composition. Cherr (1974) concluded that salinity influences the fish species composition of intertidal marshes by excluding less euryhaline forms. Lipcius and Subrahmanyam (1986) examined temporal factors influencing abundance and recruitment of *F. grandis* and *F. similis* in *Juncus* marshes at St. Marks and Wakulla. They concluded that temporal abundance patterns are regulated significantly by tidal phase, water depth, water temperature, and dissolved oxygen, but not by salinity or percent cloud cover. Killifishes apparently moved over submerged marsh surfaces at high tide and retreated to tidal creeks at low tide. However, tidal movements were modulated by water depth. Abundance at mouths of tidal creeks was correlated negatively with water depth, indicating that killifish remained in the salt marsh when its surface was inundated. Abundance of killifishes was positively correlated with water temperature and dissolved oxygen, reflecting juvenile recruitment periods during warmer seasons and increased recruitment during periods of higher dissolved oxygen.

Dissolved oxygen concentration may influence species composition within tidal creeks. Subrahmanyam (1980) compared the oxygen consumption of *Leiostomus xanthurus*, *Lagodon rhomboides*, *F. grandis*, and *F. similis* in closed aquaria. All sizes of killifish were found to consume less oxygen per gram body weight than comparable sizes of spot and pinfish. Within the normal range of oxygen concentration in the natural environment (10 to 85 mmHg), all species demonstrated distinctive respiratory patterns. Killifishes exhibited stronger metabolic conformity to oxygen tensions than the other species. Also, spot and pinfish died at a PO_2 of 24 mmHg, while the two species of killifish were able to survive at a PO_2 of 12 to 15 mmHg.

Cherr (1974) found significant changes in the fish fauna within tidal creeks in day and night samples at high and low tide (Note: data in Table 6-25 of Stout (1984) for high and low tide are transposed.) Species caught mostly during night samples include *Strongylura timucu* (95%), *S. notata* (84%), *Menidia beryllina* (85%), and *O. saurus* (80%). Species captured mostly during day samples included *Sphoeroides nephelus* (88%), *Lucania parva* (86%), and *Eucinostomus* spp. (73%). Fish caught mostly at low tide were *L. parva* (100%), *Cyprinodon variegatus* (98%), *F. grandis* (92%), *F. similis* (91%), *P. albigutta* (91%), and *E. argenteus* (85%).

Marsh residents (cyprinodonts) were most common in low tide collections. This is consistent with the findings of Subrahmanyam and Drake (1975), who observed that cyprinodont species were more abundant at low tide because they were forced to leave the marsh with falling water. Although it is risky to compare the results of different studies using different collecting gear at different marshes, based upon Cherr's (1974) observations, fish abundance data presented by Subrahmanyam and Drake (1975) and Subrahmanyam and Coultas (1980) probably represent minimum densities for many species since all samples at St. Marks and Wakulla marshes were made during daylight.

Subrahmanyam and Coultas (1980) compared the ten most common fish species in tidal creeks and marsh ponds with common estuarine species (Table 6.6) and observed distinct species assemblages within the habitats. However, some species, such as *Leiostomus xanthurus* and *Anchoa mitchilli* were found in all three habitats. They concluded that these communities are characterized by a few nonpredaceous, ubiquitous, visible dominants and a few less abundant species (Thorson, 1957).

A comprehensive study of the fishes of a regularly flooded *Spartina* marsh in North Carolina utilized a block net to funnel organisms into a trap as they exited the marsh on a falling tide (Hettler, 1989). Although the species are somewhat different, the trends observed in this Atlantic coast marsh are similar to those observed for *Juncus* marshes of northwest Florida. A total of 35 species of fish were collected, including 8 residents, 26 estuarine-dependent transients, and 5 marine visitors. Ten species accounted for 88% of the biomass and 92% of the numbers. *F. heteroclitus* ranked first in both average biomass and average numbers, followed by *Lagodon rhomboides*, *Leiostomus xanthurus*, and *C. variegatus*. Fish biomass, total number, and number of species were lowest in January and February. Numbers peaked in April, due primarily to juvenile spot, and again in July and August when the residents *F. heteroclitus*, *C. variegatus*, and *F. majalis* made up 80% of the catch.

Hettler (1989) concluded that the assessment of tidal marshes as a nursery area must include an understanding of differences in microhabitats within marshes. Further studies are required to quantify the penetration by individual species into tidal marsh habitats and determine whether edge or total area is the dominant feature controlling marsh use by nekton (Turner, 1977). Without a thorough understanding of nekton abundance and distribution in various marsh microhabitats, the value of tidal marshes for fisheries production cannot be definitively assessed and quantified.

Subrahmanyam and Coultas (1980) presented evidence that predation by fishes acts as a controlling factor in macroinvertebrate abundance. Peak densities in macroinvertebrates were observed in late winter and fall, a month before or after maxima for fishes. Subrahmanyam and Drake (1975) found that residents, migrants, and carnivorous fishes move into the marsh with high tide. Killifish spread out over the entire marsh surface and feed on macroinvertebrates and plant detritus. The diversity of the benthic community and the variation of seasonal abundances by different species of benthos make the tidal marsh habitat a rich and productive food source for the entire year.

References

Adam, P. 1990. *Salt Marsh Ecology*. Cambridge University Press, Cambridge.

Bell, S. S. 1979. Short- and long-term variation in the high marsh meiofauna community. *Estuarine Coast. Mar. Sci.* 9:331–350.

Bell, S. S. 1980. Meiofauna-macrofauna interactions in a high salt marsh habitat. *Ecol. Monogr.* 50:487–505.

Bell, S. S., M. C. Watzin, and B. C. Coull. 1978. Biogenic structure and its effect on the spatial heterogeneity of the meiofauna of a salt marsh. *J. Exp. Mar. Biol.* 35:99–107.

Bingham, F. O. 1972. The influence of environmental stimuli on the direction of movement of the supralittoral gastropod *Littorina irrorata*. *Bull. Mar. Sci.* 22:309–335.

Boesch, D. F. and R. E. Turner. 1984. Dependence of fishery species on salt marshes: roles of food and refuge. *Estuaries* 7:460–468.

Breder, C. M., Jr. 1940. The spawning of *Mugil cephalus* on the Florida west coast. *Copeia* 2:138–139.

Burbanck, W. D. 1959. The distribution of the estuarine isopod *Cyathura* sp. along the eastern coast of the United States. *Ecology* 40:507–511.

Caldwell, D. K. 1954. Additions to the known fish fauna in the vicinity of Cedar Key, Florida. *Quart. J. Fla. Acad. Sci.* 17:182–184.

Caldwell, D. K. 1955. Notes on the distribution, spawning, and growth of the spot-tailed pinfish, *Diplodus holbrooki*. *Quart. J. Fla. Acad. Sci.* 18:73 83.

Cammen, L. M. 1979. The macro-infauna of a North Carolina salt marsh. *Am. Midl. Nat.* 102:244–253.

Cammen, L. M., E. D. Seneca, and B. J. Copeland. 1974. *Animal Colonization of Salt Marshes Artificially Established on Dredge Spoil*. Sea Grant Publication SG-74-15. University of North Carolina, Chapel Hill. 67 pp.

Chabreck, R. H. 1988. *Coastal Marshes, Ecology and Wildlife Management*. University of Minnesota Press, Minneapolis.

Cherr, G. 1974. Species composition and diel variations in the ichthyofaunal community of intertidal grassbeds in the northeastern Gulf of Mexico. M.S. thesis. Florida State University, Tallahassee. 141 pp.

Chester, A. R., R. L. Ferguson, and G. W. Thayer. 1983. Environmental gradients and benthic macroinvertebrate distributions in a shallow North Atlantic estuary. *Bull. Mar. Sci.* 33:282–295.

Coates, K. A. and C. Erseus. 1986. Marine enchytraeids (Oligochaeta) of the coastal northwest Atlantic (northern and mid U.S.A.). *Zool. Scripta* 14:103–116.

Cooper, A. W. 1974. Salt marshes. Pages 55–98 in H. T. Odum, B. J. Copeland, and

E. A. MacMahan, Eds. *Coastal Ecological Systems of the United States.* Vol. 2. Conservation Foundation, Washington, D.C.

Coull, B. C. and S. S. Bell. 1979. Perspectives of marine meiofaunal ecology. Pages 189–216 in R. J. Livingston, Ed. *Ecological Processes in Coastal and Marine Systems.* Marine Science, Vol. 10. Plenum Press, New York.

Coull, B. C. and J. W. Fleeger. 1977. Long-term temporal variation and community dynamics of meiobenthic copepods. *Ecology* 58:1136–1143.

Coull, B. C., S. S. Bell, A. M. Savory, and B. W. Dudley. 1979. Zonation of meiobenthic copepods in a southeastern United States salt marsh. *Estuarine Coast. Mar. Sci.* 9:181–188.

Coultas, C. L. 1969. Some saline marsh soils in north Florida. *Proc. Soil Crop Sci. Soc. Fla.* 29:111–123.

Daiber, F. C. 1982. *Animals of the Tidal Marsh.* Van Nostrand-Reinhold, New York.

Dardeau, M. R., R. F. Modlin, W. W. Schroeder, and J. P. Stout. 1992. Estuaries. Pages 615–744 in C. T. Hackney, S. M. Adams, and W. A. Martin, Eds. *Biodiversity of Southeastern United States/Aquatic Communities.* John Wiley and Sons, New York.

Day, J. W., Jr., W. G. Smith, P. R. Wagner, and W. C. Stone. 1973. *Community Structure and Carbon Budget of a Salt Marsh and Shallow Bay Estuarine System in Louisiana.* Publication No. LSU-SG-72-04. Center for Wetland Research, Louisiana State University, Baton Rouge. 80 pp.

de la Cruz, A. A. 1974. Primary productivity of coastal marshes in Mississippi. *Gulf Res. Rep.* 4:351–356.

de la Cruz, A. A. 1980. Recent advances in our understanding of salt marsh ecology. Pages 51–65 in P. L. Fore and R. D. Peterson, Eds. *Proceedings of the Gulf of Mexico Coastal Ecosystems Workshop.* FWS/OBS-80/30. Albuquerque, New Mexico.

de la Cruz, A. A. 1981. Differences between South Atlantic and Gulf coast marshes. Pages 10–20 in R. C. Carey and P. S. Markovitz, Eds. *Proceedings of U.S. Fish and Wildlife Service Workshop on Coastal Ecosystems of the Southeastern United States, February 1980, Big Pine Key, FL.* FWS/OBS-80/59.

Durako, M. J., J. A. Browder, W. L. Kruczynski, C. B. Subrahmanyam, and R. E. Turner. 1985. Salt marsh habitat and fishery resources of Florida. Pages 189–280 in W. Seaman, Jr., Ed. *Florida Aquatic Habitat and Fishery Resources.* Florida Chapter, American Fisheries Society, Eustis, Florida.

Eleuterius, L. N. 1972. The marshes of Mississippi. *Castanea* 3:153–168.

Eleuterius, L. N. 1973. The marshes of Mississippi. Pages 147–190 in *Cooperative Gulf of Mexico Estuarine Inventory and Study, Mississippi.* Gulf Coast Research Lab., Ocean Springs, Mississippi.

Eleuterius, L. N. 1976. The distribution of *Juncus roemerianus* in the salt marshes of North America. *Chesapeake Sci.* 17:289–292.

Eleuterius, L. N. and C. K. Eleuterius. 1979. Tide levels and salt marsh zonation. *Bull. Mar. Sci.* 29(3):394–400.

Erseus, C. 1986. Marine Tubificidae (Oligochaeta) at Hutchinson Island, Florida. *Proc. Biol. Soc. Wash.* 99:286–315.

Erseus, C. 1990. The marine Tubificidae of the barrier reef ecosystem at Carrie Bow Cay, Belize, and other parts of the Caribbean Sea, with descriptions of 27 new species and revisions of *Heterodrilus, Thalassodrilides*, and *Smithsonidrilus. Zool. Scripta* 19:243–303.

Fotheringham, N. 1980. *Beachcomber's Guide to Gulf Coast Marine Life*. Gulf Publishing, Houston, Texas. 124 pp.

Frankenberg, D. and W. D. Burbanck. 1963. A comparison of physiology and ecology of the estuarine isopod *Cyathura polita* in Massachusetts and Georgia. *Biol. Bull.* 125:81–95.

Gilmore, G. and L. Trent. 1974. *Abundance of Benthic Macroinvertebrates in Natural and Altered Estuarine Areas*. NOAA Technical Report NMFS SSRF-67. U.S. National Oceanic and Atmospheric Administration, Washington, D.C. 13 pp.

Gunter, G. 1945. Studies on marine fishes of Texas. *Publ. Inst. Mar. Sci. Texas* 1: 1–190.

Hackney, C. T. and A. A. de la Cruz. 1981. Some notes on the macrofauna of an oligohaline tidal creek in Mississippi. *Bull. Mar. Sci.* 31:658–661.

Haines, E. B. 1979. Interactions between Georgia salt marshes and coastal waters: a changing paradigm. Pages 35–46 in R. J. Livingston, Ed. *Ecological Processes in Coastal and Marine Systems*. Marine Science, Vol. 10. Plenum Press, New York. 548 pp.

Haines, E. B. and C. L. Montague. 1979. Food sources of estuarine invertebrates analyzed using $^{13}C/^{12}C$ ratios. *Ecology* 60:48–56.

Hamilton, P. V. 1976. Predation on *Littorina irrorata* (Mollusca: Gastropoda) by *Callinectes sapidus* (Crustacea:Portunidae). *Bull. Mar. Sci.* 26:403–409.

Hamilton, P. V. 1977. Daily movements and visual location of plant stems by *Littorina irrorata* (Mollusca:Gastropoda). *Mar. Behav. Physiol.* 4:293–304.

Hamilton, P. V. 1978a. Adaptive visually-mediated movements of *Littorina irrorata* (Mollusca:Gastropoda) when displaced from their natural habitat. *Mar. Behav. Physiol.* 5:255–271.

Hamilton, P. V. 1978b. Intertidal distribution and long-term movements of *Littorina irrorata* (Mollusca:Gastropoda). *Mar. Biol.* 46:49–58.

Hansen, D. J. 1969. Food, growth, migration, reproduction, and abundance of pinfish, *Lagodon rhomboides*, and Atlantic croaker, *Micropogon undulatus*, near Pensacola, Florida, 1963–65. *U.S. Fish Wildl. Serv. Fish. Bull.* 68:135–146.

Harlos, D. 1976. Environmental distribution of *Melampus bidentatus* (Pulmonata) and *Cerithidea scalariformis* (Prosobranchia) in a Florida tidal marsh. M.S. thesis. Mississippi State University, Mississippi State. 66 pp.

Harrington, R. W. and E. S. Harrington. 1972. Food of female marsh killifish, *Fundulus confluentus* Goode and Bean, in Florida. *Am. Midl. Nat.* 87:492–502.

Hausman, S. A. 1936. Food and feeding activities of the salt marsh snail (*Melampus bidentatus*). *Anat. Rec.* 67:127.

Healy, B. 1975. Fauna of the salt marsh, North Bull Island, Dublin (Eire). *Proc. R. Ir. Acad. Sci. B* 75(10):225–234.

Healy, B. 1994. New species of *Marionina* (Oligochaeta, Enchytraeidae) from *Spartina* salt marshes on Sapelo Island, Georgia, U.S.A. *Proc. Biol. Soc. Wash.* 107: 164–173.

Heard, R. W. 1982. *Guide to Common Tidal Marsh Invertebrates of the Northeastern Gulf of Mexico.* MS-AL Sea Grant Consortium Publication No. MASGP-79-004.82. Ocean Springs, Mississippi.

Heip, C., M. Vincz, N. Smol, and G. Vranken. 1982. The systematics and ecology of free-living marine nematodes. *Helminthol. Abstr. Ser. B* 51:1–31.

Heip, C., M. Vincx, and G. Vranken. 1985. The ecology of marine nematodes. *Oceanogr. Mar. Biol. Annu. Rev.* 23:399–489.

Herald, E. S. and R. S. Strickland. 1949. An annotated list of fishes of Homosassa Springs, Florida. *Quart. J. Fla. Acad. Sci.* 11:99–109.

Hettler, W. F. 1989. Nekton use of regularly-flooded saltmarsh cordgrass habitat in North Carolina, U.S.A. *Mar. Ecol. Prog. Ser.* 56:111–118.

Horlick, R. G. and C. B. Subrahmanyam. 1983. Macroinvertebrate infauna of a salt marsh tidal creek. *Northeast Gulf Sci.* 6:79–89.

Humphrey, W. D. 1979. Diversity, distribution and relative abundance of benthic fauna in a Mississippi tidal marsh. Ph.D. dissertation. Mississippi State University, Mississippi State. 93 pp.

Ivester, M. S. 1978. Faunal dynamics. Part III. Pages 1–36 in L. R. Brown, A. A. de la Cruz, M. S. Ivester, and J. P. Stout, Eds. *Evaluation of the Ecological Role and Techniques for the Management of Tidal Marshes on the Mississippi and Alabama Gulf Coast.* MS-AL Sea Grant Publication No. MASGP-78-004. Ocean Springs, Mississippi.

Jordan, D. S. 1884. The fishes of Key West, Florida, with notes and descriptions. *Proc. U.S. Natl. Mus.* 7:103–150.

Keppner, E. J. 1986. New species of free-living marine nematodes (Nematoda:Enoplida) from Bay County, Florida, U.S.A. *Trans. Am. Microsc. Soc.* 105:319–337.

Keppner, E. J. 1987a. Observations on three known free-living marine nematodes of the Family Ironidae (Nematoda:Enoplida) and a description of *Thalassironus lynnae* n.sp. from northwest Florida. *Proc. Biol. Soc. Wash.* 100:1023–1035.

Keppner, E. J. 1987b. Five new and one known species of free-living marine nematodes of the Family Oncholaimidae (Nematoda:Enoplida) from northwest Florida, U.S.A. *Trans. Am. Microsc. Soc.* 106:214–231.

Keppner, E. J. 1987c. Another new species of free-living marine nematode (Nematoda:Enoplida) from northwest Florida, U.S.A. *Trans. Am. Microsc. Soc.* 106:348–353.

Keppner, E. J. 1987d. Five new species of free-living marine nematodes (Nematoda:Enoplida) from a northwest Florida, U.S.A. estuary. *Trans. Am. Microsc. Soc.* 106:333–347.

Keppner, E. J. 1988a. Six new species of free-living marine nematodes (Nematoda: Araeolaimida:Enoplida) from two estuaries in northwest Florida. *Trans. Am. Microsc. Soc.* 197:79–95.

Keppner, E. J. 1988b. *Thoonchus longisetosus* and *Oxyonchus striatus*, new species of free-living marine nematodes (Nematoda:Enoplida) from northwest Florida, U.S.A. *Proc. Biol. Soc. Wash.* 101:183–191.

Keppner, E. J. 1989. Four new species of free-living nematodes in the genus *Pareurytomina* (Nematoda:Enoplida) with observations on other members of the genus. *Proc. Biol. Soc. Wash.* 102:249–263.

Keppner, E. J. and L. A. Keppner. 1988. A redescription of *Oncholaimoides elonatus* Hopper, 1961 (Nematoda:Enoplida) with descriptions of the other two members of the genus. *Gulf Res. Rep.* 8:137–144.

Keppner, E. J. and A. C. Tarjan. 1989. Illustrated key to the genera of free-living marine nematodes of the Order Enoplida. *NOAA Tech. Rep. NMFS* 77:1–26.

Kerwin, J. A. 1972. Distribution of the salt marsh snail (*Melampus bidentatus*) (Say) in relation to marsh plants in the Poropotank River area, Virginia. *Chesapeake Sci.* 13(2):150–153.

Kilby, J. D. 1948. A preliminary report on the young striped mullet (*Mugil cephalus* Linnaeus) in two Gulf coastal areas of Florida. *Quart. J. Fla. Acad. Sci.* 2:7–23.

Kilby, J. D. 1955. The fishes of two Gulf coastal areas of Florida. *Tulane Stud. Zool.* 2:175–247.

Kraeuter, J. N. and P. L. Wolf. 1974. The relationship of marsh macroinvertebrates to salt marsh plants. Pages 449–462 in R. F. Reimold and W. H. Queen, Eds. *Ecology of Halophytes.* Academic Press, New York.

Kruczynski, W. L. and C. B. Subrahmanyam. 1978. Distribution and breeding cycle of *Cyathura polita* (Isopoda:Anthuridae) in a *Juncus roemerianus* marsh of northern Florida. *Estuaries* 1:93–100.

Kruczynski, W. L., C. B. Subrahmanyam, and S. H. Drake. 1978a. Studies on the plant community of a north Florida salt marsh. Part I. Primary production. *Bull. Mar. Sci.* 28:316–334.

Kruczynski, W. L., C. B. Subrahmanyam, and S. H. Drake. 1978b. Studies on the plant community of a north Florida salt marsh. Part II. Nutritive value and decomposition. *Bull. Mar. Sci.* 28:707–715.

Kuenzler, E. J. 1961. Structure and energy flow of a mussel population in a Georgia salt marsh. *Limnol. Oceanogr.* 6:191–204.

Kurz, H. and D. Wagner. 1957. Tidal marshes of the Gulf and Atlantic coasts of north Florida and Charleston, South Carolina. *Fla. St. Univ. Stud.* 24:1–168.

LaSalle, M. W. and L. P. Rozas. 1991. Comparing benthic macrofaunal assemblages of creekbank beds of the spikerush *Eleocharis parvula* (R&S) link and adjacent unvegetated areas in a Mississippi brackish marsh. *Wetlands* 11(2):229–244.

Leatham, W., P. Kinner, and D. Maurer. 1976. Northern range extension of the Florida marsh clam *Cyrenoidea floridana* (Superfamily Cyrenoidacea). *Nautilus* 90:93–93.

Lipcius, R. N. and C. B. Subrahmanyam. 1986. Temporal factors influencing killifish abundance and recruitment in Gulf of Mexico salt marshes. *Estuarine Coastal Shelf Sci.* 22:101–114.

Lynne, G. D., P. Conroy, and F. J. Prochaska. 1981. Economic valuation of marsh areas for marine production processes. *J. Environ. Econ. Manage.* 8:175–186.

Macnae, W. 1957a. The ecology of the plants and animals in the intertidal regions of the Zwartkops estuary near Port Elizabeth, South Africa. Part I. *J. Ecol.* 45: 113–131.

Macnae, W. 1957b. The ecology of the plants and animals in the intertidal regions of the Zwartkops estuary near Port Elizabeth, South Africa. Part II. *J. Ecol.* 45: 361–387.

Mare, M. F. 1942. A study of a marine benthic community with special reference to the micro-organisms. *J. Mar. Biol. Assoc. U.K.* 25:517–554.

McIvor, C. C. and W. E. Odum. 1986. The flume net: a quantitative method of sampling fishes and marcocrustaceans on tidal marsh surfaces. *Estuaries* 9:219–224.

Miller, J. M. and M. L. Dunn. 1980. Feeding strategies and patterns of movement in juvenile estuarine fishes. Pages 437–448 in V. S. Kennedy, Ed. *Estuarine Perspectives*. Academic Press, New York.

Mitsch, W. J. and J. G. Gosselink. 1986. *Wetlands*. Van Nostrand-Reinhold, New York.

Moody, W. D. 1950. A study of the natural history of the spotted trout, *Cynoscion nebulosus*, in the Cedar Key, Florida, area. *Quart. J. Fla. Acad. Sci.* 12:147–171.

Nakamura, E. L., J. R. Taylor, and I. K. Workman. 1980. *The Occurrence of Life Stages of Some Recreational Marine Fishes in Estuaries of the Gulf of Mexico*. NOAA Technical Memo. NMFS-SEFC-45. National Oceanic and Atmospheric Administration, SE Fisher Center, Miami, Florida.

Nichol, E. A. 1936. The ecology of a salt marsh. *J. Mar. Biol. Assoc. U.K.* 20:203–261.

Nixon, S. W. and C. A. Oviatt. 1973. Ecology of a New England salt marsh. *Ecol. Monogr.* 43:463–498.

Odum, E. P. 1971. *Fundamentals of Ecology*. W.B. Saunders, Philadelphia.

Odum, W. E. and T. G. Smith. 1981. Habitat value of coastal wetlands. Pages 30–35 in R. C. Carey and P. S. Markovitz, Eds. *Proceedings of the U.S. Fish and Wildlife Service Workshop on Coastal Ecosystems of the Southeastern United States, February 1980, Big Pine Key, FL*. FWS/OBS-80/59.

Peters, D. S., D. W. Ahrenholz, and T. R. Rice. 1979. Harvest and value of wetland associated fish and shellfish. Pages 606–617 in P. E. Greeson, J. R. Clark, and J. E. Clark, Eds. *Wetland Functions and Values: The State of Our Understanding*. American Water Resources Association, Minneapolis, Minnesota.

Pomeroy, L. R. and R. G. Wiegert. 1981. *The Ecology of a Salt Marsh*. Springer-Verlag, New York.

Provost, M. W. 1976. Tidal datum planes circumscribing salt marshes. *Bull. Mar. Sci.* 26:558–563.

Ranwell, D. S. 1972. *Ecology of Salt Marshes and Sand Dunes*. Chapman and Hall, London.

Reid, G. K., Jr. 1954. An ecological study of the Gulf of Mexico fishes in the vicinity of Cedar Key, Florida. *Bull. Mar. Sci. Gulf Caribb.* 4:1–94.

Rozas, L. P. and M. W. LaSalle. 1990. A comparison of the diets of Gulf killifish, *Fundulus grandis* Baird and Girard, entering and leaving a Mississippi brackish marsh. *Estuaries* 13:332–336.

Rufison, R. A. 1981. Substrate preferences of juvenile shrimps in estuarine habitats. *Contrib. Mar. Sci.* 24:35–52.

Russell-Hunter, W. D., M. L. Apley, and R. D. Hunter. 1972. Early life-history of *Melampus* and the significance of semilunar synchrony. *Biol. Bull.* 143:623–656.

Sanders, H. L. 1960. Benthic studies in Buzzards Bay. III. The structure of the soft bottom community. *Limnol. Oceanogr.* 5:138–153.

Sheridan, P. F. 1992. Comparative habitat utilization by estuarine macrofauna within the mangrove ecosystem of Rookery Bay, Florida. *Bull. Mar. Sci.* 50(1):21–37.

Shirley, M. A., A. H. Hines, and T. G. Wolcott. 1990. Adaptive significance of habitat selection by molting adult blue crabs *Callinectes sapidus* (Rathburn) within a subestuary of central Chesapeake Bay. *J. Exp. Mar. Biol. Ecol.* 140:107–119.

Stout, J. P. 1984. The ecology of irregularly flooded salt marshes of the northeastern Gulf of Mexico: a community profile. *U.S. Fish Wildl. Serv. Biol. Rep.* 85(7.1): 1–98.

Subrahmanyam, C. B. 1980. Oxygen consumption of estuarine fish in relation to external oxygen tension. *Comp. Biochem. Physiol.* 67A:129–133.

Subrahmanyam, C. B. and C. L. Coultas. 1980. Studies on the animal communities in two north Florida salt marshes. Part III. Seasonal fluctuations of fish and macroinvertebrates. *Bull. Mar. Sci.* 30:790–817.

Subrahmanyam, C. B. and S. H. Drake. 1975. Studies on the animal communities in two north Florida salt marshes. Part I. Fish communities. *Bull. Mar. Sci.* 25:445–465.

Subrahmanyam, C. B. and W. L. Kruczynski. 1979. Colonization of polychaetous annelids in the intertidal zone of a dredged material island. Pages 279–296 in J. F. Grassle et al., Eds. *Ecological Diversity in Theory and Practice.* Statistical Ecology Series, Vol. 6. Intern. Coop. Publ. House, Fairland, Maryland.

Subrahmanyam, C. B., W. L. Kruczynski, and S. H. Drake. 1976. Studies on the animal communities in two north Florida salt marshes. Part II. Macroinvertebrate communities. *Bull. Mar. Sci.* 26:172–195.

Sullivan, M. J. and C. A. Moncreiff. 1990. Edaphic algae are an important component of salt marsh food-webs: evidence from multiple stable isotope analyses. *Mar. Ecol. Prog. Ser.* 62:149–159.

Tagatz, M. E. 1968. Biology of the blue crab, *Callinectes sapidus* Rathburn, in the St. Johns River, Florida. *U.S. Fish Wildl. Serv. Fish. Bull.* 67(1):17–33.

Tarjan, A. C. 1980. *An Illustrated Guide to the Marine Nematodes.* Institute of Food and Agricultural Sciences, University of Florida, Gainesville.

Teal, J. M. 1962. Energy flow in the salt marsh ecosystem of Georgia. *Ecology* 43: 614–624.

Teal J. M. and W. Wieser. 1966. The distribution and ecology of nematodes in a Georgia salt marsh. *Limnol. Oceanogr.* 6:217–222.

Thayer, G. W. and J. F. Ustach. 1981. Gulf of Mexico wetlands: value, state of knowledge and research needs. in *Proceedings of a Symposium on Environmental Research Needs in the Gulf of Mexico, Key Biscayne, FL, May 1981.* National Oceanic and Atmospheric Administration, Washington, D.C.

Thayer, G. W., H. H. Stewart, W. J. Kenworthy, J. F. Ustach, and A. B. Hall. 1979. Habitat values of salt marshes, mangroves, and seagrasses for aquatic organisms. Pages 235–247 in P. E. Greeson, J. R. Clark, and J. E. Clark, Eds. *Wetland Functions and Values: The State of Our Understanding.* American Water Resources Association, Minneapolis, Minnesota.

Thomas, J. L., R. J. Zimmerman, and T. J. Minello. 1990. Abundance patterns of juvenile blue crabs (*Callinectes sapidus*) in nursery habitat of two Texas bays. *Bull. Mar. Sci.* 46:115–125.

Thorson, G. 1957. Bottom communities (sublittoral and shallow shelf). *Geol. Soc. Am. Mem. 67*, 1:461–534.

Townsend, B. C., Jr. 1956. A study of the spot, *Leiostomus xanthurus* Lacépéde, in Alligator Harbor, Florida. M.S. thesis. Florida State University, Tallahassee. 42 pp.

Turner, R. E. 1977. Intertidal vegetation and commercial shrimp yields of penaeid shrimp. *Trans. Am. Fish. Soc.* 106:41–416.

Wiegert, R. G. and B. J. Freeman. 1990. Tidal salt marshes of the southeast Atlantic coast: a community profile. *U.S. Fish Wildl. Serv. Biol. Rep.* 85(7.29):1–70.

Wieser, W. and J. Kanwisher. 1961. Ecological and physiological studies on marine nematodes from a small salt marsh near Woods Hole, Massachusetts. *Limnol. Oceanogr.* 6:262–270.

Wilder, B. J. 1979. Distribution and ecology of the salt marsh snail *Melampus bidentatus* Say in a Mississippi tidal marsh, with a brief study of salinity preference. M.S. thesis. Mississippi State University, Mississippi State. 42 pp.

Young, D. L. 1973. Studies of Florida Gulf coast salt marshes receiving thermal discharges. Pages 532–550 in J. W. Gibbons and R. R. Sharitz, Eds. *Thermal Ecology: Proceedings of a Symposium held at Augusta, GA, USA, May 3–5, 1973.*

Zilberberg, M. H. 1966. Seasonal occurrences of fishes in a coastal marsh of northwest Florida. *Publ. Inst. Mar. Sci. Texas* 11:126–134.

Zimmerman, R. J. and T. J. Minello. 1984. Densities of *Penaeus aztecus*, *P. setiferus*, and other natant macrofauna in a Texas salt marsh. *Estuaries* 7:421–433.

Zimmerman, R. J., T. J. Minello, and G. Zamora. 1982. Habitat selection by penaeid shrimp within a Texas marsh. *Am. Zool.* 22:882.

Zimmerman, R. J., T. J. Minello, D. Smith, and J. Kostera. 1990a. *The Use of Juncus and Spartina Marshes by Fisheries Species in Lavaca Bay, Texas, with Reference to the Effects of Floods.* NOAA Technical Memo. NMFS-SEFC-251. National Oceanic and Atmospheric Administration, SE Fisher Center, Miami, Florida.

Zimmerman, R. J., T. J. Minello, T. Baumer, and M. Castiglione. 1990b. Utilization of Nursery Habitats in San Antonio Bay in Relation to Annual Salinity Variation. Final Report to Texas Parks and Wildlife and the Texas Water Development Board. NMFS Galveston Laboratory, February 1990. 60 pp.

Terrestrial Arthropods

7

Jorge R. Rey and Earl D. McCoy

Salt marshes are attractive subjects for systems ecologists because of their apparent simplicity (few species of plants, with relatively sharp zonations). In monitoring pathways of energy and nutrient flow, scientists have examined many of the floral and faunal components of the marsh, including an often-overlooked one, the terrestrial arthropods (insects, arachnids, chilopods, and diplopods). Marsh arthropods have been studied as both "compartments" in energy flow schemes (Odum and Smalley, 1959; Smalley, 1959, 1960; Teal, 1959, 1962; Williams and Murdoch, 1972) and as elements in trophic webs (Kale, 1964, 1965; Marples, 1966); as a result, seasonal abundance patterns of at least a few species have been determined.

Early studies of salt marsh terrestrial arthropods centered mainly on natural history observations (Arndt, 1914; Metcalf, 1920; Metcalf and Osborne, 1920) and on control of anthropophilic pests (see Daiber [1974] for a review). More recently, abundance patterns and habitat preferences of various insects and arachnids have been investigated (Beshear, 1971; Cameron, 1972; Dukes et al., 1974; Foster and Treherne, 1975, 1976; Luxton, 1967; Rockal and Hansens, 1970; Tippins and Beshear, 1968), but these studies have generally been limited to one or a few species.

Very little is known of the species diversity and abundance patterns of whole arthropod communities. The few existing large-scale surveys reveal that these communities can vary greatly in species diversity and relative abundance over time (Barnes, 1953; Bickley and Seek, 1975; Davis, 1964; Davis and Grey, 1966; Pavoir-Smith, 1956; Payne, 1972). LaSalle and de la Cruz (1985) studied the spider fauna of *Spartina cynosuroides* and *Juncus roemerianus* marshes in Mississippi, and Palmer (1987) describes the phytophagous insect fauna associated with *Baccharis* spp. in Texas, Louisiana, and northern Mexico.

The terrestrial arthropod faunas of northwest Florida Gulf coast salt marshes have been studied in terms of their abundance and diversity (McCoy, 1977; McCoy and Rey, 1981b, 1987; Rey, 1978; Rey and McCoy, 1982, 1983, 1986), community and species-level biogeographic patterns (Rey, 1979, 1981, 1985; Rey and McCoy, 1979; Rey and Strong, 1983; Strong and Rey, 1983), factors influencing the dynamics of several insect populations (Stiling and Strong, 1982b, 1984), and relationships between population density and habitat structure and wing polymorphism in planthoppers (Homoptera: Delphacidae) (Denno and Grissel, 1979; McCoy and Rey, 1981a; Strong and Stiling, 1983; Juilianna, 1984).

Our data for Gulf coast marshes come from studies carried out in the marshes of Wakulla County, within the St. Marks National Wildlife Refuge. The study sites were on Apalachee Bay, 6.8 km of west of the town of St. Marks. Four sites were chosen for the study: (1) Wakulla Beach (WB), (2) a narrow peninsula which forms the east side of Goose Creek Bay (PE), (3) John's Island (JI), a 1.7-km^2 island approximately 450 m west of the peninsula, and (4) Patty's Island (PI) a 1.2-km^2 island approximately 900 m south of Wakulla Beach (Figure 7.1). There were 20 sampling stations established at these sites to span a range of vegetation types (large expanses of *J. roemerianus*, fringes and small clumps of *S. alterniflora*, meadows dominated by *Distichlis spicata*, and small shrubs growing on levees) and exposure (windward vs. leeward). Primary collections were made with heavy-duty muslin sweep nets, augmented with visual and suction (D-Vac) collections. We computed two diversity indices, the reciprocal of Simpson's Index (SIR) and simple species number (S), to monitor the possible range of diversity changes; SIR is most sensitive to common species, whereas S is most sensitive to rare species (Hill, 1973). We also computed a dominance index (D1 = number of individuals in the most abundant species/total number of individuals in the sample).

During the course of the study, we collected 25,245 individuals representing approximately 585 morphospecies in 20 arthropod orders. Visual collections yielded 13 additional species (primarily odonates, lepidopterans, and aculeate hymenopterans) and suction collections yielded 14 additional species (mainly parasitic hymenopterans and dolichopodid dipterans).

Tables 7.1 and 7.2 list the arthropod species recorded from the St. Marks marshes. Specific identifications for some groups are more complete than others. This situation reflects the difficulty in performing certain specific determinations, the unwillingness of some taxonomic specialists to examine the material, and the absence of large, taxonomically useful series of specimens for many of our taxa (some being represented by only one or a few individuals).

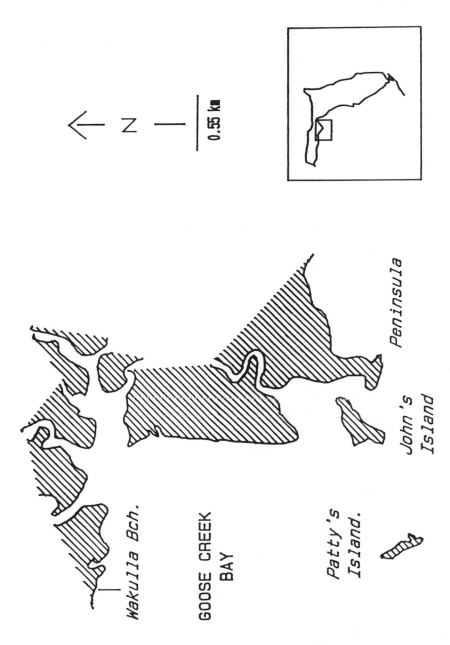

Figure 7.1 Map of the study area showing the sampling sites.

Table 7.1 List of Terrestrial Arthropods Collected in Northwest Florida Salt Marshes

Taxon	SF	JM	DM	HS
Collembola				
Entomobryidae *Willowsia plantani* Nicolet				x
Poduridae *Xenylla grisea* Axelson	x	x		
Sminthoridae *Sminthurus dorsalis* Banks	x	x		
Orthoptera				
Mantidae *Stagmomantis* sp.	x	x	x	x
Tettigoniidae *Conocephalus nigropleuroides* (Fox)	x	x	x	x
Conocephalus sp.	x	x	x	
Amblycorypha sp.		x		x
Acrididae *Neoconocephalus triops* (L)	x		x	
Neoconocephalus sp.	x		x	x
Paroxya sp.		x		x
Melanoplus sp.				x
Leptysma sp.		x		
Orphuela sp.		x		x
Melanoplus sp.				x
Schistocerca gregaria	x	x		x
Gryllidae *Anaxipha* sp.		x		
Oecanthus quadripunctatus Beutenmuller				x
Cycloptilum sp.	x	x		
Hapithus agitator Uhler	x	x		
Dermaptera				
Forficulidae *Doru taeniatum* (Dohrn)	x			
Psocoptera				
Ectopsocidae *Ectopsocopsis cryptomeriae* (Enderlein)	x			x
Psocidae *Indiopsocus insulanus* Chapman	x			
Caeciliidae *Caecilius* sp. nr *manteri* (Summerman)	x	x		
Thysanoptera				
Thripidae *Frankliniella bispinosa* (Morgan)	x	x	x	
Frankliniella sp. 2	x	x	x	x
Chirothrips mexicanus Crawford	x		x	x
Phloeothripidae *Bolothrips glivipes* (Hood)	x	x	x	x
Bolothrips sp. 2		x		
Bolothrips sp. 3			x	
Karnyothrips melalecus (Bgn.)	x	x	x	x

Table 7.1 (continued)   List of Terrestrial Arthropods Collected in Northwest Florida Salt Marshes

Taxon		SF	JM	DM	HS
			Zone[a]		
	Adraneothrips sp.	x	x	x	
	Haplothrips graminis (Hood)	x		x	
Hemiptera					
Anthocoridae	*Orius insidiosus* (Say)	x			x
	Gen. et sp. indet. 1	x			x
Miridae	*Taylorilygus pallidulus* (Blandchard)		x	x	x
	Trigonotylus doddi (Dist.)	x			
	Trigonotylus uhleri (Reuter)	x	x	x	
Reduviidae	*Zelus cervicalis* Stal			x	
	Doldina interjungens Bergroth			x	
Nabidae	*Nabis capsiformis* (Germar)	x	x	x	
Lygaeidae	*Cymoninus notabilus* (Dist.)		x		
	Cymus virescens (F.)	x	x		
	Ischnodemus badius Van Duzee	x		x	
	Nysius raphanus Howard			x	
	Pachybrachius vincta (Say)	x	x	x	x
	Ptochiomera nodosa Say				x
	Gen. et sp. indet. 1				x
	Gen. et sp. indet. 2				x
	Gen. et sp. indet. 3	x	x		
	Gen. et sp. indet. 4				x
Coreidae	*Leptoglossus ashmeadi* Heideman		x		
	Leptoglossus phyllopus (Linnaeus)				x
Pentatomidae	*Rhytidolomia saucia* (Say)			x	
	Thyanta custator (F.)		x		
	Thyanta casta Stal			x	
Homoptera					
Membracidae	*Micrutalis calva* (Say)			x	x
	Idioderma virescens Van Duzee	x	x	x	x
	Spissistilus festinus (Say)	x			x
Cicadellidae	*Balclutha* sp. 1				x
	Balclutha sp. 2			x	
	Carneocephala floridana (Ball)			x	
	Draeculacephala portola (Ball)	x	x	x	x
	Empoasca sp.	x		x	x
	Erythroneura sp.		x		

Table 7.1 (continued)   List of Terrestrial Arthropods Collected in Northwest Florida Salt Marshes

	Taxon	SF	JM	DM	HS
	Graminella nigrifrons (Forbes)	x	x	x	x
	Graminella villica (Crumb)	x			x
	Oncometopia orbona (F)			x	
	Pendarus fumidus (Osborn)			x	
	Sanctanus sp.			x	x
	Spangbergiella sp.	x		x	x
	Stirellus bicolor (Van Duzee)				x
	Stragania robusta (Uhler)				x
	Tinobregmus vittatus (Van Duzee)				x
	nr. *Deltocephalus*				x
	Gen. et sp. indet. 1			x	
Cercopidae	*Clastoptera* sp.				x
	Prosapia bicincta (Say)	x			
Delphacidae	*Chloriona slossonae* (Ball)			x	
	Delphacodes spp.[b]	x	x	x	x
	Keyflana hasta Beamer	x	x		
	Liburniella ornata (Stal)		x	x	
	Megamelanus bicolor Ball		x	x	
	Megamelus trifidus Beamer	x	x	x	
	Neomegamelanus dorsalis (Metcalf)			x	
	Neomegamelanus elongatus (Ball)	x		x	
	Pissonotus albovenosus (Osborn)	x		x	x
	Prokelesia marginata (Osborn)	x	x	x	x
	Sogatella kolophon meridiana (Beamer)	x	x		x
	Stobaera pallida Osborn				x
Cixiidae	*Myndus enonatus* Van Duzee			x	
Issidae	*Bruchomorpha* sp.			x	
Psyllidae	*Trioza magnoliae* (Ashmead)		x		
	Paratrioza sp.	x	x	x	x
	Subfamily *Triozinae*				x
Aphidae	*Dactynotus* sp. 1			x	x
	Dactynotus sp. 2				x
	Aphis sp.	x			
Pseudococcidae	*Pseudococcus* sp.				x
	Rhizoecus maritimus (Cockerell)			x	
Coleoptera					
Carabidae	*Platynus* sp.	x			

Table 7.1 (continued) List of Terrestrial Arthropods Collected in Northwest Florida Salt Marshes

	Taxon	SF	JM	DM	HS
				Zone[a]	
	Selenophorus sp.			x	
	Tachys sp.				x
Dytiscidae	*Copelatus caelatipennis* Young	x			
Staphylinidae	*Eudectoides crassicornis* (LeC.) (?)			x	
	Heterota plumbea (Waterhouse) (?)	x			
	Stenus sp.	x			
	Gen. et sp. indet. 1	x			
Silphidae	*Necrodes surinamensis* (Fab.)	x		x	
Helodidae	*Prionocyphon* sp.				x
Scarabaeidae	*Euphoria fulgida* (Fab.)				x
Elateridae	*Conoderus difformis* Fleut.	x			
	Glyphonyx sp.			x	x
Dermestidae	*Dermestes caninus* (Germ.)	x			
Malachiidae	*Attalus* sp.		x		
	Collops sp.	x	x	x	x
	Temnopsophus bimaculatus Horn (?)	x	x	x	x
Nitidulidae	*Carpophilus* sp.nov. [nr. *dimidatus* (Fab.)][c]	x		x	
Cryptophagidae	Gen. et sp. indet. 1				x
Phalacriidae	*Phalacrus* sp.	x		x	x
Coccinellidae	*Cycloneda sanguinea* (L.)				x
	Diomus floridanus (Mulsant)	x	x		
	Diomus terminatus (Say)		x		
	Hippodamia convergens Guer.	x	x		x
	Microweisea sp. (?)	x			
	Olla abdominalis (Say)	x			x
Lathridiidae	*Melanophthalma* sp.	x	x		x
Alleculidae	*Hymenorus densus* LeC.				x
Oedemeridae	*Hypasclera dorsalis* (Hels.)				x
	Oxycopis thoracica (Fab.)				x
Mordellidae	*Mordellistena* sp.	x	x	x	
	Mordellistena splendens Smith[d]	x			
Anthicidae	*Anthicus* sp. 1	x			
	Anthicus sp. 2	x			
Cerambycidae	*Lypsimena fuscata* LeC.				x
	Spalacopsis stolata Neuman		x		
Chrysomelidae	*Altica amoena* Horn	x	x	x	x
	Chaetocnema sp.	x			
	Cryptocephalus pumilus Haldeman	x			x

Table 7.1 (continued)   **List of Terrestrial Arthropods Collected in Northwest Florida Salt Marshes**

Taxon		Zone[a]			
		SF	JM	DM	HS
	Diabrotica undecipunctata Barber				x
	Disonycha pennsylvanica (Illiger)			x	
	Erynephala maritima (LeC.)			x	
	Lexiphanes saponatus (Fab.)	x			
	Monoxia batesia Blatchley	x	x	x	x
	Ophraella integra LeC.	x	x	x	x
	Trirhabda bacharidis (Weber)			x	x
Amthribidae	*Ormiscus saltator* LeC.	x			x
Curculionidae	*Lissorhoptrus lacustris* Koschel	x			
	Notolomus basalis LeC.	x			x
	Pachylobius picivorous Germ.	x			
Scolytidae	*Pityophthorus* sp. (?)	x			
	Stephanoderes sp. (?)			x	x
Languriidae	*Languria taedata* LeConte[d]	x			
Strepsiptera					
Elenchidae	*Elenchus koebelei* (Pierce)	x			
Neuroptera					
Chrysopidae	*Chrysops cubana* Hog.	x	x	x	x
	Chrysops rufilabris Burns	x			
Diptera					
Tipulidae	*Limonia floridana* (Osten Sacken)		x	x	x
	Gen. sp. indet.	x			
Culicidae	*Aedes taeniorhynchus* (Wiedemann)		x	x	x
	Anopheles atropos Dyar & Knab	x			
	Aedes sollicitans (Walker)	x			
	Anopheles bradleyi King	x	x		
	Culex salinarius Coquillett	x			
Ceratopogonidae	*Culicoides furens* (Poey)	x	x	x	x
	Culicoides melleus (Coquillett)			x	
	Culicoides mississipiensis (Hoffman)	x	x	x	x
	Dasyhelea grisea (Coquillett)	x	x		
	Dasyhelea sp. 2	x	x		
	Dasyhelea sp. 3	x	x	x	x
	Dasyhelea sp. 4	x		x	
	Forcipomyia sp.	x	x		x
	Atrichopogon sp.	x	x	x	x

Table 7.1 (continued) List of Terrestrial Arthropods Collected in Northwest Florida Salt Marshes

	Taxon	Zone[a]			
		SF	JM	DM	HS
	Bezzia sp.	x	x		
	Stilobezzia beckae Wirth			x	x
Chironomidae	*Orthocladius* sp.	x	x	x	x
	Microtendipes anticus (Walker)		x		
	Gen. sp. indet.		x	x	
Sciaridae	*Bradysia* sp.				x
Scatopsidae	Gen. sp. indet.		x		
Stratiomyidae	*Brachycara slossonae* (Johnson)		x	x	x
	Sargus sp.	x			
Tabanidae	*Tabanus nigrovittatus* (Macquart)		x	x	x
	Chrysops fulginosus Wiedemann		x		
	Chrysops atlanticus Pechuman	x			
Asilidae	*Laphystia litoralis* Curran	x			
	Laphystia sp. 2	x	x		x
Empididae	Gen. sp. indet.				x
Dolichopodidae	*Thrypticus violaceus* Van Duzee[d]	x			
	Pelastoneurus lamellatus Loew	x	x	x	x
	Paracleius alternans (Lowe)	x	x	x	
Cecidomyiidae	*Calamomyia alterniflorae* Gagne[d]	x			
Pipunculidae	*Tomosaryella* sp.	x	x	x	x
Otitidae	*Chaetopsis apicalis* Johnson	x	x	x	x
	Chaetopsis fulvifrons (Macquart)	x	x	x	x
	Chaetopsis aenea (Wiedemann)	x	x	x	
	Gen. sp. indet.	x		x	x
Tephritidae	*Tephritis stigmatica* (Coquillett)	x	x	x	x
Lauxanidae	Gen. sp. indet.			x	
Lonchaeidae	*Lonchaea* sp.			x	x
Canacidae	*Canace snodgrassii* Coquillett	x		x	x
	Gen. sp. indet.	x		x	x
Ephydridae	*Dimecoenia fuscifemus* Steyskal	x	x	x	
	Dimecoenia spinosa (Loew)	x	x		
	Dimecoenia austrina (Coquillett)	x	x	x	x
	Cressonomyia hinei (Cresson)	x		x	x
	Polytrichophora agens Cresson	x	x	x	
	Polytrichophora sp. 2		x		
	Notiphila bispinosa Cresson	x	x	x	x
	Clanoneurum americanum Cresson	x	x	x	

Table 7.1 (continued)   List of Terrestrial Arthropods Collected in Northwest Florida Salt Marshes

	Taxon	SF	JM	DM	HS
	Hydrellia valida Loew	x	x	x	x
	Psilopa flavida Coquillett			x	
	Psilopa sp. 2			x	
	Letopsilopa sp. 1			x	x
	Letopsilopa sp. 2			x	
	Ceropsilopa costalis Wirth	x		x	x
	Creopsilopa sp.			x	
	Parydra unituberculata Loew	x	x	x	x
	Discocerina sp.	x		x	x
Drosophilidae	Gen. sp. indet.	x			
Chloropidae	*Rhopalopterum carbonarium* (Loew)	x	x	x	x
	Apallates dissidens (Tucker)	x	x	x	
	Apallates particeps (Becker)	x	x	x	
	Monochaetoscinella nigricornis (Loew)	x	x	x	x
	Liohippelates pusio (Loew)			x	x
	Liohippelates bicolor Coquillett	x			
	Incertella sp. 1	x	x	x	x
	Incertella insularis (Malloch)	x	x		x
	Paractecephala aristalis (Coquillett)	x			
Agromyzidae	*Cerodontha* sp.	x			
Aucaligasteridae	*Stenomicra* sp.		x		
	Gen. sp. indet.	x			
Anthomyiidae	*Hylemya* sp.	x	x	x	
Muscidae	*Stomoxys calcitrans* (L.)	x	x		x
	Coenosia sp. 1			x	
	Coenosia sp. 2	x	x	x	x
	Coenosia sp. 3		x		
	Mydaea sp.	x	x	x	
	Lispe albitarsis Stein	x		x	
	Lispe sp.	x			x
	Phaonia sp. 1	x			x
	Phaonia sp. 2	x	x	x	x
	Phaoniinae gen. sp. indet.		x	x	x
	Musca domestica L.	x			
	Gen. sp. indet.	x	x		
	Gen. sp. indet.		x		
Sarcophagidae	*Sarcophaga johnsoni* Aldrich	x		x	
	Tricharaea simplex (Aldrich)	x			

Table 7.1 (continued) List of Terrestrial Arthropods Collected in Northwest Florida Salt Marshes

		Zone[a]			
	Taxon	SF	JM	DM	HS
Hymenoptera[e]					
Formicidae	*Aphaenogaster miamiana* (Wheeler)				x
	Camponotus impressus (Roger)	x	x	x	x
	Cardiocondyla nuda Forel			x	x
	Conomyrma flavopectus (Smith)		x		x
	Crematogaster clara Mayr	x	x	x	x
	Monomorium viridum (Brown)			x	
	Paratrechina longicornis (Laetrille)		x		
	Pseudomyrmex brunneus (Smith)				x
	Pseudomyrmex pallidus Smith			x	x
	Solenopsis invicta (Buren)		x		
	Gen.sp. 1		x		
	Gen.sp. 2 (male reproductive)	x	x	x	
	Gen.sp. 3 (male reproductive)				x
	Gen.sp. 4 (female reproductive)	x	x		
	Gen.sp. 5 (female reproductive)	x			
Vespidae	*Didineis* sp.	x			
	Liris argentata (Beauvois)	x			
Halictidae	*Ceratina dupla* (Say)	x			
Megachilidae	*Megachile* sp.	x			
Apidae	*Dialictus* sp.	x			
Pseudoscorpiones					
Cheliferidae	*Paisochelifer* sp.	x	x	x	
Chernetidae	*Parachernes littoralis* Muchmore	x	x	x	x
	Gen. sp. nov. [nr. *Parachernes*][f]			x	x
Acarina					
Cymbaeremaeidae	*Scapheremaeus* sp.		x		
Oripodidae	*Oripoda* sp.		x	x	
Ceratozetidae	*Antarctozetes* sp.		x	x	
Bdellidae	*Cyta latirositris* (Hermann)	x			
	Gen. et sp. indet.			x	
	Bdellodes sp.	x	x	x	
Erythraeidae	Gen. et sp. indet.				
Caligonellidae	*Neognathus* sp.		x		x
Uropodidae	?*Cilliba* sp.	x			
Phytoseiidae	*Amblyseius* sp.	x	x	x	x

Table 7.1 (continued)   **List of Terrestrial Arthropods Collected in Northwest Florida Salt Marshes**

Taxon		SF	JM	DM	HS
			Zone[a]		
Euzeroconidae	*Euzercon* sp.		x	x	
	Proprioseiopsis lepidus (Chant)		x		
Laelapidae	*Pseudoparasitus* sp.				x
Trombidiidae	Gen. et sp. indet.	x			
	Hypoapsis sp.			x	
	Ololaelaps sp.	x		x	
Passalozendae	*Passalozetes* sp.				x
Galumnidae	Gen. et sp. indet				x
Oribateidae	Gen. et sp. indet.	x	x		
Tarsonemidae	*Ogmotarsonemis erepsis* Lindquist[d]	x			
Diplopoda					
Pyrogodesmidae	*Psochodesmus crescentis*		x		
Polyxenidae	*Polyxenus* sp. 1			x	x
Araneae					
Araneidae	*Acanthapeira venusta* (Banks)	x	x	x	
	Neoscona pratensis (Hentz)	x	x	x	x
	Argiope trifasciata (Forskal)		x		
	Agriope sp.		x		
	Singa eugeni Levi			x	
	Cyclosa sp.			x	
	Nephila clavipes (L.)		x	x	
	Gen. et sp. indet.	x			
Lycosidae	*Pardosa* sp.	x	x		
	Pirata sp.	x			
	Lycosa sp.		x		
Oxyopidae	*Oxyopes* sp. nr. *salticus* (Hentz)			x	x
	Peucetia viridans Hentz				x
Micryphantidae	*Grammonota trivittata* (Banks)	x	x	x	x
	Ceraticelus sp.			x	
Thomisidae	*Misumenops bellulus* Banks				x
	Misumenops sp.	x	x	x	x
	Tibellus sp. nr. *chamberlini* (Gertsch.)		x		
	Tibellus sp.		x		
	Xysticus sp.				x
	Philodromus sp.				x

Table 7.1 (continued) **List of Terrestrial Arthropods Collected in Northwest Florida Salt Marshes**

Taxon		SF	JM	DM	HS
Dictynidae	*Dictyna altamita* (Gertsch. & Davis)	x	x	x	
Tetragnathidae	*Tetragnatha branda* Levi	x	x	x	x
	Tetragnatha laboriosa Hentz	x	x	x	x
	Tetragnatha sp.	x	x	x	x
Theridiidae	*Dipoena* sp.		x		
Linyphiidae	*Florinda coccinea* (Hentz)	x	x		
Gnaphosidae	*Sergiolus ocellatus* (Walck.)	x	x		x
	Gen. et sp. indet.	x			
Clubionidae	*Clubiona littoralis* Banks	x		x	
	Clubiona sp.	x	x		
	Gen. et sp. indet.	x		x	x
Salticidae	*Hentzia palmarum* (Hentz)	x	x	x	
	Hentzia sp.			x	x
	Metaphidippus galathea (Walck.)		x		
	Metaphidippus sp.	x			
	Marpissa bina (Hentz)	x	x		
	Marpissa pikei (Peckham)	x	x	x	
	Marapissa wallacei Barnes	x		x	x
	Marpissa sp.	x			
	Eris marginata (Walck.)		x	x	x
	Phidippus sp.		x		
	Synemosyna sp. nr. *Petrunkevitchi* (Chapin)		x		
	Synageles noxiosa (Hentz)		x		

[a] SF = *Spartina alterniflora* fringe, JM = *Juncus roemerianus* marsh, DM = *Distichlis spicata* meadows, HS = halophytic shrub zone.

[b] *D. propinqua* (Fieber), *D. detecta* (Van Duzee), and *D. puella* (Van Duzee).

[c] Det. W. A. Connell (University of Delaware).

[d] Record added by Stiling and Strong (1981, 1982).

[e] Aculeate Hymenoptera only. See Table 7.2 for nonaculeate Hymenoptera.

[f] Det. W. B. Muchmore (University of Rochester).

Table 7.2   List of Nonaculeate Hymenoptera Collected at St. Marks

Family	Genus	No. species	Species names
Braconidae	*Apanteles*	3	
	Bracon	3	
	Chelonus	1	
	Heterospilus	2	
	Microplitus	1	
	Myosoma	1	
	Opius[a]	1	
	Phanerotoma	1	
	Triapsis	1	
	Indet.	1	
Ichneumonidae	*Casinaria*	1	
	Charitope	1	
	Diadegma	1	*D. pattoni* Ashmead
	Diaparsis	1	
	Enytus	1	
	Isdromus	1	
	Itoplectis	1	*I. conquistator* (Say)
	Labena	1	*L. grallator* (Say)
	Phaeogenes	1	
	Pseuderipternus	1	
	Temelucha	1	
	Tricholabus	1	*T. mitchelli* Heinr.
	Tromatobia	1	*T. variabilis* (Hgn.)
Mymaridae	*Anagrus*	1	*A. delicatus* Dozier[a]
	Gonatocerus	3	*G. dolichocerus*
			sp. nov. 1
			sp. nov. 2
	Polynema	1	
Eulophidae	*Achrysocharella*	1	
	Aprostocetus	6	
	Chryosonotomyia	1	
	Cirrospilus	2	
	Elechertus	2	
	Elasmus	1	*E. setosiscutellatus* Crawford
	Euderus	3	
	Galeopsomyia	3	
	Horsimenus	1	*H. floridanus* (Ashmead)
	Notanisomorpha	1	
	Pnigalio	1	
	Sympiesis	1	*S.* nr. *stigmatipennis* (Ashmead)
	Tetrastichus	5	
Encyrtidae	*Meromizobia*	1	
	Indet.	8	

Table 7.2 (continued) List of Nonaculeate Hymenoptera Collected at St. Marks

Family	Genus	No. species	Species names
Eupelmidae	*Eupelmus*	2	
	Indet.	1	
Pteromalidae	nr. *Aggelma*	1	
	Callitula	2	
	Catolaccus	1	
	Crytogaster	1	
	Halticoptera	1	
	?Merismus	1	
	Neocatalaccus	1	
	Pachyneuron	1	
	Pteromalus	1	
	Trichomalopsis	2	
	Zatropsis	2	
Eurytomidae	*Eurytoma*	1	
	Rileya	1	
	Tenuipetiolus	1	
Chalcidae	*Brachymeria*	1	
	Haltichella	1	
	Spilochalcis	2	*S. maria* (Riley)
			S. side Walker
Charipidae	*Alloxysta*	6	
Eucolidae	*Ganapsis*	1	
	Hexacola	1	
	Kleiodotoma	2	
	Rhoptromeris	1	
Ceraphronidae	*Ceraphron*	1	
Diapriidae	Indet.	2	
Scelionidae	*Macrotelia*	1	
	Opisthacantha	1	
	Telenomus	5	
	Trimorus	1	
	Trissoleus	1	
	Tritelia	1	
	Indet.	4	
Platygasteridae	*Inostema*	1	
	Platygaster	3	
	Synopea	1	
Bethylidae	*Cephalonomia*	1	*C. hyalinipennis* Ashmead
	Gonozius	2	*G. floridanus* (Ashmead)
	Laelius	1	*L. pedatus* Say
	Scerodermus	1	*S. macrogaster* (Ashmead)
Dryinidae	*Dicondylus*	1	*D. americanus* (Perkins)
	Pseudogonatopus	1	*P. sjoestedti* (Kieffer)

[a] Indicates record added by Stiling and Strong (1981, 1982).

For the sake of brevity, many of the uncommon morphospecies not yet identified beyond the family level are not included in our species lists. Notes on some of the orders are presented below and a discussion of the abundance and diversity patterns in the next section. As we have not compared our specimens with those of others, seven species reported by other investigators and not identified from our collections are listed as separate species; the sources of these records are identified in the tables. Only our data were used in our discussions of abundance and diversity patterns in these wetlands.

Approximately 57% of the Homoptera and 50% of the Hemiptera identified to species have also been reported from other North American marshes (Bickley and Seek, 1975; Cameron, 1972; Davis, 1978; Davis and Grey, 1966; Kale, 1964; Lane, 1969; Marples, 1966). In total, about 83% of the Homoptera and 72% of the Hemiptera have at least congeneric species recorded from other North American marshes (Rey and McCoy, 1982).

The numbers of species of Coleoptera recorded from St. Marks compare well with those reported from other marshes in North America (McCoy and Rey, 1981b). Davis (1978) listed nearly 80 species from North Carolina, but included many species from habitats such as marsh ponds and drier coastal habitats not sampled by us. When such species are removed from consideration, Davis's records are comparable in size to ours. Bickley and Seek (1975) report about 140 species of beetles from Maryland marshes, but they include data from brackish marshes, which were not considered in our study.

The composition of the Diptera at St. Marks resembles that of other North American marshes more closely than other orders that we have examined, at least at the generic level (Rey and McCoy, 1986). The composition at the family level is also similar. The five most common families in Georgia, North Carolina, and South Carolina (excluding the Tabanidae; Davis, 1964; Marples, 1966) are Dolichopodidae (17.7% of the total number of species), Chloropidae (15.4%), Ephydridae (14.0%), Ceratopogonidae (7.3%), and Culicidae (6.0%). At St. Marks, the five most common families are Ephydridae (19.6%), Muscidae (15.0%), Ceratopogonidae (12.8%), Chloropidae (10.3%), and Culicidae (5.7%).

Because of the difficulty in identifying nonaculeate Hymenoptera without large series and detailed information concerning hosts, we had only moderate success in obtaining determinations for this group beyond the genus or family level. For this reason, we present a separate table (Table 7.2) with the taxonomic information for this group. Data for the aculeate Hymenoptera are included in Table 7.1. The ants *Camponotus impressus* and *Crematogaster clara* were the only aculeate hymenopterans that were abundant and widespread.

Stiling and Strong (1981, 1982a, 1982b; see below) have added some information about nonaculeate Hymenoptera that occur in these marshes. They reported that *Sympiesis* sp. (Eulophidae), *Pteromalus* sp. (Pteromalidae), and *Opius* sp. (Braconidae) are important parasites of the fly *Hydrellia valida* (Ephydridae). They also reported that *Anagrus delicatus* (Mymaridae) parasitizes the eggs of the planthopper *Prokelisia marginata* (Delphacidae) and that *Macrotelia surfacei* (Scelionidae) is an egg parasite of the grasshopper *Orchelimum fidicinium* (Tettigoniidae).

Abundance and Diversity

The average numbers of individuals and species collected in the various vegetation types are shown in Figure 7.2. Two major abundance peaks are evident: one in March or April and another in late summer or early fall.

Over 95% of the individuals captured during the study belonged to seven arthropod orders (Thysanoptera, Hemiptera, Homoptera, Coleoptera, Diptera, Acarina, and Araneida). The relative abundance of the different orders, however, were not consistent among vegetation types (Figure 7.3). For example, Coleoptera was the most abundant order found in the shrub zone, comprising 31.2% of all the individuals captured. In the other three vegetation types, however, this order never comprised more than 3% of the total. A majority of the beetles collected in the shrubs belong to three species of the family Chrysomelidae (*Monoxia batesia*, *Ophraella integra*, and *Trirhabda baccharidis*; Table 7.1) which became abundant during the spring and summer. Diptera was the most abundant order in the *Distichlis spicata* meadows and the *Spartina alterniflora* fringes (34.7 and 56.1%, respectively) and also the most abundant overall (29.8%). It was third in abundance in the expanses of *Juncus roemerianus* (13%) and fourth in the shrub zone (11.3%), but the D-Vac samples indicate that some flies may be much more abundant around the bases of shrubs, an area not well sampled with sweep nets. Homoptera was the second most abundant order in all vegetation types. The relative abundance of this order was as follows: *D. spicata* meadows (34.2%), shrub zone (23.0%), *S. alterniflora* fringes (17.5%), *J. roemerianus* expanses (17.2%). Overall, the order comprised 22.5% of all individuals collected. Thysanoptera was somewhat less abundant than the previous three orders (less than 10% of the total for any vegetation type), but individual species generated large population sizes at some locations, and some species were moderately abundant throughout the study area. The spiders and mites were both common and ubiquitous throughout the marsh; however, only a few species of mites built up large population sizes. Mites were particularly abundant in the *J. roemerianus*, where they accounted for over 50% of the individuals collected (22.7% in all vegetation types). The spiders accounted for 33.3% of the total number of individuals collected in the *J. roemerianus* but less than 10% in other vegetation types. All other orders of insects and arach-

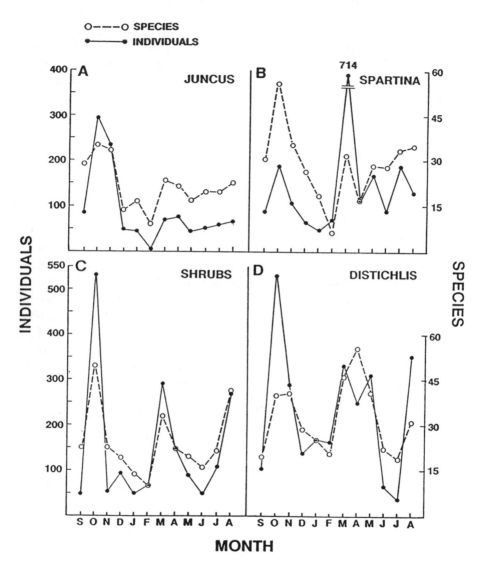

Figure 7.2 Numbers of species and individuals captured in the different habitat types. Points plotted are means for all stations within a given habitat.

nids were much less abundant, although some, like Orthoptera, had ubiquitous species.

Several groups of terrestrial arthropods were underrepresented in these samples. Most Lepidoptera, Odonata, and some aculeate Hymenoptera are adept at fleeing from the sweep nets and were, therefore, rarely captured. Horseflies

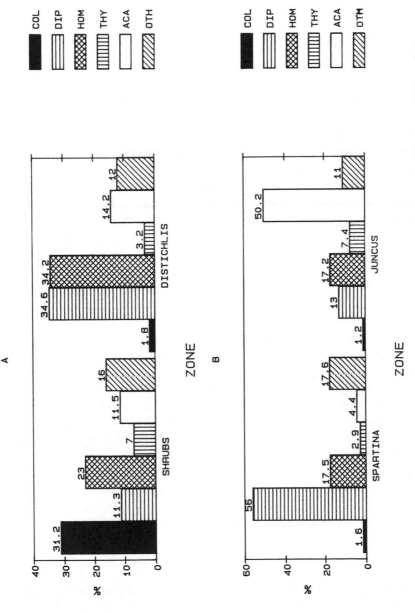

Figure 7.3 Relative abundance of the more common orders in the different vegetation zones. COL = Coleoptera, DIP = Diptera, HOM = Homoptera, THY = Thysanoptera, ACA = Acarina, OTH = others.

and deerflies (Diptera: Tabanidae) likewise can escape sweep nets by flight. Most of the soil-inhabiting arthropods were not captured at all by sweeping and rarely by suction sampling.

Seasonal Patterns

Juncus

At leeward *J. roemerianus* stations, numbers of species ranged from 5 in February to 30 in April at PI, from 6 in February to 36 in September at JI, and from 7 in February to 46 in October and August at PE. Thus, while each station had about the same number of species during winter, peaks in numbers of species in spring and fall displayed an increasing trend proceeding from the far island (PI) to the near island (JI) and finally to the mainland (PE). A similar pattern existed for the windward stations: the range in species richness for PI (9 to 24 species) was less than that of JI (9 to 31 species) which, in turn, is less than that of PE (11 to 35 species) and WB (7 to 37 species). In general, windward stations appeared less variable in species richness than leeward stations.

Peaks in species richness usually occurred at two times of the year: in the fall (August to November) and in the spring (March to May), depending upon the site. Peaks in numbers of species were uniformly accompanied by peaks in numbers of individuals; however, the relative magnitudes of each did not always coincide (i.e., relatively large seasonal increases in numbers of species were sometimes accompanied by only small increases in numbers of individuals and vice versa).

Little regularity was evident in patterns of D1 and SIR for the *J. roemerianus* stations. For example, the peak in numbers of species in March at JI leeward was accompanied by a relatively low value of D1 (0.17), while the peak at PE leeward in the same month was accompanied by a relatively high value of D1 (0.81). Such inconsistencies stem from the apparently haphazard occurrence of large numbers of individuals of several species of mites. This conclusion is corroborated by the fact that in general, when insects alone are considered, peaks in numbers of species are accompanied by relatively low values of D1 and relatively high values of SIR and vice versa.

Spartina

The leeward *S. alterniflora* stations were similar in the numbers of species captured in various months throughout the year except the station at JI leeward, which displayed very low and fairly constant species numbers throughout the year (Figure 7.2), probably because of the generally poor condition of the habitat there; the wide spacing and short stature of the individual plants, together with the narrowness of the entire fringe, is noteworthy. JI and PE windward were tremendously variable, displaying the largest numbers of species recorded on a single sampling occasion at any station. We believe that these very large peaks are due to mass invasions by transients (presumably wind-aided) to these locations.

At *S. alterniflora* stations, peaks in numbers of species were generally accompanied by peaks in numbers of individuals, the same pattern found at the *J. roemerianus* stations. However, in contrast to *J. roemerianus* stations, species peaks were not generally accompanied by low values of D1. Very high values of D1 at PI leeward (0.70) and JI leeward (0.66) accompanied the fall peaks in numbers of species, due to the presence of large numbers of individuals of a species of chloropidfly (Diptera: Chloropidae). These high values of D1 coincided with low values of SIR, as expected. At other times of the year, the values of D1 and SIR remained relatively constant.

Distichlis

The patterns of numbers of species and individuals sampled at stations located within *Distichlis spicata* meadows were quite similar to those described above for the *J. roemerianus* stations, except that numbers of species and individuals were generally greater and more variable than in the *J. roemerianus*. Two factors likely account for these large numbers: (1) *D. spicata* meadows have more species of plants than do the other habitats, although most of these species occur in low abundances and (2) the habitat is much denser than the others and thus provides much greater cover and habitat space per unit area than the others. Again, as in the *J. roemerianus*, peaks in species richness occurred in both spring and fall and were accompanied by low values of D1 and relatively high values of SIR when insects alone are considered.

Shrubs

Diversity and abundance patterns in the shrubs were not as distinct as in the other habitats, except that there were large peaks in species richness in the fall. In general, these peaks also were accompanied by peaks in numbers of individuals, low values of D1, and high values of SIR. There was, however, a peak in numbers of individuals in the spring which was accompanied by high values of D1. This peak was due to the increases in the populations of the three chrysomelid species mentioned above.

Overall

We attribute the spring and fall peaks in species richness to a flush of growth of marsh plants after winter dormancy and the fall peak to the large amounts of standing plant biomass available then. New resources of fruits appear at both times. This reasoning is similar to that of Cameron (1972) and Davis (1964), who showed that marsh plant biomass was positively correlated with insect diversity. Attack on seeds by flies (mostly Ottitidae) and lepidopteran larvae was observed to be seasonally heavy, especially on *S. alterniflora* seeds. We estimated the number of *S. alterniflora* seeds damaged by seed parasites to vary from 10% to more than 30% depending upon the site. The seed utilizers were not solely responsible for the increase in numbers of species that coincided with fruiting, but also a group of associated arthropods. For example, flies inhabiting the seeds were themselves attacked by parasitoids. Although we were not able to estimate the rates of parasitization, emerging wasps were often captured, still within the seeds. We also observed that several species of spiders constructed webs between the *S. alterniflora* fruiting spikes during the fall, perhaps as a response to the activity about this resource. Clearly, similar sequences must accompany fruiting of other marsh plants.

There was a degree of randomness in the distribution of arthropods among the study sites, which becomes evident when species similarities are compared. An interesting pattern, the same for all four habitats, was evident when we compared the species complements within stations (through time) and within habitat types (across all stations) using Jaccard's Similarity Index. At individual stations, similarity over time was generally low (<0.25). However, between stations within the same habitat type, values were often higher, except during the spring and fall peaks when Jaccard Similarity was also very low. In the winter, the similarity values rose to as high as 0.60 (note however that no test of significance is available for the Jaccard Index). All of this indicates that species lists constructed based on the results of one sampling occasion are largely unpredictable, except in winter when a small group of hardy species persists at many stations. These same species account for much of the similarity existing among stations at other times of the year and also exhibit some of the largest population sizes.

By comparison with other habitats, the numbers of species of terrestrial arthropods in salt marshes are low (Table 7.3). Data from Davis (1964) and Barnes (1953) indicate that there are about 380 to 430 species of insects and spiders in the salt marshes around Beaufort, North Carolina. Evans and Murdoch (1968) reported 1000 to 1200 species from a single old field in Michigan; Wheeler (1973a, 1973b) and Pimentel and Wheeler (1973) collected more than 800 species from an alfalfa field in New York, and both Janzen (1973a, 1973b) and Janzen and Schoener (1968) captured more than 500 spe-

Table 7.3 Comparison of the Results of Several Salt Marsh Arthropod Studies that Utilized the Sweep-Netting Method

Study	Location	Sample size	No. of species	No. of individuals
Davis (1964)	Carteret County, North Carolina	8,500	124	13,000
		33,120	400	40,000+
Marples (1966)	Sapelo Island, Georgia	—	—	—
		960	60	12,000
Lane (1969)	San Francisco Bay, California	—	—	—
		120	67	4,000
This study	Wakulla County, Florida	4,680	284	4,500
		16,800	514	18,000

Note: Top figures indicate the results of sweeps in *Spartina alterniflora* and the bottom the total for all habitats sampled.

cies of insects in only 2000 sweeps of tropical American vegetation. Since the arthropods recorded from salt marsh plants usually include all species captured by general sampling techniques, such as sweep-netting, there is no indication of how many species are transients (but see Rey [1981] and Rey and Strong [1983]); thus, the number of true salt marsh associates is likely even smaller than reported.

The number of species taken by Davis (1964) and Barnes (1953) in North Carolina is only 30 to 50 species (about 10%) less than we found at St. Marks (Table 7.3). The figures from North Carolina are the result of over 33,000 sweeps taken over 250 m^2, while our figures are the result of 15,600 sweeps (in habitats comparable to those in North Carolina) over a few square kilometers. Although species richness is similar between the two areas, North Carolina marshes support far greater numbers of individuals than do the St. Marks marshes. Thus, as with the studies of Marples (1966) in Georgia and Lane (1969) in California (see below), a given sweep sampling effort in North Carolina yielded high numbers of individuals (say 500 to 1000) but usually about 10 to 30 species, while in our samples there were rarely more than 200 individuals (especially if the Acarina are excluded) but often 30 to 60 species. In the North Carolina marshes, species also tended to show a stronger habitat preference than they do in northwest Florida (distributions of species among habitats are much more even in northwest Florida, indicating a greater sharing [over time] of species between habitats). More rare species were found at St. Marks than at sites in North Carolina. While about 40% of all arthropods captured in our collections were considered rare (fewer than five individuals captured), less than 25% of those captured by Davis can be considered rare using the same criterion.

The differences between North Carolina and northwest Florida marshes listed above may all be due to a single causative factor: the greater patchiness of the northwest Florida salt marshes. In the marshes covered by this study, all habitats save the *J. roemerianus* expanses are composed of small, largely isolated patches. In contrast, the San Francisco Bay marshes studied by Lane (1969) and Cameron (1972) comprised a few habitat types covering hundreds of acres each. Lane, collecting by sweep netting, captured only 43 species, but nearly 4000 individuals, in 6 months of sampling (a total of only 120 sweeps). Marples (1966) worked in a *S. alterniflora* marsh that covered more than 50 ha; he collected tens of thousands of individuals in a year's sampling. Finally, McMahan et al. (1972), working in a large *Spartina alterniflora* marsh in North Carolina, captured 126 species of insects and over 6000 individuals with a D-Vac suction sampler. Data from Davis (1964) indicate that most of the habitats at his sites were also extensive.

The presence of small, isolated patches in northwest Florida salt marshes, coupled with the seasonal availability of some resources (Hansen et al., 1976; Mendlessohn and Marcellus, 1976; Odum and Smalley, 1959; personal observation), does not allow the accumulation of large numbers of individuals of any species within the patches. The small absolute population sizes and the small size of habitat patches make movement between patches frequent. Since patch invasion is probably a highly stochastic event, patches are generally very different from one another in species composition at any one time.

The habitat that occurs in northwest Florida in large stands, *J. roemerianus*, also has relatively more species of arthropods than it does in North Carolina. We believe that this difference is because the *J. roemerianus* marsh derives many species from other habitats. Population sizes within the *J. roemerianus* do not appear to be large in northwest Florida; in fact, they are no larger than those reported from North Carolina *J. roemerianus* marshes (Davis, 1964). The poor quality of *J. roemerianus* as a food source (Williams and Murdoch, 1972) likely accounts for the generally small arthropod populations found in *J. roemerianus* marshes in all areas.

On the local level (the few square kilometers covered by our study), patchiness likely accounts for the high diversities; however, it does not account for the observation that identical sampling efforts would likely yield more species in the northwest Florida marshes than in others that have been studied. If these marshes do contain more species than those in Georgia, North Carolina, and California, several factors may be responsible. First, most of the increased diversity in local marshes can be attributed principally to two orders of insects, Diptera and Hymenoptera. Perhaps the large numbers of flies are a consequence of the presence of extensive *J. roemerianus* stands. Since these stands are flooded less often than the *S. alterniflora* stands found in east coast marshes, detritus build-up is greater in northwest Florida, and species of flies that pass their larval lives in decomposing plant material are provided with large amounts of suitable habitat. The numerous species of parasitoid wasps present in our study sites may also be a result of the large *J. roemerianus* stands. We have observed that many herbivores, particularly cicadellid and delphacid homopterans, that are rare as adults in the *J. roemerianus* (they are most common in the *S. alterniflora* fringe), are commonly found in the *J. roemerianus* as nymphs. These juveniles occur mainly around the bases of the plants and are collected in large numbers only by suction sampling. The concentration of juveniles, coupled with the lack of flooding in these habitats, may expose the juvenile homopterans and dipterans to increased levels of parasitization (see also Stiling and Strong, 1982a, 1982b). This process may account, at least partially, for the large number of species of parasitoid hymenopterans in the local marshes. Second, marshes are much more extensive along the Gulf coast from Texas to Florida than along the Atlantic coast from Florida to Delaware (Eleuterius, 1976). We do not know, however, if marsh arthropod diversities are relatively high all along the Gulf coast.

A latitudinal gradient in diversity (Fischer, 1960; Pianka, 1966) may also contribute to the high diversity of the Florida marshes. While it appears that North Carolina marshes are more diverse than those in Georgia, habitats other than *S. alterniflora* have not been well studied in Georgia marshes. A test of the importance of latitude could be performed by comparing our data with those derived from a comparable sampling program carried out along the east coast of Florida. Some marshes there lie at about the same latitude as those near Wakulla Beach but differ in structure, there being large expanses of *S. alterniflora*, particularly in the upper St. Johns River area (Kurz and Wagner, 1954; personal observation).

Several factors may account for the lower number of individuals per sampling effort collected in northwest Florida marshes than in Georgia and North Carolina. As previously mentioned, the small size of many of the habitat patches in northwest Florida marshes may prevent many species from building large populations (Raupp and Denno, 1979). Predators and parasites also could be important in controlling marsh arthropod populations. The abundance of certain hymenopteran parasitoids has already been mentioned. In addition, strepsipterans were common parasites of the family Delphacidae. Exact estimates of parasitization were not obtained, but rates often appeared to be high. The main invertebrate predators in the marsh were spiders, which have been shown to exert considerable predation pressure on marsh insects in Massachusetts (Vince and Valiela, 1981). Hemipteran (mainly Reduviidae) and coleopteran (mainly

Coccinellidae) predators were also abundant at certain times. The main verte-brate predators were fish and birds. Kale (1964, 1965) has shown that insects comprise an important fraction of the long-billed marsh wren's diet. In addi-tion, the seaside sparrow, sharp-tailed sparrow, swamp sparrow, song sparrow, great kingbird, and eastern kingbird, all common in our study area, are known to depend heavily on insects for food, whereas terns, herons, rails, and many other shorebirds are known to include insects in their diet occasionally (Bent, 1963; N. Wamer, personal communication).

During periods of inundation, many species of fish penetrate into the marsh and feed on insects, both under and above water. Subrahmanyan and Drake (1975) found 55 species of fish, many of which are known to be arthropod predators, in tidal marshes of northwest Florida. Several times we have ob-served small fish jumping out of the water and capturing insects resting on the vegetation. A single specimen of *Fundulus* sp. was followed for 5 min and was observed to make nine successful strikes in that short period of time. The extent to which fish feed on eggs and immature and adult arthropods in the marsh is not known, but fish predation pressure on arthropods can be considerable (Hoy and Reed, 1971).

In a related study (Rey and McCoy, 1979), we examined the arthropod colonization patterns of the *Spartina* fringes in more detail. We uncovered high species "turnover" rates due to the large number of transient species colonizing these habitats. We also found that barriers to dispersal restricted colonization by predators/parasites to a significantly greater extent than the colonization rates of herbivores, but we found no time lags in the predator/parasite accumu-lation curves when compared to those of herbivores.

Ecological Biogeography

The patchy nature of many of the habitats in these northwest Florida marshes means that some of these patches may be considered "habitat islands" in a sea of dissimilar habitats. In addition, many real islands, varying in size from only a few square meters to several hectares, exist in the area. The smaller of these islands usually consist of monocultures of *S. alterniflora* growing on oyster or sand bars at various distances from shore. Rey (1979, 1981, 1985) took advan-tage of this situation to test several of the assumptions and predictions of the equilibrium theory of island biogeography (MacArthur and Wilson, 1967) us-ing the terrestrial arthropod faunas of *S. alterniflora* islands.

The experimental system consisted of eight *S. alterniflora* islets, ranging in size from 56 to 1023 m^2 and in distance from the mainland from 29 to 1752

m. Six of the islands were defaunated using methyl bromide, and their patterns of colonization, immigration, and extinction were monitored during a year. Two islands and a 2145-m^2 *S. alterniflora* stand on the mainland were not defaunated and served as controls. Caging experiments in four additional islands helped to discriminate between resident and transient species, while dispersal traps were used to examine the movements of arthropods throughout the study area. More complete descriptions of the experiments are given by Rey (1981). Although a detailed discussion of the results is beyond the scope of this chapter, a summary of the major findings is presented below.

Using very conservative criteria (Rey, 1981), 56 species of arthropods were determined to be resident in the *S. alterniflora* islands. Recolonization of the defaunated islands was rapid; the defaunated islands attained species richness corresponding to predefaunation levels in about 20 weeks and maintained similar numbers of species during the rest of the study. Island-wide immigration and extinction rates were high during the first 20 weeks following defaunation and then dropped to low but still measurable levels (3.1 to 9.4 spp. per year) after that. In general, the relationships between immigration and extinction rates and species numbers were in the direction predicted by the MacArthur–Wilson theory (positive for extinction and negative for immigration), but there were significant departures from the predictions and considerable variability (Rey and Strong, 1983). As predicted by the model, equilibrium species numbers were positively correlated with area and island extinction rates were negatively correlated with area, but no relationship between immigration rates and various isolation measures was evident.

The turnover observed on the *S. alterniflora* archipelago was intermediate between the totally homogeneous (all species with equal turnover rates) and the totally heterogeneous (a small subset of species turn over frequently and the rest not at all). Some species immigrated and went extinct more frequently than others, but all species in the pool contributed to the archipelago-wide extinction and immigration rates. Turnover thus resulted in observable changes in the species composition of the various islands through time. The evidence suggests that turnover in these islands was real and not a result of sampling error (Rey, 1985).

Alary Polymorphism in Marsh Fulgoroids

The production of different wing morphs is a common phenomenon in fulgoroid homopterans. Habitat stability is one of the factors that appears to control the production of long-winged (macropterous) vs. short-winged and flightless (brachypterous) individuals (Denno, 1976, 1978). Macropterous individuals are able to escape from unsuitable or deteriorating habitats, but brachypterous morphs have a higher fecundity and earlier oviposition than macropters (Tsai et al., 1964; Kisimoto, 1965; Nasu, 1969).

Denno and Grissel (1979) examined the patterns of wing polymorphism in populations of the *S. alterniflora*-associated planthopper *Prokelisia marginata* (Delphacidae) from marshes along the Atlantic coast of North America and from Cedar Key, Florida and found that the proportion of brachypters in Cedar Key was much higher than in Atlantic marshes. They attributed this difference to the fact that in Atlantic marshes, streamside *S. alterniflora* stands were seasonally the most "desirable" because of superior nutritive value, greater abundance of food, and freedom from parasites, but became unsuitable during winter because litter and standing vegetation disappeared due to seasonal mortality and tidal action. The advantage to switching between the always-available high marsh stands and the streamside stands, they reasoned, provides the impetus for the production of fully winged individuals at these sites. At Cedar Key, however, all *S. alterniflora* stands persisted throughout the year, and there was little difference in structure between streamside and high marsh stands. Thus, in those marshes, habitat switching was of little advantage and brachypters predominated.

We compared the Cedar Key and Atlantic coast polymorphism data with our data from St. Marks marshes (McCoy and Rey, 1981a) and also found a higher proportion of brachypters in several species of delphacids than reported by Denno and Grissel (1979) and Denno (1976, 1978) from Atlantic coast marshes. We concluded, however, that the reasons for the predominance of brachypters at St. Marks is not due to habitat uniformity (as in Cedar Key), but is a result of the fact that resources at St. Marks are patchy, isolated, and, more importantly, disturbances, in the form of flooding tides and seasonal vegetation diebacks, are correlated in time between all habitat patches. Frequent dispersal may thus be counterproductive since a dispersing delphacid not only has very little chance of finding a better quality patch than the one it just left, but it also runs the risk of being blown out to sea or ending up in the wrong habitat. Furthermore, as discussed above, the habitat distribution and characteristics at St. Marks are not conducive to the generation of large population sizes, a factor that has also been shown to be important in the production of macropterous individuals in marsh delphacids. For example, Strong and Stiling (1983) were able to increase the proportion of macropters produced by *P. marginata* by artificially increasing population densities on caged *S. alterniflora* islands at St. Marks. Juilianna (1984), however, was unable to replicate these results in *S. alterniflora* marshes in Tampa Bay. He attributed the different results to a higher brachypter–macropter switching threshold in Tampa Bay marshes than in St. Marks due to the uniform and permanent nature of the former.

Additional Ecological Work

Stiling and Strong (1984) studied the effects of experimental density manipulations in guilds of stem-boring and leaf-mining insects on *S. alterniflora* in the St. Marks marshes. The insects involved were *Thrypticus violaceous* (Diptera: Dolichopodidae), *Calamomyia alterniflorae* (Diptera: Cecidomyiidae), *Mordellistena splendens* (Coleoptera: Mordellidae), *Languria taedata* (Coleoptera: Languriidae), and *Chilo demotellus* (Lepidoptera: Pyralidae), all of which overwinter inside the *S. alterniflora* stems. Through laboratory experiments, Stiling and Strong were able to demonstrate severe competition between species inside the stems and established the following competitive hierarchy: (1) *Chilo demotellus*, (2) *Languria taedata*, (3) *Mordellistena splendens*, (4) *Thrypticus violaceous*, (5) *Calamomyia alterniflorae*. They hypothesized that the low densities of the lesser competitors in certain areas were a result of competition.

To test their hypothesis in the field, Stiling and Strong (1984) erected exclusion cages to prevent colonization of *S. alterniflora* stems already containing *T. violaceous* by two of the most competent predators (*L. taedata* and *M. splendens*). They also prevented colonization of some stems by *M. splendens* by clipping the flowering heads before this insect, which only oviposits in that location, colonized the plants. They showed that in both experiments, the larval densities of *T. violaceus* increased dramatically. They point out that competition in the stem-boring insect guild associated with *S. alterniflora* may be facilitated by the hollow stems of this plant, which promote interspecific encounters. Stiling and Strong's results contrast with those of other stem-boring guilds in plants with solid piths, where interspecific competition was not demonstrated (Rathcke, 1976).

In other experiments, Stiling et al. (1991) showed that nitrogen fertilization of mainland *S. alterniflora* plots produced a significant increase in populations of the grasshopper *Orchelimum fidicinium* (Tettigoniidae), but not populations of *Prokelisia marginata* (Delphacidae). In offshore islands, where grasshoppers were very rare, *P. marginata* populations increased after nitrogen fertilization. Stiling and Strong (1984) attribute these results to the fact that in the mainland plots, grasshopper feeding damage was so severe that it prevented population increases by delphacids in response to fertilization. In the absence of grasshopper feeding damage, delphacids were able to take advantage of the increased foliar nitrogen content.

Conclusion

Salt marshes indeed seem to be relatively simple systems, which facilitates the study of the organisms that reside there (see Denno and Roderick, 1989). Our own studies of salt marsh terrestrial arthropods, as well as those of other researchers, indicate that the distribution of resources in space and time influences the biologies of these organisms to a great extent. We have proposed in northwest Florida, for example, that plant spacing and stature influence species richness; that buildup of detritus under the most common plant species promotes high densities of certain insect species, which in turn attract a diversity of parasites and other associates; that patchiness of other plant species modifies the species composition of arthropods and even dictates ecological and evolutionary adjustments of life histories; and that periodic flushes of resources allow the accrual of additional individuals and species.

We believe the structure of salt marshes to also be important in ways that have yet to be well investigated. For instance, we suggest that where *S. alterniflora* occurs in small isolated stands, the resident arthropods will be exposed to relatively higher predation rates than when this plant covers vast expanses. The small stands are probably more "open" to birds and fishes, and differences in predation rates may interact with factors such as dispersal, which we have discussed in some detail above. The relative simplicity of the salt marsh system makes it amenable to creative experiments and manipulations to address important problems such as this one.

References

Arndt, C. H. 1914. Some insects of the between tides zones. *Proc. Ind. Acad. Sci.* 1914:323–336.

Barnes, R. D. 1953. The ecological distribution of spiders in non-forest maritime communities at Beaufort, North Carolina. *Ecol. Monogr.* 23:315–337.

Bent, A. C. 1963. *Life Histories of North American Marsh Birds.* Dover, New York. 392 pp.

Beshear, R. J. 1971. Thrips collected from *Spartina* spp. from Cumberland Island, Georgia. *J. Ga. Entomol. Soc.* 6:243–246.

Bickley, W. E. and T. R. Seek. 1975. *Insects in Four Maryland Marshes.* University of Maryland Agriculture Experiment Station Misc. Publ. 870. 27 pp.

Cameron, G. N. 1972. Analysis of insect trophic diversity in two salt marsh communities. *Ecology* 53:58–73.

Daiber, F. C. 1974. Salt marsh plants and future coastal salt marshes in relation to animals. Pages 475–510 in R. J. Reimold and W. H. Queen, Eds. *Ecology of Halophytes.* Academic Press, New York.

Davis, L. V. 1964. Insects in the herbaceous strata of salt marshes in the Beaufort, North Carolina area. Ph.D. dissertation. Duke University, Durham, North Carolina.

Davis, L. V. 1978. Class Insecta. Pages 186–220 in R. G. Zingmark, Ed. *An Annotated Checklist of the Biota of the Coastal Zone of South Carolina.* University of South Carolina Press, Columbia.

Davis, L. V. and I. W. Grey. 1966. Zonal and seasonal distribution of insects in North Carolina salt marshes. *Ecol. Monogr.* 36:275–295.

Denno, R. F. 1976. Ecological significance of wing polymorphism in Fulgoroidea which inhabit tidal salt marshes. *Ecol. Entomol.* 1:257–266.

Denno, R. F. 1978. The optimum population strategy for planthoppers (Homoptera: Delphacidae) in stable marsh habitats. *Can. Entomol.* 110:135–142.

Denno, R. F. and E. E. Grissel. 1979. The adaptiveness of wing dimorphism in the salt marsh-inhabiting planthopper *Prokelisia marginata* (Homoptera: Delphacidae). *Ecology* 60:221–236.

Denno, R. F. and G. K. Roderick. 1989. Influence of patch size, vegetation texture, and host plant architecture on the diversity, abundance, and life history styles of sap-feeding herbivores. Pages 169–196 in S. S. Bell, E. D. McCoy, and H. R. Mushinsky, Eds. *Habitat Structure: The Physical Arrangement of Objects in Space.* Chapman and Hall, London.

Dukes, J. C., T. D. Edwards, and R. C Axtell. 1974. Associations of Tabanidae (Diptera) larvae with plant species in salt marshes, Carteret County, North Carolina. *Environ. Entomol.* 3:280–286.

Eleuterius, L. N. 1976. The distribution of *Juncus roemerianus* in the salt marshes of North America. *Chesapeake Sci.* 17:282–292.

Evans, F. C. and W. W. Murdoch. 1968. Taxonomic composition, trophic structure, and seasonal occurrence in a grassland insect community. *J. Anim. Ecol.* 37:259–274.

Fisher, A. G. 1960. Latitudinal variations in organic diversity. *Evolution* 14:64–81.

Foster, W. A. and J. E. Treherne. 1975. The distribution of an intertidal aphid, *Pemphigus trehernei,* on marine salt marshes. *Oecologia* 21:141–155.

Foster, W. A. and J. E. Treherne. 1976. The effect of tidal submergence on an intertidal aphid *Pemphigus trehernei* Foster. *J. Anim. Ecol.* 37:259–274.

Hansen, D. J., P. Dayanandan, P. B. Kaufman, and J. D. Brotherson. 1976. Ecological adaptations of salt marsh grass, *Distichlis spicata* (Graminea), and environmental factors affecting its growth and distribution. *Am. J. Bot.* 63:635–650.

Hill, M. O. 1973. Diversity and evenness: a unifying notation and its consequences. *Ecology* 54:427–432.

Hoy, J. B. and D. E. Reed. 1971. The efficacy of mosquitofish for the control of *Culex tarsalis.* *Mosquito News* 31:567–572.

Janzen, D. H. 1973a. Sweep samples of tropical foliage insects: description of study sites with data on species abundances and size distributions. *Ecology* 54:659–686.

Janzen, D. H. 1973b. Sweep samples of tropical foliage insects: effects of seasons, vegetation types, elevation, time of day, and insularity. *Ecology* 54:687–708.

Janzen, D. H. and T. W. Schoener. 1968. Differences in insect abundance and diversity between wetter and drier sites during a tropical dry season. *Ecology* 49:96–110.

Juilianna, J. 1984. Wing polymorphism and habitat quality in a species of delphacid planthopper, *Prokelisia marginata* (Homoptera: Delphacidae). M.S. thesis. University of South Florida, Tampa.

Kale, H. W. 1964. Food of the long-billed marsh wren, *Telmatodytes palustris griseus*, in the salt marshes of Sapelo Island, Georgia. *The Oriole* 29:47–61.

Kale, H. W. 1965. *Ecology and Bioenergetics of the Long-Billed Marsh Wren in Georgia Salt Marshes.* Publ. Nutall Orn. Club #5, Cambridge, Massachusetts.

Kisimoto, R. 1965. Studies on the polymorphism and its role-playing in the population growth of the brown planthopper, *Nilaparvata lugens* Stal. *Bull Shikoku Agric. Exp. Sta.* 13:1–106.

Kurz, H. and K. Wagner. 1954. *Tidal Marshes of the Gulf and Atlantic Coasts of Northern Florida and Charleston S.C.* Florida State University Studies #24. 168 pp.

Lane, R. S. 1969. The insect fauna of a coastal salt marsh. M.A. thesis. San Francisco State College, San Francisco, California. 78 pp.

LaSalle, M. W. and A. A. de la Cruz. 1985. Seasonal abundance and diversity of spiders in two intertidal marsh plant communities. *Estuaries* 8:381–393.

Luxton, M. 1967. The ecology of salt marsh Acarina. *J. Anim. Ecol.* 36:257–277.

MacArthur, R. H. and E. O. Wilson. 1967. *The Theory of Island Biogeography.* Monogr. in Pop. Biology #1. Princeton University Press, Princeton, New Jersey.

Marples, T. G. 1966. A radionuclide tracer study of arthropod food chains in a *Spartina* salt marsh ecosystem. *Ecology* 47:270–277.

McCoy, E. D. 1977. Diversity of terrestrial arthropods in northwest Florida salt marshes. Ph.D. dissertation. Florida State University, Tallahassee. 104 pp.

McCoy, E. D. and J. R. Rey. 1981a. Patterns of abundance, distribution, and alary polymorphism among the salt marsh Delphacidae (Homoptera: Fulgoroidea) of northwest Florida. *Ecol. Entomol.* 6:285–291.

McCoy, E. D. and J. R. Rey. 1981b. Terrestrial arthropods of northwest Florida salt marshes: Coleoptera. *Fla. Entomol.* 64:405–411.

McCoy, E. D. and J. R. Rey. 1987. Terrestrial arthropods of northwest Florida salt marshes: Hymenoptera (Insecta). *Fla. Entomol.* 70:90–97.

McMahan, E. A., R. L. Knight, and A. R. Camp. 1972. A comparison of microarthropod populations in sewage-exposed and sewage-free *Spartina* salt marshes. *Environ. Entomol.* 1:244–252.

Mendelssohn, I. A. and K. L. Marcellus. 1976. Angiosperm production of three Virginia marshes in various salinity and soil nutrient regimes. *Chesapeake Sci.* 17:15–23.

Metcalf, Z. P. 1920. Some ecological aspects of the tidal zone of the North Carolina coast. *Ecology* 1:193–197.

Metcalf, Z. P. and H. Osborne. 1920. Some observations of the insects of the between-tide zone of the North Carolina coast. *Ann. Entomol. Soc. Am.* 13:108–120.

Nasu, S. 1969. *The Virus Diseases of the Rice Plant.* Johns Hopkins Press, Baltimore.

Odum, E. P. and A. E. Smalley. 1959. Comparison of population energy flow of a herbivorous and a deposit-feeding invertebrate in a salt marsh ecosystem. *Proc. Natl. Acad. Sci. U.S.A.* 45:617–622.

Palmer, W. A. 1987. The phytophagous insect fauna associated with *Baccharis halimifolia* L. and *B. neglecta* Britton in Texas, Louisiana, and northern Mexico. *Proc. Entomol. Soc. Wash.* 89:185–199.

Pavoir-Smith, K. 1956. The biotic community of a salt meadow in New Zealand. *Trans. R. Soc. N.Z.* 83:525–554.

Payne, K. 1972. A survey of the *Spartina*-feeding insects in Poole Harbor, Dorset. *Entomol. Mon. Mag.* 108:66–79.

Pianka, E. 1966. Latitudinal gradients in species diversity: a review of concepts. *Am. Nat.* 100:33–46.

Pimentel, D. and A. G. Wheeler. 1973. Species diversity of arthropods in the alfalfa community. *Environ. Entomol.* 2:659–668.

Rathcke, B. J. 1976. Competition and coexistence within a guild of herbivorous insects. *Ecology* 57:76–88.

Raupp, M. J. and R. F. Denno. 1979. The influence of patch size on a guild of sap-feeding insects that inhabit the salt marsh grass *Spartina patens*. *Environ. Entomol.* 8:412–417.

Rey, J. R. 1978. Abundance patterns of terrestrial arthropods in north Florida salt marshes. M.S. thesis. Florida State University, Tallahassee. 187 pp.

Rey, J. R. 1979. Colonization, turnover, and equilibrium of arthropods on *Spartina alterniflora* islands in northwest Florida. Ph.D. dissertation, Florida State University, Tallahassee. 171 pp.

Rey, J. R. 1981. Ecological biogeography of arthropods on *Spartina* islands in northwest Florida. *Ecol. Monogr.* 51:237–265.

Rey, J. R. 1985. Insular ecology of salt marsh arthropods: species level patterns. *J. Biogeogr.* 12:97–107.

Rey, J. R. and E. D. McCoy. 1979. Application of island biogeographic theory to the pests of cultivated crops. *Environ. Entomol.* 8:577–582.

Rey, J. R. and E. D. McCoy. 1982. Terrestrial arthropods of northwest Florida salt marshes: Hemiptera and Homoptera (Insecta). *Fla. Entomol.* 65:241–248.

Rey, J. R. and E. D. McCoy. 1983. Terrestrial arthropods of northwest Florida salt marshes: Araneae and Pseudoscorpiones (Arachnida). *Fla. Entomol.* 66:497–503.

Rey, J. R. and E. D. McCoy. 1986. Terrestrial arthropods of northwest Florida salt marshes: Diptera (Insecta). *Fla. Entomol.* 69:197–205.

Rey, J. R. and D. R. Strong. 1983. Immigration and extinction of salt marsh arthropods on islands: an experimental study. *Oikos* 41:396–401.

Rockal, E. G. and E. J. Hansens. 1970. Distribution of larval horseflies and deerflies (Diptera: Tabanidae) of a New Jersey salt marsh. *Ann. Entomol. Soc. Am.* 63: 681–684.

Smalley, A. E. 1959. The growth cycle of *Spartina* and its relation to the insect populations in the marsh. Pages 96–100 in *Proceedings of the Salt Marsh Conference*. University of Georgia, Athens.

Smalley, A. E. 1960. Energy flow of a salt marsh grasshopper population. *Ecology* 41:672–677.

Stiling, P. D. and D. R. Strong. 1981. A leaf miner (Diptera: Ephydridae) and its parasitoids on *Spartina alterniflora* in northwest Florida. *Fla. Entomol.* 64: 468–471.

Stiling, P. D. and D. R. Strong. 1982a. Parasitoids of the planthopper *Prokelisia marginata* (Homoptera: Delphacidae). *Fla. Entomol.* 65:191–192.

Stiling, P. D. and D. R. Strong. 1982b. Egg density and the intensity of parasitism in *Prokelisia marginata* (Homoptera: Delphacidae). *Ecology* 63:1630–1635.

Stiling, P. D. and D. R. Strong. 1984. Experimental density manipulation of stem-boring insects: some evidence for interspecific competition. *Ecology* 65:1683–1685.

Stiling, P. D., B. V. Brodbeck, and D. R. Strong. 1991. Population increases of planthoppers on fertilized saltmarsh cordgrass may be prevented by grasshopper feeding. *Fla. Entomol.* 74:88–97.

Strong, D. R. and J. R. Rey. 1983. Testing for MacArthur–Wilson equilibrium with the arthropods of the miniature *Spartina* archipelago at Oyster Bay, Florida. *Am. Zool.* 22:355–360.

Strong, D. R. and P. D. Stiling. 1983. Wing dimorphism changed by experimental density manipulation in a planthopper (*Prokelisia marginata*, Homoptera: Delphacidae). *Ecology* 64:206–209.

Subrahmanyam, C. B. and S. Drake. 1975. Studies on the animal communities in two north Florida salt marshes. *Bull. Mar. Sci.* 25:445–465.

Teal, J. M. 1959. Energy flow in the salt marsh ecosystem. Pages 101–107 in *Proceedings of the Salt Marsh Conference*. University of Georgia, Athens.

Teal, J. M. 1962. Energy flow in the salt marsh ecosystem of Georgia. *Ecology* 43: 614–624.

Tippins, H. H. and R. J. Beshear. 1968. Scale insects (Homoptera: Coccoidea) from grasses in Georgia. *J. Ga. Entomol. Soc.* 3:134–136.

Tsai, P., F. Hwang, W. Feng, Y. Fu, and Q. Dong. 1964. Studies on the *Delphacodes striatella* Fallen (Homoptera, Delphacidae) in north China. *Acta Entomol. Sinica* 13:552–571.

Vince, S. W. and I. Valiela. 1981. An experimental study of the structure of herbivorous insect communities in a salt marsh. *Ecology* 62:1662–1678.

Wheeler, A. G. 1973a. Studies on the arthropod fauna of alfalfa. IV. Species associated with the crown. *Can. Entomol.* 105:353–366.

Wheeler, A. G. 1973b. Studies on the arthropod fauna of alfalfa. V. Spiders (Araneida). *Can. Entomol.* 105:425–432.

Williams, R. B. and M.-B. Murdoch. 1972. Compartmental analysis of the production of *Juncus roemerianus* in a North Carolina salt marsh. *Chesapeake Sci.* 13:69–79.

EMERGY Evaluation of Florida Salt Marsh and Its Contribution to Economic Wealth

8

Howard T. Odum and Douglas A. Hornbeck

In this chapter, data on salt marshes typical of northwest Florida were used to estimate the contributions of marsh production and storage to real wealth using EMERGY (spelled with an "M"), a scientific measure of environmental work (Odum, 1986, 1988). Then, for perspective, the part of the regional economic buying power due to the EMERGY of marsh productivity is estimated by proportion. The contribution of marshes is evaluated for Cedar Key, Florida, and its surroundings, Levy County (Figure 8.1).

Money is paid only to people, not to nature. As money circulates, it facilitates human purchase of real wealth such as fuels, food, houses, books, etc. (Figure 8.2). The real wealth comes from environmental resources, some renewable (e.g., sun, wind, and waves) and some nonrenewable (in the short run) (e.g., oil, minerals, and virgin forest wood). To measure the real wealth, a scientific measure was proposed in 1967 (Odum, 1967, 1971), subsequently revised to include geologic processes, and given the name EMERGY in 1983.

Definitions

Solar EMERGY

In Figure 8.2, environmental production (salt marsh example) is connected to an economic use (economic production) process. Money, shown with a dashed

209

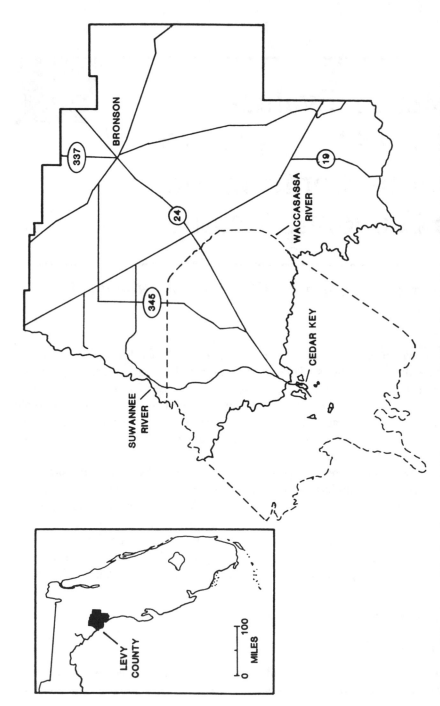

Figure 8.1　Area of land and water used for EMERGY evaluation of Levy County (encircled by the dashed line).

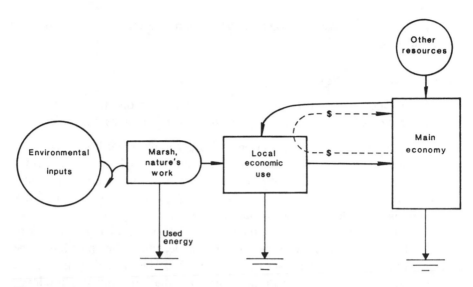

Figure 8.2 Energy diagram of an environmental system and its economic interface. Money flow is shown with a dashed line.

line, circulates as a counter current to the real items it purchases (example: fish landed by fishermen). Money is only paid for the human services. Services of nature come from many kinds of energy inputs to the salt marsh, including solar energy, wind energy, tidal energy, and the Gibbs free energy of rain and fresh water relative to salt water. The contribution from the salt marsh is measured by expressing all its inputs in the energy of one kind, the solar energy required to make each of the other kinds of energy input. The previously processed solar energy required for an input is its solar EMERGY.

> Solar EMERGY is defined as the solar energy previously required for all the inputs necessary to make a product and is expressed as solar emjoules.

EMERGY measures both the products of nature and the goods and services of the human economy on a similar basis (energy of one type required). Decisions that maximize the EMERGY of the combined economy of humans and nature lead to more real wealth, more buying power for the circulating currency, and the highest standard of living. Maximum EMERGY use comes from using the local environment and the purchased inputs in a symbiotic way.

Joules of sunlight, waves, oil, electricity, zooplankton, whales, and human beings cannot be compared as equivalent until they are expressed in equivalents of one kind of EMERGY required. Here, as in previous papers, all energy flows or storages that are to be compared are represented in solar EMERGY (solar energy required to make the product). Calculations are facilitated by tables of solar transformity.

Solar Transformity

> Solar transformity is the solar energy required for one unit of another kind of energy and is expressed as solar emjoules per joule.

Tables of solar transformities (solar emjoules per joule) have been developed from previous EMERGY evaluations that sum the required inputs to various processes in nature and in the economy. Where data on materials are more easily calculated in grams or human services in dollars, we use solar EMERGY per gram and solar EMERGY per dollar to compute the solar EMERGY.

Transformities express the position a product has in the energy hierarchy of nature. The more solar EMERGY required per joule, the higher position is the product. Products of higher transformity are scarce because they take more resources to make. Higher transformity products that are regularly produced have higher effects in use, which justifies their large resource requirement. Higher transformity items require larger territories of support and have larger areas of influence. Higher transformity items tend to be larger, with longer replacement times.

Tables of transformities have been assembled from prior EMERGY evaluations of environmental systems. The EMERGY evaluations of marsh in this chapter generate a new set of solar transformities for marsh components and processes. Transformities of principal components and processes of the marsh show the position of these elements in the natural energy hierarchy of the marsh and the biosphere.

Macroeconomic Dollar Value for Public Policy Decisions

> Macroeconomic dollar value of a product is defined as the dollars of the gross economic product based on the EMERGY of that product. Macroeconomic value is that proportion of the dollars of gross economic product that the total EMERGY is of the total annual EMERGY.

Microeconomic Market Value vs. EMERGY-Based Macroeconomic Value

Since money is not paid to nature for its work processes, its contribution is not measured by payments of money. When environmental resources are abundant, they contribute most to wealth and a high standard of living, but this is when they are valued least by costs, prices, and market values. When resources (such as fisheries) become scarce, their costs and prices go up, so that as much of the real wealth of the economy is going into obtaining them as is yielded to the economy from nature. This is when the contribution to real wealth is least but the economic values are highest. Thus, market values and real wealth measured by EMERGY are inverse. It is incorrect to use market value to measure the contribution to wealth of environmental resources or the impacts on environment.

Previous Evaluations of Marsh

A number of preliminary evaluations have been made of the value of marshes using energy and productivity data. Leitch and Ekstrom (1989) provide an annotated entrance to a large body of literature. Lynne et al. (1981) and Bell (1989) use economic methods to evaluate marsh, all of which directly and indirectly evaluate the flow of money to people using marshes or their products. Lynne et al. (1981) used the labor substitution that would produce the same dollars of fishing income to evaluate the economic value of fishes based on marsh–estuarine nurseries. Bell (1989) extended these methods, finding $35.61 to be the economic value of the fish contributed by 1 acre of coastal wetland. (See further discussion by Bell and Lynne in Chapter 9.) Bergstrom et al. (1990) estimated annual economic value of recreational use of wetlands in Louisiana. Some of the previous estimates of value are given in Table 8.1.

Starting in 1967 and 1971, the authors and associates, while measuring the production processes of ecosystems, used measures of nature's work to evaluate the contribution of ecosystems to the economy. Several of these methods were compared by Gosselink et al. (1974). Wetlands were evaluated in 1978. Farber and Costanza (1987), followed by Turner et al. (1988), evaluated marshes using one of our methods: multiplying marsh production energy by an estimated coal equivalent per unit energy and then dividing by the coal equivalents per dollar for the U.S. economy. These results were compared with market values (Table 8.1).

In another approach, concepts were defined more rigorously in 1983, introducing the terms EMERGY and transformity for previous terms embodied by energy and energy transformational quotient. This clarification was ambiguous because of the use of these terms by others with different meanings (Odum, 1986, 1988, 1996).

Considerable confusion in environmental evaluation comes from mixing the two different kinds of measures: one the value to humans individually and their businesses and the other the value to the public economy as a whole. The measures in this chapter evaluate nature's work and human work on a similar

Table 8.1 Comparison of Estimates of Salt Marsh Value

Items	Annual $/ha/year
Economic value (market values, willingness to pay)	
Goldstein (1971) conversion to agriculture	0
Gosselink et al. (1974); fishery sales	
Georgia	270
Florida	188
Louisiana	120
Reimold et al. (1980); duck support	100
Farber and Costanza (1987)	85
Bell (1989)	89
Bergstrom et al. (1990); recreation in Louisiana marshes	89.4
EMERGY-derived "macroeconomic value"	
Odum (1978a)	500
Farber and Costanza (1987)	1560
Turner et al. (1988)	889
Florida west coast, Tables 8.2 and 8.3	
Spartina marsh	660
Juncus marsh	743

basis, and the dollar value of the product to the public economy is then inferred from the proportional part of the economy contributed by these previous works. There is no intent to use the EMERGY-derived value in place of microeconomic market values.

Methods

For evaluating salt marshes, analyses were made at two scales: (1) for a hectare of typical marsh and (2) for an overview of a coastal region in which salt marshes were contributing (Cedar Key, Levy County, Florida).

Energy Systems Diagram

For EMERGY analysis, a systems diagram was developed first to organize knowledge of the main parts and processes to be evaluated. The salt marsh diagram is Figure 8.3 and the coastal economy diagram is Figure 8.4. A more detailed systems diagram showing more parts and processes of the marsh is given in the Introduction.

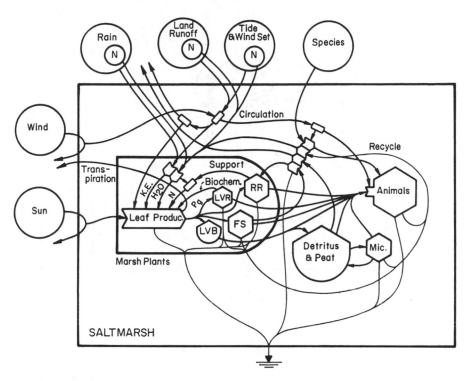

Figure 8.3   Energy systems diagram of a salt marsh showing the input sources evaluated in Table 8.2, storages evaluated in Table 8.3, and the pathways with transformity determinations in Table 8.4. Abbreviations: Biochem., flow of biochemical substances; FS, flowers and seeds; H_2O, water; K.E., kinetic energy of water motion; LVB, leaf biomass; LVR, leaf respiration; Mic., microbes; N, nutrients; Pg, gross photosynthesis; RR, runners and roots.

Procedure for Computing an EMERGY Analysis Table

Next, an EMERGY analysis table was prepared evaluating the main input sources and other flows of interest. The first example is Table 8.2, an evaluation of annual EMERGY flows of the marsh system drawn in Figure 8.3. Table 8.3 is the EMERGY evaluation of the main stored quantities of the marsh, those components that have replacement times longer than a year.

Each line item in the EMERGY table is a process or storage in an environmental system. The table includes the processes of nature and those of the economy. EMERGY analysis puts both the works of nature and those of humans on a common basis.

The first column indicates the number of the footnote where each calculation is shown. The next column lists the items. The next column gives the raw

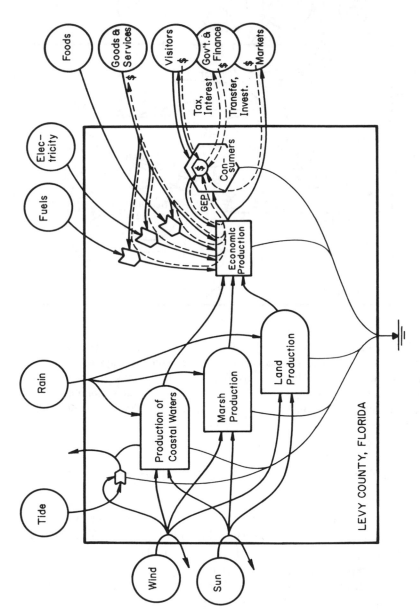

Figure 8.4 Energy systems diagram of the regional system of environment and nature in Figure 8.1. Dashed lines are flows of money.

Table 8.2 Annual EMERGY of Inputs to Salt Marsh in Northwest Florida (Figure 8.3)[a]

Note Item, units	Data (joules, J)	Solar transformity (sej/J)	Solar EMERGY E14 (sej/year)	Macroeconomic 1990 $/year[b]
Spartina marsh:				
1 Direct sun, J	5.95 E13	1	0.6	30
2 Tidal absorption, J	1.41 E10	23,564	3.3	165
3 Transpiration,[c] J	5.48 E10	18,000	9.9	495
4 Total (omitting #1)[d]			13.2	660
Juncus marsh:				
5 Direct sun, J	5.95 E13	1	0.6	30
6 Tidal absorption, J	0.7 E10	23,564	1.65	83
7 Transpiration,[c] J	7.3 E10	18,000	13.2	660
8 Total (omitting #5)[d]			14.85	743

[a] 1 ha.

[b] 1990 U.S. $ obtained by dividing annual EMERGY values by 2.0 E12 sej/1990 $.

[c] Marsh transpiration, which is the fresh water used, includes that flowing in from land runoff and that exchanging in from the seaside of the marsh. Transpiration integrates the combined energy inputs of insolation, wind energy absorbed, vapor pressure gradients in air, and freshwater activity in the estuarine waters.

[d] Totals were corrected for double counting of sunlight which contributes to the world weather system that brings rain (separate evaluation not counted).

1 Gainesville insolation with 10% albedo: $(1.58\ E6\ \text{kcal/m}^2/\text{year}) (1 - 0.10) (1\ E4\ \text{m}^2/\text{ha}) (4186\ \text{J/kcal}) = 5.95\ E13\ \text{J/ha/year}$.

2 Half of 80-cm tide absorbed: $(0.50) (1.0\ E4\ \text{m}^2) (0.5 \times 0.8\ \text{m}) (1025\ \text{kg/m}^3) (9.8\ \text{m/sec}^2) (705\ \text{tides/year}) = 1.41\ E10\ \text{J/year}$.

3 Marsh transpiration: 3.0 mm/day ? for 365 days = 1095 mm $(1.095\ \text{m}) (1.0\ E4\ \text{m}^2) (1\ E6\ \text{g/m}^3) (5\ \text{J/g}) = 5.475\ E10\ \text{J/year}$.

4 Sum of #2 and #3.

5 Same as #1.

6 One-fourth of 80-cm tide absorbed: $(0.25) (1.0\ E4\ \text{m}^2) (0.5 \times 0.8\ \text{m}) (1025\ \text{kg/m}^3) (9.8\ \text{m/sec}^2) (705\ \text{tides/year}) = 0.7\ E10\ \text{J/year}$.

7 Marsh transpiration: 4.0 mm/day ? for 365 days = 1460 mm $(1.46\ \text{m}) (1.0\ E4\ \text{m}^2) (1\ E6\ \text{g/m}^3) (5\ \text{J/g}) = 7.3\ E10\ \text{J/year}$.

8 Sum of #6 and #7.

Table 8.3 Evaluation of EMERGY Stored in Typical Salt Marsh of Northwest Florida[a,b]

Note	Item, units	Data (joules, J)	Solar transformity[c] (sej/J)	Solar EMERGY E14 (sej)	Em dollar × 1000[d]
Spartina marsh[b]					
1	Aboveground live biomass	0.78 E11	8,461	6.6	0.33
2	Aboveground dead biomass	0.51 E11	12,941	11.7	0.66
3	Detritus and peat, 1.14 m	496.00 E11	8,000	3,968.0	198.40
4	Total			3,986.3	199.39
Juncus marsh[b]					
5	Aboveground live biomass	1.86 E11	8,010	14.9	0.75
6	Underground marsh biomass	7.70 E11	9,867	76.0	3.80
7	Aboveground dead biomass	1.17 E11	10,132	11.9	0.60
8	Peaty sediment, 1.0 m	338.00 E11	13,185	4,456.5	222.83
9	Total			4,559.3	227.98
10	Fishes, crabs, shrimp	17.7 E6	2.7 E8	47.8	2.39
11	Tidal channel structure	9.8 E6	1.68 E9	164.6	8.23

[a] 1 ha.

[b] Hornbeck and Odum, Chapter 8 in this book.

[c] Table 8.4.

[d] 1990 U.S. $ obtained by dividing annual EMERGY values by 2.0 E12 sej/1990 $.

1 Aboveground live biomass: 463 dry g/m^2 (463 g/m^2) (1 E4 m^2) (4 kcal/g) (4186 J/kcal) = 7.75 E10 J/m^2.

2 Aboveground dead biomass: 307 dry g/m^2 (307 g/m^2) (4 kcal/g) (4186 J/kcal) (1/year) (1 E4 m^2/ha) = 5.1 E10 J/ha.

3 Energy in detritus and peat with data on organic carbon fraction in 4 levels (1.14-m total) of *Spartina* soil in Apalachee Bay (Coultas and Gross, 1975): (0.16 C × 0.18 m) + (0.136 C × 0.46 m) + (0.135 C × 0.25 m) + (0.116 C × 0.25 m)/1.14 m = 0.135 C; (0.135 C) (1.14 m^3/m^2) (0.7 E6 dry g/m^3) (11 kcal/g C) (4186 J/kcal) (1 E4 m^2/ha) = 4.96 E13 J/ha.

4 Sum of #1–3.

5 Aboveground live biomass: 1111 dry g/m^2 (1111 g/m^2) (4 kcal/g) (4186 J/kcal) (1/year) (1 E4 m^2/ha) = 1.86 E11 J/ha.

6 Underground biomass: 4600 dry g/m^2 (Kruczynski et al., 1978): (4600 g/m^2) (4 kcal/g) (4186 J/kcal) (1 E4 m^2/ha) = 7.7 E11 J/ha.

7 Aboveground dead biomass: 880 dry g/m^2 (880 g/m^2) (4 kcal/g) (4186 J/kcal) (1/year) (1 E4 m^2/ha) = 1.17 E11 J/ha.

8 Peaty sediment, 1-m thick: 10.5% organic carbon (Coultas and Calhoun, 1976) (0.105) (1 E4 m^3/ha) (0.7 E6 g/m^3) (11 kcal/g C) (4186 J/kcal) = 3.38 E13 J/ha.

9 Sum of #5–8.

Table 8.3 (continued)   Evaluation of EMERGY Stored in Typical Salt Marsh of Northwest Florida[a,b]

10 Larger nekton in tidal channels: 5.3 g preserved wt/m^2 of channels (Homer, 1977); channels occupying 10% of the marsh area: (5.3 g/m^2) (0.1) (1 E4 m^3/ha) (0.2 dry) (4 kcal/g) (4186 J/kcal) = 17.7 E6 J.

11 Tidal channels covering 10% ? of area; work of excavation: (0.1) (1 E4 m^2/ha) (1 m) (1000 kg/m^3) (0.5 m) (9.8 m/sec^2) = 9.8 E6 J/ha.

data in joules, grams, or dollars. The fourth column gives the solar transformities from previous studies (the solar emjoules per joule, solar emjoules per gram, or solar emjoules per dollar). In the fifth column, the raw data in column 4 are multiplied by the transformities in column 5 to obtain the solar EMERGY values. Finally, some perspective is provided by expressing the EMERGY as the part of the gross economic product due to that wealth. This is done by dividing the solar EMERGY values in column 5 by the ratio of solar EMERGY per dollar previously determined by analysis of the state economy for a particular year. As explained in a later section, EMERGY-based evaluation of dollars due to a resource should not be confused with or substituted for regular economic market values.

Some of the line items in the table may include other items in the table, but care has been taken to avoid double counting in the lines which are totals. By evaluating the source inputs, crossing the boundaries of the hectare of marsh, we obtain the total EMERGY, which is operating the marsh ecosystem. After minor corrections to prevent double counting, the solar EMERGY inputs used by the ecosystem are summed. For the *Spartina* marsh found closest to open water and inundated by tide and wind-set more often, 50% of the tidal energy was estimated to be absorbed. For the *Juncus* marsh, usually further from the open water and inundated less often, 25% of the tidal energy was estimated to be absorbed.

After the calculations are made, various items are summed or ratios calculated to provide insights on the contributions of the environment to the wealth of the economy. In this chapter, the solar EMERGY contributions of the salt marshes were related to the total economy of Cedar Key and Levy County, Florida.

Procedure for Calculating a Table of Solar Transformities for Marshes

Having evaluated the total annual EMERGY basis for salt marshes, solar transformities were obtained by dividing the proportion of the total marsh

EMERGY budget by the energy of the component process or product produced for the same time. For some marsh products, such as total gross production, it is appropriate to use the total annual EMERGY budget. For some components of the marsh, it is appropriate to use only a part of the total annual EMERGY budget, namely that part of the energy at the same hierarchical level due to the component. The transformities are as good as the data on energy flow through the marsh ecosystem food web, usually represented in an evaluated energy diagram or equivalent table. Details of assumptions and calculations were given in the footnotes to the transformity table.

Procedure for EMERGY Evaluation of a Coastal Region

The main inputs to the combined economy of nature and humans were recognized in diagramming. Figure 8.4 represents three environmental areas (coastal waters, salt marshes, and terrestrial ecosystems) from Figure 8.1 and the monied economy. An EMERGY analysis table was prepared including main solar EMERGY inputs from the three environmental areas.

For the purchased inputs, goods and services purchased from outside the county were estimated by comparing the proportions of income for various sectors with the average proportions to support normal consumers. That income for a sector in greater proportion than normal consumer economy was inferred to be from export.

The solar EMERGY budget for Florida was previously estimated (Odum et al., 1993), and from this a solar EMERGY/GNP $ ratio was estimated for 1983 as 5.4 E12 sej/$, more than twice that for the United States as a whole in 1983 (2.4 E12 sej/$) (Odum et al., 1987). The money circulating in for exports was assumed to purchase goods and services from outside (Figure 8.4). These dollar figures multiplied by the Floridian EMERGY/$ ratio provided an estimate of the EMERGY of the goods and services, including human services, exported. The purchases made from outside bring in goods and services, which are evaluated using the EMERGY/$ ratio of the U.S. The solar EMERGY equivalent of the fuels and electricity itself (different from the services involved) was obtained from data on their use and multiplied by appropriate solar transformities to obtain solar EMERGY uses.

Results

EMERGY analysis results include Tables 8.2 to 8.4 for salt marsh and Table 8.5 for the coastal economy.

Table 8.4 **Solar Transformities of Components and Processes of Salt Marsh**

Note	Item	Solar EMERGY (sej)	Energy (J)	Solar transformity[a,b] (sej/J)
	PROCESSES, replacement times <1 year			
1	Solar insolation	—	—	1
	Spartina:		13.2 E14	
2	Gross photosynthesis		2.85 E11	4,280
3	Day net photosynthesis		1.47 E11	8,979
4	24-hr net photosynthesis		2.62 E11	5,038
5	Aboveground live biomass		0.78 E11	8,461
6	Peaty sediment		1.65 E11	8,000
7	Aboveground dead biomass		0.51 E11	12,941
8	Annual detritus production		2.5 E10	52,800
9	Detritus export		1.26 E10	104,761
	Juncus:		14.9 E14	
10	Gross photosynthesis		6.1 E11	2,442
11	Day net photosynthesis		3.9 E11	3,820
12	24-hr net photosynthesis		1.91 E11	7,801
13	Aboveground live biomass		1.86 E11	8,010
14	Underground organic matter		1.54 E11	9,867
15	Aboveground dead biomass		1.47 E11	10,132
16	Detritus production		1.17 E11	12,735
17	Peaty sediment		1.13 E11	13,185
18	Fishes, crabs, shrimp	3.8 E15	2.74 E8	1.38 E7
19	Tidal channel landform	1.65 E16	9.8 E6	1.68 E9

[a] Quotient of two previous columns (solar EMERGY/energy).

[b] Solar transformities based on world energy flows (Odum et al., 1983, 1987): average wind, 623; rain over continents (chemical purity), 18,000; tide absorbed, 23,664; land runoff, 48,459.

1 Unity by definition.

2–8 EMERGY, 13.2 E14 sej/ha/year from Table 8.2.

2 Energy in gross production: (2.33 g C/m^2/day) (8 kcal/g C) (4186 J/kcal) (365 days/year) (1 E4 m^2/ha) = 2.85 E11 J/ha/year.

3 Energy in day net production: (1.21 g C/m^2/day) (8 kcal/g C) (4186 J/kcal) (365 days/year) (1 E4 m^2/ha) = 1.47 E11 J/ha/year.

4 Energy in 24-hr net photosynthesis: (2.15 g C/m^2/day) (8 kcal/g C) (4186 J/kcal) (365 days/year) (1 E4 m^2/ha) = 2.62 E11 J/ha/year.

5 Aboveground live biomass: 463 dry g/m^2, replacement time 1 year (463 g/m^2) (4 kcal/g) (4186 J/kcal) (1/year) (1 E4 m^2/ha) = 7.75 E10 J/ha/year.

Table 8.4 (continued)   Solar Transformities of Components and Processes
of Salt Marsh

6   Peaty sediment, 13.5% organic carbon in 1.14 m; see footnote #4 in Table 8.3, replacement time 300 years? (0.135 C) (1.14 m^3/m^2) (0.7 E6 dry g/m^3) (11 kcal/g C) (4186 J/kcal) (1 E4 m^2/ha)/300 year = 1.65 E11 J/ha/year.

7   Aboveground dead biomass: 307 dry g/m^2, replacement time 1 year (307 g/m^2) (4 kcal/g(4186 J/kcal) (1/year) (1 E4 m^2/ha) = 5.1 E10 J/ha/year.

8   Energy in detritus formation (0.15 g $C/m^2/day$) (11 kcal/g C) (4186 J/kcal) (365 days/year) (1 E4 m^2/ha) = 2.5 E10 J/ha/year.

9   Energy in detritus export assuming half of that formed (0.075 g $C/m^2/day$) (11 kcal/g C) (4186 J/kcal) (365 days/year) (1 E4 m^2/ha) = 1.26 E10 J/ha/year.

10–17   EMERGY, 14.9 E14 sej/ha/year from Table 8.2.

10   Energy in gross production: (5 g $C/m^2/day$) (8 kcal/g C) (4186 J/kcal) (365 days/year) (1 E4 m^2/ha) = 6.11 E11 J/ha/year.

11   Energy in day net production: (3.2 g $C/m^2/day$) (8 kcal/g C) (4186 J/kcal) (365 days/year) (1 E4 m^2/ha) = 3.9 E11 J/ha/year.

12   Energy in 24-hr net photosynthesis: (1.57 g $C/m^2/day$) (8 kcal/g C) (4186 J/kcal) (365 days/year) (1 E4 m^2/ha) = 1.91 E11 J/ha/year.

13   Aboveground live biomass: 1111 dry g/m^2, replacement time 1 year (1111 g/m^2) (4 kcal/g) (4186 J/kcal) (1/year) (1 E4 m^2/ha) = 1.86 E11 J/ha/year.

14   Underground biomass: 4600 dry g/m^2 (Kruczynski et al., 1978); replacement time 5 years? (4600 g/m^2) (4 kcal/g) (4186 J/kcal) (1 E4 m^2/ha)/(5 year) = 3.8 E12 J/ha/year.

15   Aboveground dead biomass: 880 dry g/m^2, replacement time 1 year (880 g/m^2) (4 kcal/g) (4186 J/kcal) (1/year) (1 E4 m^2/ha) = 1.47 E11 J/ha/year.

16   Energy in underground and detritus annual net production: (0.7 g $C/m^2/day$) (11 kcal/g C) (4186 J/kcal) (365 days/year) (1 E4 m^2/ha) = 1.17 E11 J/ha/year.

17   Peaty sediment, 1 m thick: 10.5% organic carbon in 1.0 m, replacement time 300 years? (0.105 C) (1.0 m^3/m^2) (0.7 E6 dry g/m^3) (11 kcal/g C) (4186 J/kcal) (1 E4 m^2/ha)/300 year = 1.128 E11 J/ha/year.

18   Fish, crabs, and shrimp in and out of the marsh, 8.2 g preserved wt/m^2; 24-month replacement; food chain based on 5 $g/m^2/day$ organic matter at start of consumer chain (Homer, 1977; Hall et al., 1986); transformity of base organic matter assumed 10,000 sej/J. EMERGY supporting the animals: (5 $g/m^2/day$) (365 days/year) (5 kcal/g) (4186 J/kcal) (1 E4 m^2/ha) (10,000 sej/J) = 3.8 E15 sej/m^2/year. Energy: (8.2 $g/m^2/year$) (0.2 dry) (4 kcal/g) (4186 J/kcal) (1 E4 m^2/ha) = 1.37 E8 J/ha/year.

19   Tidal channel landform if generated in 100 years (sedimentation and erosion). Half of tidal energy per marsh flooding-draining area (#2 in Table 8.2). EMERGY: (0.5) (100 year) (3.3 E14 sej/ha/year) = 1.65 E16 sej/ha. Energy of displacing mud, 1 m deep, 10% of area, 1000 kg/m^3 mud displaced, center of gravity 0.5 m: (0.1) (1 E4 m^2/ha) (1 m mean depth) (1000 kg/m^3) (0.5 m) (9.8 m/sec^2) = 9.8 E6 J/ha.

Table 8.5 Annual EMERGY of the Coastal Economy of Levy County, Florida (Figure 8.1)[a]

Note	Item, units	Data	Solar EMERGY/unit (sej/unit)	Solar EMERGY E18 (sej/year)	Macroeconomic 1990 $ (million $/year)[b]
	Coastal waters:				
1	Direct sun, J	4.44 E18	1	4.4	2.2
2	Tidal absorption, J	1.16 E15	23,564	27.3	13.6
3	Rain, J	4.1 E15	18,000	73.8	36.9
4	Runoff	6.0 E14	18,000	10.8	5.4
5	Total[c]			111.9	55.9
	Salt marshes:				
6	Direct sun, J	0.57 E18	1	0.57	0.29
7	Tidal absorption, J	2.69 E14	23,564	6.3	3.2
8	Transpiration, J	5.20 E14	18,000	9.4	4.7
9	Total[c]			15.7	7.9
	Land environment:				
10	Direct sun, J	3.57 E18	1	3.6	1.8
11	Transpiration, J	3.0 E15	18,000	54.0	27.0
12	Total[c]			57.6	28.8
13	Total of all environmental areas			185.2	92.6
	Human economy (1974):				
14	Electricity use, J	2.62 E13	1.59 E5	4.2	2.1
15	Gasoline use, J	1.49 E14	6.6 E4	98.3	49.2
16	Natural gas use, J	2.63 E12	4.8 E4	0.13	0.07
17	Goods and services, $	1.78 E6	5.8 E12	10.3	5.15
18	Total of the economy			112.9	56.52
19	Total environment and economy			298.1	149.12
20	**Fisheries landed (1974)**				
	Mullet, J	7.5 E11	5 E5?	0.37	0.19
	Game fish, J	3.1 E11	30 E6?	9.1	4.55
	Blue crabs, J	11.5 E11	4 E6?	4.65	2.3
	Total			14.1	7.0
	Fishery sales, 1974 $	5.63 E5	5.5 E12	3.1	1.55

[a] Areas: coastal waters, 8.2 E8 m²; marsh, 9.47 E7 m²; land, 6.0 E8 m².
[b] 1990 U.S. $ obtained by dividing annual EMERGY values by 2.0 E12 sej/1990 $.
[c] Totals were corrected for double counting by omitting sunlight, which contributes to the world weather system that brings rain.

Table 8.5 (continued) Annual EMERGY of the Coastal Economy of
Levy County, Florida (Figure 8.1)[a]

1   Gainesville insolation, 1.58 E6 kcal/m²/year, with 18% albedo: (1.58 E6 kcal/m²/year) (1 − 0.18) (8.2 E8) (4186 J/kcal) = 4.45 E18 J/year.

2   Tidal absorption in coastal waters as 50% of tidal energy, 80-cm mean tide with half of amplitude for center of gravity: (8.2 E8 m²) (0.5 × 0.8 m) (1025 kg/m³) (9.8 m/sec²) (0.5) (705 tides/year) = 1.16 E15 J/year.

3   Rain: 1.0 m ?; 5 J Gibbs free energy per gram relative to sea water (1.0 m/year) (8.2 E8 m²) (1 E6 G/m³) (5 J/g) = 4.1 E15.

4   Runoff estimated as 20% of land rain in #11: (0.2 m) (6.0 E8 m²) (1 E6 g/m³) (5 J/g) = 6.0 E14 J/year.

5   Sum of #2–4.

6   Gainesville insolation with 10% albedo: (1.58 E6 kcal/m²/year) (1 − 0.10) (9.5 E7 m²) (4186 J/kcal) = 5.67 E17 J/year.

7   All of 80-cm tide absorbed: (9.5 E7 m²) (0.5 × 0.8 m) (1025 kg/m³) (9.8 m/sec²) (705 tides/year) = 2.68 E14 J/year.

8   Marsh transpiration: 3.0 mm/day for 365 days = 1095 mm (1.095 m) (9.5 E7 m²) (1 E6 g/m³) (5 J/g) = 5.20 E14 J/year.

9   Sum of #6 and #7.

10 Gainesville insolation with 10% albedo: (1.58 E6 kcal/m²/year) (1 − 0.10) (6.0 E8) (4186 J/kcal) = 3.57 E18 J/year.

11 Transpiration as 80% of rainfall (1.0 m): (1.0 m) (6.0 E8 m²) (1 E6 g/m³) (5 J/g) = 3.0 E15 J/year.

12 Sum of #5, #9, and #12.

14 Electricity use: (83 E6 kWh/year/Levy County) (0.0876 area people/total people) (7.27 E6 kWh/year) (860 kcal/kWh) (4186 J/kcal) = 2.62 E13 J/year.

15 Gasoline: (11.2 E6 gal/year/county) (0.0876 population fraction) = 9.8 E5 gal/year; (9.8 E5 gal/year) (36,225 kcal/gal) (4186 J/kcal) = 1.49 E 14 J/year.

16 Natural gas per person, 1.87 E6 Btu/year (Florida Department of Administration, 1978): (1.87 E6 Btu/year) (1338 people) (1054 J/Btu) = 2.63 E12 J/year.

17 Money for purchases: the difference between transfer funds plus sales minus taxes: ($0.725 E6 + 2.278 E6 − 0.43 E6) = 2.573 E6 $/year. Goods and services estimated by subtracting prices paid for fuels from total money for purchase: (2.573 E6 − 0.786 E6) = 1.787 E6 $/year.

18 Sum of #14–17.

19 Sum of #12 and #18.

20 Catch ratios for Levy County in the period 1970–1972 from Prochaska and Cato (1975) were multiplied by total catches (Mathis et al., 1978). Mullet: 301 E6 fresh wt/year; 10% dry organic: (301 E6 g/year) (0.1) (6 kcal/g) (4186 J/kcal) = 7.5 E11 J/year. Game fish: 150 E6 fresh wt/year; 10% organic dry: (150 E6 g/year) (0.1) (5 kcal/g) (4186 J/kcal) = 3.1 E11 J/year. Blue crabs: 687 E6 g fresh wt/year; 8% organic dry: (687 E6 g/year) (0.08) (5 kcal/g) (4186 J/kcal) = 1.15 E12 J/year. Fishery sales multiplied by 5.5 sej/$ in 1974.

Solar Transformities for Salt Marsh Ecosystems

Table 8.4 presents solar transformities of some principal component processes and populations of the marsh, which were calculated by dividing their solar EMERGY share by their energy flow. The more successive transformations and stages of storage, the higher the transformity, as more and more resources are required per unit of energy transformed. Thus, transformities increase from sun to photosynthate to organic biomass to detritus to fishes, etc.

EMERGY Overview of Levy County

Table 8.5 is the overview EMERGY evaluation of the coastal county. Comparison of the total EMERGY from the marsh to the total for the county provides perspective on the importance of the marsh: about 9% of the environmental contributions and 5% of the total wealth of the county for 1974. With the coastal waters (grass flat, mud flat, and plankton ecosystems) and land ecosystems (forests, freshwater wetlands, successional ecosystems) included, environmental contributions as a whole were 62% of the annual county EMERGY value (7.9 million 1990 U.S. $) to the coastal economy.

Macroeconomic Value of a Hectare of Salt Marsh

In the last column of Tables 8.2 and 8.3, the annual EMERGY production of the marsh and other components of wealth was divided by the solar EMERGY/$ ratio of Florida to obtain the macroeconomic $ values. Direct studies of people making economic use of the marsh in fishing, for example, were not made, but a value of $89 in Table 8.1 may be appropriate and could be added.

Macroeconomic contributions of *Spartina* and *Juncus* marshes in Table 8.2—660 $/year/ha (264 $/year/acre) and 743 $/year/ha (297 $/year/acre)—were larger than was obtained from economic evaluations that measure only the services of humans (Table 8.1).

For Levy County as a whole, the macroeconomic value was 7.9 million 1990 U.S. $. In other words, the portion of the gross economic product due to environmental work supporting the economy, mostly indirectly, was $7.9 million/year.

EMERGY Investment Ratio Measure of Environmental Matching

In addition to the direct EMERGY contributions to the economy of nature and humans, environmental resources are the basis for attracting matching EMERGY from economic investment. Marshes and their fisheries attract fishing, tourists, human settlements for retirees, etc. Purchased EMERGY comes into a local economy as the electricity, fuels, goods and services, etc. (Figures 8.2 and 8.4). The ratio of the purchased EMERGY to the environmental EMERGY is called the EMERGY investment ratio. For Florida as a whole, as for the United States as a whole, the typical regional ratio is 7.0.

From Table 8.5 in Levy County in 1974, however, the ratio of purchased EMERGY to free environmental energy was 112.9/185.2 = 0.61, a lower value than the present state and national average. This means that economic activities in the Cedar Key area had more than usual environmental support, providing more opportunity for environmental waste absorption, aesthetics, and other natural services which in cities require payment and higher tax. Areas with lower investment ratio may be expected to continue economic growth relative to those with higher ratios. Certainly, Cedar Key has exhibited economic growth since 1974.

Ratio of Services to Environmental Contributions

Our EMERGY analysis procedure includes line items for regular economic values, and these are expressed in solar EMERGY units for comparison with the environmental EMERGY contributions. Measuring the degree of development and environmental loading is the ratio of human services to the environmental contribution, both expressed in EMERGY units (or macroeconomic $).

For example, from Table 8.1, $89/ha/year economic contribution (services) may be divided by $700/ha/year macroeconomic value from environment, which results in 0.127 (12.7%). Studies by Stellar (1976) and Sell (1977) had much higher human loading to mangroves in the Marco Island area of Florida.

Regional EMERGY Imports Attracted by Environmental Contributions

In one sense, the attracted EMERGY flows depend on the environmental ones, and thus all the EMERGY wealth can be attributed to the presence of the environmental resource in the sense that potential for economic matching is destroyed in proportion as the environmental resources are eliminated.

In this sense, a hectare of marsh had the macroeconomic value of its direct annual EMERGY contribution (Table 8.2), plus the economic matching according to the regional investment ratio at Cedar Key in 1974 (0.61), for a total of $12.7 million/year. If this value is divided by the area of marsh (9470 ha), the macroeconomic value that results is $1341 (1990 $)/ha/year.

The potential for growth due to the environmental value of the marsh in Levy County is the direct free contribution ($7.9 million) times the state EMERGY investment ratio (7.0), which is $55.3 million (1990 $)/year. Dividing by the area of marsh (9470 ha), the potential public value that depends on marshes is $5839/ha/year.

Public Policy, EMERGY Value, and Economic Value

Officials responsible for purchase of environmental lands have voiced concerns that EMERGY-based evaluation might be used for land appraisal in determining money to be paid to owners. This is a misconception. Market values concern the contributions of humans and their businesses, which are small values, because human services involved in utilizing environmental products and services are generally small compared to the large work in the environmental products and services themselves. It is wrong to pay individuals for the work of nature.

On the other hand, it is appropriate to consider the EMERGY-based macroeconomic dollars in deciding what the marsh is worth to the public good as compared with proposals for alternative use. Tax reductions so as to preserve marshes may be appropriately judged as some fraction of the macroeconomic dollar value.

A new trend in law, the public trust doctrine, is giving priority to the rights of the public to environmental values over rights to private property, and EMERGY evaluations are a means for quantitative implementation. One of our calculations was used to settle a court case on mangrove damage in south Florida in 1990.

As described by Johnston (1990), the public trust doctrine is

> an ancient, but rapidly expanding judicially-created doctrine that says: the public has an interest akin to an easement, which predates all private ownership for the protection of navigation, commerce, wildlife habitat, and kindred interests.

Mitigation

Laws and policies for environmental regulation now concern mitigation, the substitution of one environmental value for another. The EMERGY and transformity tables in this chapter provide ready means for judging equivalent values in salt marshes for comparison with other areas of nature and the economy. A set of similar tables for the main 10 to 20 types of ecosystem in Florida needs to be constructed. These and some procedural guidelines will constitute a mitigation manual.

Data Needs

Whereas EMERGY evaluations were herewith completed for marshes and a coastal county, a number of approximations and assumptions were made. The results can be made more precise with further measurements or location of other data. Among the data needs are time for deposition of 1 m of marsh–peaty soil, annual export of detritus and food extracted by nekton, rates of transpiration, and percentages of tidal energy absorbed in various coastal zones. With more data, some changes in transformities are to be expected, but probably not order of magnitude changes.

References

Bell, F. 1989. *Application of Wetland Valuation Theory to Florida Fisheries.* Florida Sea Grant Publications SGR-95. University of Florida, Gainesville. p. 14.

Bergstrom, J. C., J. R. Stoll, J. P. Titre, and V. L. Wright. 1990. Economic value of wetlands-based recreation. *Ecol. Econ.* 2:129–147.

Coultas, C. L. and F. G. Calhoun. 1976. Properties of some tidal marsh soils of Florida. *Soil Sci. Soc. Am. J.* 40(1):72–76.

Coultas, C. L. and E. Gross. 1975. Distribution and properties of some tidal marsh soils of Apalachee Bay, Florida. *Soil Sci. Soc. Am. Proc.* 39(5):914–919.

Farber, S. C. and R. Costanza. 1987. The economic value of wetlands systems. *J. Environ. Manage.* 24:41–51.

Florida Department of Administration. 1978. *Forecasts of Energy Consumption in Florida.* State Energy Office, Tallahassee.

Goldstein, J. H. 1971. Competition for wetlands in the midwest, an economic analysis. *Resources for the Future.* Johns Hopkins Press, Baltimore. 105 pp.

Gosselink, J. G., E. P. Odum, and R. M. Pope. 1974. *The Value of Tidal Marsh.* LSU-SG-74-03. Center for Wetland Resources, Louisiana State University, Baton Rouge. 29 pp.

Hall, C., C. J. Cleveland, and R. Kaufmann. 1986. *Energy and Resource Quality.* John Wiley and Sons, New York. 577 pp.

Homer, M. 1977. Pages 259–267 in *Thermal Ecology II.* Energy Research and Development Administration, Division of Technology Information, Oak Ridge, Tennessee.

Johnson, R. W. 1990. The public trust doctrine and Alaska Oil. Pages 1–22 No. 8.2 in Appendix M of Alaska Oil Spill Commission (Vol. III), Sea Grant Legal Research Report by T. Bader, R. Johnson, Z. Plater, and A. Rieser.

Kruczynski, W. L., C. B. Subrahmanyam, and S. H. Drake. 1978. Studies on the plant community of a north Florida salt marsh. Part I. Primary production. *Bull. Mar. Sci.* 28(2):316–334.

Leitch, J. A. and B. L. Ekstrom. 1989. *Wetland Economics and Assessment, An Annotated Bibliography.* Garland Publishing, New York. 194 pp.

Lynne, G. D., P. Conroy, and F. J. Prochaska. 1981. Economic valuation of marsh areas for marine production processes. *J. Environ. Econ. Manage.* 8:175–186.

Mathis, K., J. C. Cato, R. T. Degner, P. D. Landrum, and F. J. Prochaska. 1978. *Commercial Fishing Activity and Facility Needs in Florida: Dixie, Levy and Taylor Counties.* Department of Food and Resource Economics, University of Florida, Gainesville.

Odum, H. T. 1967. Energetics of world food production. Pages 55–94 in *Problems of World Food Supply.* President's Science Advisory Committee Report, Vol. 3. White House, Washington, D.C.

Odum, H. T. 1971. *Environment, Power and Society.* John Wiley and Sons, New York. 331 pp.

Odum, H. T. 1978a. Value of wetlands as domestic ecosystems. Pages 9–18 in *Proceedings of the National Wetland Protection Symposium.* Stock #024-010-00504-1. U.S. Department of Interior, Fish and Wildlife Service, U.S. Government Printing Office, Washington, D.C.

Odum, H. T. 1978b. Principle of environmental energy matching for estimating potential economic value, a rebuttal. *Coastal Zone Manage. J.* 5(1):247–249.

Odum, H. T. 1986. Emergy in ecosystems. Pages 337–369 in N. Polunin, Ed. *Ecosystem Theory and Application.* John Wiley and Sons, New York.

Odum, H. T. 1988. Self organization, transformity, and information. *Science* 242: 1132–1139.

Odum, H. T. 1996. *Environmental Accounting,* EMERGY *and Environmental Decision Making.* John Wiley and Sons, New York, 370 pp.

Odum, H. T. and E. C. Odum, Eds. 1983. *Energy Analysis Overview of Nations.* Working Paper WP-83-82. International Institute for Applied Systems Analysis, Laxenburg, Austria. 467 pp.

Odum, H. T., E. C. Odum, and M. Blissett, Eds. 1987. *Ecology and Economy: Emergy Analysis and Public Policy in Texas.* Lyndon B. Johnson School of Public Affairs and Texas Department of Agriculture, Policy Research Publication #78. University of Texas, Austin. 178 pp.

Odum, H. T., E. C. Odum, and M. T. Brown. 1993. *Environment and Society in Florida.* Center for Environmental Policies, University of Florida, Gainesville, 446 pp.

Prochaska, F. J. and J. C. Cato. 1975. *Landings, Values and Prices in Commercial Fishing for the Florida West Coast.* Marine Advisory Bulletin SUSF-SG-75-003. University of Florida Cooperative Extension Service, Gainesville.

Reimold, R. J., J. H. Phillips, and M. A. Hardistey. 1980. Sociocultural values of wetlands. Pages 79–89 in V. S. Kennedy, Ed. *Estuarine Perspectives.* Academic Press, New York.

Sell, M. G. 1977. Modelling the response of mangrove ecosystems to herbicide spraying, hurricanes, nutrient enrichment, and economic development. Ph.D. dissertation. Environmental Engineering Sciences, University of Florida, Gainesville. 412 pp.

Stellar, D. L. 1976. An energy evaluation of residential development alternatives in mangroves. M.S. thesis. Environmental Engineering Sciences, University of Florida, Gainesville. 124 pp.

Turner, M. G., E. P. Odum, R. Costanza, and T. M. Springer. 1988. *Environ. Manage.* 12(2):209–217.

Marginal Value of Coastal Wetlands to Florida's Commercial Fisheries

9

Frederick W. Bell and Gary D. Lynne

Saltwater or marine wetlands are ecologically linked to nearly 92% of the value of Florida's commercial fish harvest (Bell, 1989). The link arises because of two basic functions in supporting fish populations. First, wetlands provide grazing food and detritus material (i.e., decomposed matter). The rich source of vitamins and proteins in detritus may be especially important (Heald and Odum, 1970). Second, Boesch and Turner (1984) have argued that food support is secondary to the wetland function as a refuge from predators. Wetlands appear to be a critical factor in determining the magnitude of the biomass and consequently the level of the commercial fishery harvest.

The purpose of this chapter is to demonstrate the marginal-value-productivity approach as one method for estimating the dollar value of an acre of marsh land. The value estimates suggest the important economic role of Florida's Gulf coast wetlands (including the Panhandle) to the commercial fisheries in this area. First, problems in estimating a value for the services from saltwater wetlands are discussed. Generally, poorly functioning wetlands markets (i.e., there are not sellers of wetland services because, in part, those who benefit may be unwilling to pay for such services) and a large number of wetlands services make it difficult, at best, to estimate dollar value. Second, a brief review of previous valuation literature is provided, highlighting the character of the controversy over valuation methodologies. Next, marginal productivity theory is explained as one framework for guiding the value estimating process. Lastly, the empirical estimates of wetland values arising because of the Florida commercial fishery industry are presented and discussed.

Wetlands Markets, Wetland Values, and Market Failure

In the main, the market process determines land use and allocation in the United States. A particular area of land enters a specific use determined by the highest bidder. Also, generally the land market works reasonably well at meeting many social objectives. Sometimes, however, the market process breaks down or fails to meet such objectives, which sometimes happens in the case of wetlands. Consider the following scenario.

One could inherit an undeveloped acre of salt marsh. This "swamp land" may have an alternative use such as residential or commercial development, assuming the owner has the legal right to convert the inherited swamp. Several problems may emerge, all related to the fact that unaltered wetlands perform many useful functions that may be lost if development takes place, but for which the owner may not generally receive any payment.

If an additional acre of wetland increases fishery production, for example, marsh service is worth something to the commercial fishery industry. The owner could profit from charging the fishery for the service. Here is the rub! How is the present owner of the inherited additional acre of wetland to collect for the service provided by said land to individuals in the fishery? The market process will not generally work to help the owner capture a share of the benefits (rent, services) enjoyed in the fishery. Basically, there is an information transfer problem in that the value information cannot easily get from the fishers to the marsh owner. Similarly, at least some consumers may also place a high value on fish products coming from a natural wetland system and thus be willing to pay more for fish that depend on such a system. Again, the market may fail in that the consumer could not transfer the value information down to the fishers through the fish market. The inability of the wetland owner to obtain payment from services provided to both fish consumers and the fishery (and, as we will see, from a large number of other beneficiaries of wetland services) is called "market failure" by economists. At a minimum, the laws (and rules, regulations) would have to be changed to facilitate such payments to our hypothetical inheritor.

Under market failure, we would also not expect that individuals in the fishery would enter into contracts to rent wetlands, or otherwise hire the services of wetlands, for a number of reasons. First, individual fishery firms will find it difficult to quantify the increased profits received due to an additional acre of wetland. Thus, the appropriate payment to an owner could not be easily determined. Second, wetlands vary in quality, making benefits (i.e., profits) to the individual even more difficult to quantify. Third, there may be hundreds of wetland owners: who would individual fishery firms pay for the alleged service? Fourth, the sheer costs (transactions costs) for individuals in the fishery to negotiate with hundreds of wetland owners could be immense. Fifth, no provision exists in current law to facilitate a marsh rental market. That is, it is not clear that a fishery firm could in fact rent the services from our hypothetical owner, largely because the law does not specifically define what services our owner actually can claim as privately owned. Because of such problems, a rental market would also likely fail to provide the socially desired quantity of wetlands, at least until the law could be changed.

The net result may be that our hypothetical new owner will convert the land to condominiums or other developments. With the market failing (due mainly to problems in underlying laws) to help reveal the true social value of wetlands, we could expect a level of conversion to exceed the socially efficient amount.

One route often followed in the face of such problems is to regulate current wetland owners. In effect, the importance of the other values is imbedded in regulatory law in order to affect behavior relative to wetland use. A zoning regulation, for example, may restrict or even prohibit commercial development. In effect, the regulation says (implicitly) that other values are more important than the economic values that would come with the development. The regulation changes the property right. While still owning the land, others in society would now determine who benefits and who pays the costs of any use and what the use will be. The relationship between the owner and the rest of society has now changed. The new law or regulation changes the "choice sets" (Bromley, 1989) of the owner.

Increasingly, wetlands have been moved from the private into the public domain and, thus, from private ownership with very few limits to common property ownership with more extensive limits. In the extreme, owners may be prohibited (e.g., in dredge and fill regulations) from converting wetlands to any other use. Increasingly, society has taken steps to express other values (nonmonetary) through the vehicle of government (usually regulation). Wetlands have come to be viewed as public goods.

Changes in property rights suggest value shifts. A new zoning regulation which limits commercial development to preserve wetlands suggests the natural service is valued more highly than the commercial development. In this setting, economic analysis involves (1) a values analysis, focusing on improving understanding and explaining the important values being served by some law, rule, or regulation and (2) examining the effects of the law, rule, or regulation on economic values by estimating the opportunity losses from reducing the level of economic activity and retaining the natural system intact. The marginal productivity approach can also be useful in estimating such opportunity costs. Such costs are really estimates of the willingness-to-pay in order to use wetlands in commercial development.

The problem of estimating value is further compounded by the difficulty in measuring the wide range and variety of services emanating from wetlands. Wetlands serve as a fish and shellfish habitat (Table 9.1), which benefits the commercial fishery. Wetlands also provide year-round habitats for resident birds and serve as breeding grounds. Saltwater marshes along the Atlantic coast are used for nesting by birds such as black duck and marsh hawks. Wetlands are important to many birds, ranging from waterfowl and shorebirds to songbirds. Such areas also produce furbearers, like muskrats, beavers, and nutria.

Wetlands provide an environmental service with considerable economic value, called environmental quality. Wetlands help maintain water quality or improve degraded waters by (1) removing nutrients, (2) processing chemical and organic waste, and (3) reducing sediment loads of water. The Brillion Marsh in Wisconsin, for example, has been used to process domestic sewage since 1923. Wetlands also reduce turbidity of flood waters, which is especially important for aquatic life and for reducing siltation of ports, harbors, rivers, and reservoirs. Marshes also reduce biochemical oxygen demand, thereby increasing oxygen levels. Wetlands may also be regarded as the farmland for plants

Table 9.1   List of Major Wetland Values

Fish and wildlife values	Socioeconomic values
Fish and shellfish habitat	Flood control
Waterfowl and other bird habitat	Wave damage protection
Furbearer and other wildlife habitat	Erosion control
	Groundwater recharge and water supply
Environmental quality values	Timber and other natural products
Water quality maintenance	Energy source (peat)
Pollution filter	Livestock grazing
Sediment removal	Fishing and shellfishing
Oxygen production	Hunting and trapping
Nutrient recycling	Recreation
Chemical and nutrient absorption	Aesthetics
Aquatic productivity	Education and scientific research
Microclimate regulator	
World climate (ozone layer)	

Source: U.S. Fish and Wildlife Service (1984b).

that are particularly good converters of solar energy to vegetable biomass. While direct grazing of wetlands plants may be generally limited, a major food value is reached upon death when plants fragment to form detritus. Shrimp and mullet eat detritus or graze upon the bacteria, fungi, diatoms, and protozoa growing on its surface (see Crow and Macdonald, 1979).

A wetland may also temporarily store flood waters, thereby protecting downstream property owners from flood damage. Wetlands slowly release the water downstream. Under storm conditions, wetlands may also buffer the land from wave damage. Mangrove swamps are so efficient in buffering that the Federal Insurance Administration's regulations state that insured communities shall prohibit mangrove destruction or lose federal flood insurance. Wetland vegetation can reduce shoreline erosion by (1) increasing the durability of the sediment through binding with its roots, (2) damping waves through friction, and (3) reducing current velocity through friction. Coastal wetlands may also be an indirect source of drinking water by serving to protect coastal freshwater supplies. Many recreational activities such as hunting and fishing also take place in and around wetlands.

The wide array of services and problems in measuring services causes value estimating to be a complex effort. Despite the difficulties, the increasing demand for wetlands has stimulated intense interest in estimating values. Further knowledge and estimates of value help in comparing the benefits of wetland preservation (i.e., providing public goods) with the benefits of wetland development (i.e., allowing private owners to prevail) and thus improve the public decision process.

Importantly, the methods for valuing wetlands are affected by the failure (or total lack of) markets for the wide array of wetlands services. Said somewhat differently, *the concept of economic value presumes the existence of a market process to help reveal the value of a good or service through time; therefore, quantifying economic value is difficult without well-functioning markets.* In fact, without such markets, or in poorly functioning markets, the exact economic value cannot be known. The best we can do is to estimate the probable willingness-to-pay as if markets exist for the wetland service(s) in question. The marginal productivity approach demonstrated in this chapter represents one way to arrive at such a willingness-to-pay estimate. We now turn to a discussion of other valuation approaches that have been used.

The Literature on Wetland Valuation

Two general features of the literature need emphasis. First, the literature on techniques of wetland valuation has been controversial. Considerable debate has gone before, and continues, regarding the best way to value a wetland. Second, the term "value" and the idea of "valuation" are open to various interpretations. One of the more controversial approaches has been that offered by Gosselink, Odum, and Pope (1974) (hereinafter referred to as GOP), who used what they term a "life-support" valuation methodology.

GOP added the market value of the products of wetlands (e.g., ex-vessel value of estuarine-dependent species, sales [gross, total] value of furbearing animals) and divided by the number of wetland acres to arrive at an estimate of annual dollar value (R). GOP then used the income capitalization formula V = R/i, where V = the present value of an acre of land and i = interest or discount rate. The value of an acre of wetland to the fishery may be $2000, for example, if R = $100 and i = 5%.

GOP state "...all of the value of the fishery harvest is here imputed to the marsh, imputing nothing to capital, labor and management" (p. 5). GOP found that the three estuarine functions of (1) fishery habitat, (2) aquaculture, and (3) wastewater treatment alone generate a present value per acre of $81,500 using a discount rate of 5%. This suggests an annual return into perpetuity of $81,500 × 0.05 = $4075/acre. The GOP procedure has been used in other studies such as that of Raphael and Jaworski (1979).

The GOP study has also been used extensively in policy discussion. The estimates can be found in the 1974 U.S. Council on Environmental Quality report. GOP also claim their estimates have been widely used as a basis for formulating laws (see Odum, 1973). Even though used extensively, the method is not without problems.

Shabman and Batie (1978) have argued the method to be in error by the very fact that the whole market value of the fishery is attributed to the wetlands. The method assumes that the value of capital and labor in the fishing industry is zero (i.e., free goods). Yet, in reality, if the fishery were to experience immediate economic collapse, capital and labor would likely be able to earn positive returns in other (next best) employment. To truly calculate economic value of the wetland, we need to calculate the difference in value with and without the fishery.

A with and without analysis recognizes that capital and labor resources could be used in other economic activity to generate some value, albeit less than in the fishery. Because GOP did not use with and without analysis, we tend to agree with Shabman and Batie, who state, "As such, GOP's estimates of the value of marsh services are, at best, inaccurate. At worst, these inaccurate estimates may capture the focus of policy debate, and hinder rather than improve, the resource management process for coastal wetlands" (p. 244).

Tihansky and Meade (1976) have suggested a major alteration of the GOP procedure discussed above. They suggest using net (as opposed to gross or total) economic rent. Net economic rent is defined as the difference between the value of the catch and the market cost of explicit factors of production (i.e., capital equipment, labor, and returns to management). GOP estimates overstate the value per acre by the amount of the capitalized value of the explicit factors of production.

If the ocean fishery stocks were privately owned, such a residual rent could be collected by the owner for the use of the wetlands service. With the reality being that ocean fishery stocks are often virtually open access (no property) resources, residual rents will accrue to individuals who put capital and labor into fishing effort. As a result, these rents tend to get dissipated through excessive entry into the fishery. With unrestricted entry, net rents would tend toward zero in the long run. The value of the wetland, then, would also tend toward zero.

Another framework for estimating the value of wetlands has been developed by Odum (1977). To quote Odum (1977, p. 11):

> Understanding the value of wetlands for their partnership with human activity cannot be done with traditional economic valuations, because wetlands do not receive and deliver money for their outputs and inputs. Evaluating their contribution, however, can be done with energy units which are the common denominator of useful work both in economic exchanges and environmental exchanges.

Odum proposes that we calculate the energy of ecosystems, under the guidance of what has come to be called the energy theory of value. Rather than using dollar values as the common unit for measuring value, all values are measured in energy units. Estuarine ecosystems provide many services, as discussed earlier. Wetlands use energy to produce the services. For renewable resources, the sun provides energy, which Odum calls a "free externality." All such services are converted to energy units and added to give the total energy value of a wetland: energy becomes the common denominator for measuring the values of all the wetlands services.

The calculated energy value for an acre of wetland is then converted to dollar value by dividing gross national product (GNP) by kilocalories consumed in the United States. The logic is that because energy generates all economic value, it should be possible to calculate the value of each energy unit by measuring the amount of kilocalories used to generate the GNP and dividing to get an average dollar value per energy unit. The formula becomes:

$$\$/\text{acre/year} = \text{kcal/acre/year} \times \frac{\text{Gross National Product}}{\text{National Energy Consumption}}$$

In 1974, GOP calculated the dollar value per kilocalorie to be 10^{-4}/kcal. Multiplying 10^{-4} times the energy flowing through an acre of wetlands equals $4163 per year. Capitalized at 5%, an acre of wetlands would be valued at $83,260.

One must engage in fairly detailed studies to determine the kcal/acre/year. Hoffman (1986) indicates a wetland productivity of 7.08×10 kcal/acre/year, whereas GOP indicate a range of 2.08 (high marshes) to 6.24×10 kcal/acre/year (well-irrigated low marshes). This calculation was for net primary production (NPP). If NPP were increased by 25% and multiplied by 1850 kcal, gross kcal/acre/year would be 3.99×10 kcal/acre/year. The energy productivity numbers used in Odum's equation seem to vary considerably. This variation may be a cause for concern.

Shabman and Batie (1978) argue that the imputation of all GNP to caloric use is fallacious. The resulting system of energy-based prices suggests that all goods and services, as well as inputs such as labor, machinery, and raw materials, are merely transformed energy. This suggests a one-commodity world (reducible calories), and the only constraint on production is energy supply. While the fundamental role of energy is clear (i.e., the earth and human systems are driven by energy), we, like Shabman and Batie, also see the need to account for the role of humans.

We take an anthropocentric view: we consider energy as one input among many involved in producing fish. Production may increase due to new knowledge embodied in technologies produced by human innovation. Human labor and management, as well as human-created capital resources (while admittedly composed of energy; see Costanza, 1980), all affect what can be produced. The human value system, and the choices and actions arising as a result of that value system, are seemingly missing from the energy theory of value.

Another way to approach the matter is to argue that the energy theory of value does not consider consumer demand. It appears to fail to recognize the systematic relation between consumer demand and energy use in that a relative rise in energy prices (not GNP) will shift consumer demand to less energy-using goods and services. Ignoring consumer demand is tantamount to ignoring the human element in valuation.

An approach that does include human values involves land price analysis. A land parcel's sale price is determined, in part, by the human benefit related to some wetland service. Parcels near natural wetlands that offer open space or recreational opportunities will usually command higher market prices than land parcels that do not have these locational advantages (see Batie and Mabbs-Zeno [1985] for an analysis of wetland location on development values).

A variant of this approach involves using the appraised value, which is generally based on the market price of similar property sold in the same area in the recent past. State agencies most often use appraisals when purchasing wetlands. The fundamental problem is the basis of the appraisal. For example, a filled wetland may represent a locational service when it allows a commercial or industrial operation to be near a market, transportation facility, or raw material source. A drained wetland may be a direct input into production, as when a wetland is converted to producing agricultural crops. Developed wetlands may also provide services used directly by final consumers when development allows location of residential and recreational home sites. If development of wetlands is permitted on parcels used for appraisals, then the purchase price by state agencies will be based upon developed wetland values with a partial or complete loss of natural services. The method does not generally account for the broad array of wetland values arising from the various services.

Another method suggests the replacement cost as the economic value of the wetland. Destroyed or developed wetlands can sometimes be transformed back to the original or near original state where natural services are restored. Also, a wetland can be created where one never existed to offset destruction of wetlands elsewhere. Such restoration is called mitigation. The cost of replacing a wetland becomes an estimate of the minimum economic value.

According to Hoffman (1986), mitigation and restoration techniques are applied in Tampa Bay waters today. The average price of planting mangroves is $10,000 per acre, which includes the cost of the plants and labor. An acre would require about 5000 mangrove trees (about 12 in. in height and placed 36 in. apart). The replaced acre of small mangrove trees by no means can be considered identical to a mature, functioning mangrove stand. We must add the interest on the $10,000 (i.e., opportunity cost of money) over 20 years, the minimum amount of time it may take (assuming it works at all; Hoffman [1986]) to produce a functioning stand of mangroves. At an interest rate of 8%, the replacement value would be $46,610/acre ($10,000 × [1 + 0.08]20).

The replacement cost method has limits because it does not measure the direct natural services of wetlands, which may be more crucial to a development vs. preservation decision. If one could quantify the values of services from natural wetlands, however, then such values could be compared to replacement cost. For example, if government dollars were used to fund the replacement, restoration, or mitigation, one could compute the rate of return on such investment. Replacement value of wetlands is probably important for forming social policy pertaining to government investment in man-made wetlands. Unfortunately, the replacement method does not help in quantifying the contribution of services in wetlands to value.

An opportunity cost of preservation method has been suggested by Batie and Mabbs-Zeno (1985). They argue: "Given the current inability to provide economic values of natural wetlands areas, one strategy for wetlands management would be the adoption of management guidelines which stress wetland preservation, unless the expected value of foregone development is 'unacceptably' large" (p. 1).

The Virginia Wetlands Act of 1972 allows permits to be issued for wetland alteration only when the anticipated public and private benefit of the proposed activity exceeds the anticipated public and private detriment. Batie and Mabbs-Zeno examined a 1500-acre subdivision called Captain's Cove on Virginia's Atlantic coast. For a development plan which fills a given amount of wetlands, gross private costs were subtracted from gross returns to find net returns. Although wetlands are cheaper to acquire (about 20% of the value of uplands), the cost of development is more because of dredge and fill operations. In fact, the development of wetlands is nearly five times more expensive than upland development. This cost is offset by the revenue obtained from wetland development near shorefront or canalfront. The net revenue for the development of nonlevied wetlands was greatest near the shorefront, a lesser amount for canalfront, and a much smaller amount for uplands. Marginal net returns reached a peak between the 14th and 20th converted wetland acre, because all possible shorefront lots had been included at that level of development. If 58 acres of wetlands are developed, total development benefits would be $31.9 million.

The $31.9 million represents the opportunity cost of permit denial to the development of various wetlands. The preserved wetland would have to be worth at least as much, if economic return is a concern. Batie and Mabbs-Zeno (1985) state: "...it is likely that the permit would be allowed unless very substantial negative externalities from development were identified" (p. 8). The question is: what is the value lost by such negative externalities (i.e., from not developing the wetlands)? The method demonstrates the opportunity cost of not developing wetlands, which may be valuable information for social policy decision making.

The Marginal Productivity Model

We now turn to the marginal productivity model as the method demonstrated in this chapter. The marginal productivity method quantifies the contribution of a wetland system to some product sold in a market, while recognizing that humans also contribute to the product, say, fish. The marginal productivity method estimates the share of the total (sales, gross) fish market value attributable to the marsh.

Ideal application of the marginal productivity method requires detailed knowledge of the specific linkages between wetlands and particular natural services in a primitive state. The lack of such knowledge serves as an impediment to estimating wetland values (Walker, 1974). Such technical linkages can generally be represented in the concept of a production function. In the broadest sense, the combination of labor and capital (i.e., fishing effort) of fish harvesters with fish habitat services determines fish caught. Consider the following linear production function:

$$Q_i = c_0 + c_1E_i + c_2W_i \qquad (9.1)$$

where Q_i = harvest (pounds) of the ith fishery species, E_i = fishing effort applied to the ith species (e.g., standardized days), W_i = wetlands (acres) affording services to the ith species, and $c_1 > 0$ and $c_2 > 0$.

The marginal productivity of 1 acre of wetland services is $(oQ/oW) = c_2$. Assume that $c_2 = 10$. This means that an increase (decrease) in 1 acre of wetlands will increase (decrease) the catch by 10 lb, everything else held constant. To arrive at an estimate of the value of one more acre, we ask what consumers are willing to pay for that additional 10 lb of fish. Assuming that the additional catch does not depress price (because it is a very small addition to the entire market), the marginal value product (MVP) may be calculated or $(P \times MP) = MVP$, where P is the ex-vessel price of a pound of fish. Assuming that P = \$2.00, then for MP = 10 lb, the MVP = \$20. The \$20 represents the marginal fish value for the ith species from adding (or eliminating) 1 acre of wetlands. Assume further that this MVP occurs annually, into perpetuity. Applying the capitalization formula, the current value of the marginal acre becomes \$20 ÷ 0.08, or \$250 (an 8% discount rate is used as an example).

Empirical Application of the Marginal Productivity Method: West Coast of Florida

Production function models were estimated for Florida's west coast commercial fisheries. (See the equations in the Appendix.) To empirically estimate the models, the analyst needs observations on catch, fishing effort, and wetlands (e.g., salt marsh). We thought it would be a relatively easy task to collect data on catch and fishing effort for a variety of estuarine-dependent species. This was not the case. The need for a relatively long time series on catch and effort became a significant problem. Fishery management plans developed by the Gulf Fishery Management Council were helpful, but did not include such estuarine-dependent species as blue crab and oysters. In conjunction with the National Marine Fisheries Service (NMFS) Southeast Fisheries Center, we were able to develop nine estuarine-dependent species that had relatively long time-series data on catch and excellent to rudimentary measures of fishing effort.

Another problem associated with such analysis is that some of the species

Table 9.2   Selected Estuarine-Dependent Species Used for Production
Function Estimation for West Coast of Florida, 1984

Species	Pounds	Ex-vessel value ($)
Blue crab	12,912,367	3,196,927
Stone crab	3,944,911	7,343,707
Grouper	9,462,566	13,037,260
Red snapper	2,602,408	5,545,600
Oyster	6,621,595	7,300,689
Spiny lobster	5,961,388	14,929,673
Shrimp	26,449,115	47,591,405
Black mullet	17,903,879	4,680,803
Subtotal	85,858,229	103,626,064
Grand total of estuarine dependent	110,759,538	114,704,971
Percent of grand total	77.5	90.3

Source: U.S. Department of Commerce (1987); Harris et al. (1983).

have a significant recreational component. No long time-series data on recreational catch and effort are available for grouper, red snapper, and black mullet, which do have significant recreational angler participation. This problem may not impact the estimates for the shellfish as much as for the finfish species.

The eight commercial species selected for time-series estimation are shown in Table 9.2, along with catch and ex-vessel value for the west coast of Florida. The criteria for selection include (1) the species be estuarine dependent, (2) some measure of fishing effort be available, and (3) the species be a significant component of the total catch (value and/or weight). Of the total estuarine-dependent catch, our eight species constituted 78% by weight and 90% by value (Table 9.2).

We had to consider the entire west coast of Florida, including the Panhandle as a whole, rather than selected counties or areas due to a salt marsh acreage time-series data set that cannot be separated into subcoastal regions. The original research design called for a subcoastal region approach for species that are not evenly distributed along the Florida coast. Blue crab, grouper, red snapper, and black mullet are widely distributed along Florida's west coast, whereas oysters (in Franklin County), spiny lobsters (Monroe County), stone crab (Collier County), and shrimp (Monroe) are not as evenly distributed. For the eight species, catch and fishing effort over the period 1950s through 1980s were collected. The following is a synopsis of such data by species:

Species	Measure of fishing effort	Area of catch	Years catch effort
1. Blue crab	Pots	Counties	1950–1984
2. Stone crab	Pots	Counties	1950–1981
3. Spiny lobster	Traps	Counties	1950–1985
4. Red snapper	Hand lines	Counties	1950–1981
5. Grouper	Hand lines	Counties	1950–1981
6. Oyster	Tongs	Counties	1950–1981
7. Shrimp	Days/trips	NMFS Grids	1957–1985
8. Black mullet	Gill nets	Counties	1952–1976

Of the eight species, the shrimp analysis has less theoretical problems based upon the criticisms of Hoffman (1986) on the use of catch and effort data in conjunction with wetland changes. The shrimp effort series has been developed for biological analysis and is thought to be of excellent quality. The catch is by area of the Gulf of Mexico rather than landed catch by county. The fishing effort series is really a proxy for the remaining seven species, although such proxies have been used for policy analysis by the Gulf of Mexico Fishery Management Council (e.g., stone crab, snapper–grouper, and spiny lobster).

According to the National Wetland Inventory (U.S. Fish and Wildlife Service, 1984b), Florida has 11,854,822 acres of wetlands (i.e., vegetated and nonvegetated) or 31.6% of the total area (i.e., land plus water) of the state. Only 12.6% or 1,494,910 acres of wetlands is salt and brackish and of importance to this study. Of the 12.6% estuarine wetlands, 46% is forested (includes half of the forested/emergent mix) and 28% is emergent, with the remainder being nonvegetated (26%). The estuarine intertidal forested wetlands are mangrove (red and white) swamps and to a lesser extent buttonwood and salt bush. The estuarine intertidal emergent wetlands, commonly referred to as salt marsh, consist predominantly of smooth cordgrass and are flooded at least once daily by the tides. There are 674,214 acres of mangrove and 383,317 acres of salt-water marsh in Florida.

It is important to distinguish between mangroves and saltwater marsh since primary productivity differs. The west coast of Florida has over twice as much mangrove acreage as saltwater marsh. From the U.S. Fish and Wildlife Service (1983, 1984a, 1984b), the following trends are given for estuarine wetlands in Florida:

Wetland type	1953 acres	1973 acres	Change acres	Percent change
Mangroves	571,270	548,974	–22,296	–3.9
Saltwater marsh	381,950	342,007	–39,942	–10.5
Total	953,220	890,981	–62,238	–6.5

Between 1953 and 1973, the state of Florida lost 3.9% of its forested (i.e., mangrove areas) wetlands, but 10.5% of its saltwater marsh was lost largely due to residential development and agricultural filling. According to the U.S. Fish and Wildlife Service (1984a, 1984b), estuarine wetland losses have been greatest in five states: California, Florida, Louisiana, New Jersey, and Texas. Figure 9.1 shows the coastal wetland loss estimated by Gosselink and Baumann (1980). Over the period 1954 to 1974, coastal wetlands declined at 0.5% per year. Using the above data, it would appear that Florida's loss in estuarine wetland was only 0.33% per year. We would thus also expect only modest loss in fish landings and relatively low marginal values.

The only time-series wetland data available for Florida was developed for the west coast over the period 1952 to 1976 by Conroy (1979). The only information available on changes in coastal acreage is a series of aerial photographs of Florida taken by the Soil Conservation Service at irregular time intervals. For purposes of estimation, marsh land was defined as follows (Conroy, 1979, p. 62):

> Marsh lands are areas where the water table is at or above the land surface for a significant part of the year, with alternate submergence and subaerial exposure during tidal cycles. The hydrologic regime is such that the herbaceous vegetation is usually established, although alluvial and tidal flats may be nonvegetated. Also included were small bays and estuaries up to 1.5 miles in width connecting the outer limits of the headlands. River inlets were included up to a recognizable tree line or transitional zone. Small bays and inlets were excluded if: 1) there was no significant marshland surrounding the bay or inlet, 2) the bay or inlet occurred in a highly developed urban center.

Salt marsh acreage declined by only 2.7% over the period 1952 to 1976 or 0.11% a year. Such a small decline might make one skeptical of such data. Yet, as discussed, the U.S. Fish and Wildlife Service shows a decline of only 0.5% a year for the entire state over the period 1953 to 1973. Since the east coast of Florida developed more rapidly than the west coast, the Conroy (1979) time series may not be unreasonable. Also, Lynne et al. (1981) statistically linked the decline in wetland acreage in this time series with the blue crab catch.

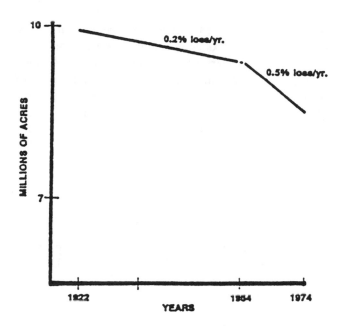

Figure 9.1 Rate of coastal wetlands loss from 1922 to 1974 in the coterminous United States (from Gosselink and Baumann, 1980). Estimates include both estuarine and tidal freshwater wetland losses.

The same time series on wetland acreage change is used here. Data on catch and fishing effort for the eight estuarine-dependent species discussed above were obtained from the NMFS (U.S. Department of Commerce, 1987).

Based on the statistical analysis of these data (see Appendix), the following marginal value products (MVP) of saltwater marsh were obtained:

Species	MVP of salt marsh (at the mean) (in $)
1. Blue crab	0.35
2. Stone crab	0.34
3. Spiny lobster	1.25
4. Red snapper	10.09
5. Grouper	1.47
6. Oyster	0.65
7. Black mullet	3.24
8. Shrimp	10.09
Total	27.49

Note: All values are in 1984 dollars.

The species selected for study constitute 90% of the estuarine-dependent catch on the west coast of Florida. Production function estimation was not possible for the remaining species primarily because of the lack of fishing effort data. We increased the sum of the estimated MVP accordingly, or ($27.48/0.90) = $30.44. This calculation is merely a linear extrapolation to the total estuarine-dependent species. Therefore, the ninth category of all other estuarine-dependent species is valued at $2.96. The capitalized value (CV) of the salt marsh for all nine categories of commercial fisheries is

$$CV = \frac{\sum_{i=1}^{9} MVP}{i}$$

The following discount rates will be used for illustrative purposes, including 8.625%, which is currently used by the US. Army Corps of Engineers for water resource projects. Also note that in Expression 9.2, we have eight species and a residual category, making nine MVPs. The CVs are as follows ($30.43/i):

Discount rate (i) (%)	CV per acre of salt marsh ($)
5	608
6	507
7	435
8.625	353
9	338
10	304

At the ex-vessel level, a salt marsh acre is worth $353 as an asset in support of the commercial fishing industry on the west coast of Florida (using the Corps of Engineers' discount rate of 8.625%).

It may be instructive and helpful to compare the marginal productivity approach with the GOP method discussed above even though the implied value systems are quite different. Recall that the GOP method involves dividing the ex-vessel value of the catch by salt marsh and discounting the result into perpetuity. Since the values of the marginal productivities are calculated at the mean, we shall use the mean value of 508,325 acres salt marsh over the period 1952 to 1976. The GOP results would be as follows:

$$\frac{\text{Ex-vessel value}}{\text{Marsh acreage}} = \frac{\$114,704,921}{508,325} = \$226 \tag{9.2}$$

Using i = 0.08265, the CV of $226 becomes $2621 per acre derived from commercial fisheries. This is 7.4 times the $353 value derived from the marginal productivity theory. In the marginal productivity approach, the difference ($2621 − 353 = $2268) is due to the fishing effort by humans, who also contributed (and, some may argue, deserve) a share of the value of the catch. Humans get no share in the GOP approach; the wetland claims all the value.

Conclusions

Saltwater wetlands provide food and refuge services for fish species important to the commercial fishery in Florida. Payment for such services cannot be captured by the owners of the wetlands. With wetland owners unable to obtain payment for services provided, an incentive is created to convert wetlands into uses such as residential and commercial development. Such economic activity destroys the natural functions of wetlands and the economic values of natural services are lost. As a result, the public sector has become ever more involved in regulating wetland systems. The decision to expand regulation, however, requires sound wetland value information.

The marginal productivity method is offered here as a way of taking into account the contribution of both wetlands and the fishing effort of humans in creating a commercial, marketable fishery catch. The catch is valued at what consumers are willing to pay. For the west coast of Florida, an additional acre of saltwater marsh added $30.44 per year. Capitalized at 8.625%, the asset support value of an acre of saltwater marsh was estimated at $353.

Further research should be conducted to improve marginal productivity value estimates herein, as well as to apply the method to other services of the marsh. Research should also be expanded into improving understanding of the important values not represented in such dollar estimates, with special focus on the values imbedded in the laws, rules, and regulations that affect wetland use. We expect that the total dollar value of a wetland system will be greatly affected by the values imbedded in such institutions. The various values important to wetland use and management need to be systematically identified and discussed.

Appendix on Derivation of the Marginal Product of an Acre of Wetlands

To derive the actual production or yield function used to calculate the marginal products of wetlands, a biological growth model must be specified. Following the work of Schaefer (1954), let us hypothesize that, without intervention by man, a wild stock of a specific species will grow in a logistic manner:*

$$B(t) = \frac{B^*}{1 + be^{-at}}$$

(A9.1)

where $B(t)$ = biomass in year t and B^* = maximum potential biomass under various constraints (i.e., habitat) while a and b are parameters. If we differentiate Expression A9.1 with respect to time, we have

$$\frac{dB(t)}{dt} = aB(t)\left[1 - \left(\frac{B(t)}{B^*}\right)\right]$$

(A9.2)

Expression A9.2 is a parabolic function. That is, the absolute increments to the stock increase to a maximum and then decline. Multiplying Expression A9.2 through by 1/B, we can obtain the rate of change in the stock as a function of stock size, or

---

* The authors are the first to recognize that the Schaefer yield curve which employs a logistic growth function is but one of many that explain the catch–effort relationship. In earlier reviews of this work, some have questioned the application of the Schaefer model to species such as shrimp, which many say cannot be overfished. Unfortunately, this is far from a settled question. Weintz (1987) has reviewed the controversy over the overfishing of shrimp and states: "...after interviewing scores of experts and reviewing countless pages of reports and documents, I noticed one factor kept emerging. The primary reason for the decline in Florida's shrimp population is excessive commercial shrimping" (p. 26). According to Weintz's research, "...there is a growing body of opinion that says you can indeed overfish shrimp populations if you are not careful" (p. 27). Thus, the simple Schaefer model may be a reasonable approximation even for shrimp. It is also argued that crab and oyster recruitment is not stock dependent. As Fox (1970) expressed it, "...the study of fishery dynamics...may be grouped into two basic approaches. Dynamic pool models, such as those of Beverton and Holt (1957), attempt to describe an exploited population in terms of its parameters—recruitment, growth, and natural mortality—in detail. Surplus-yield models, such as those of Graham (1935) and Schaefer (1954), on the other hand, combine the effects of recruitment, growth, and natural mortality into a common function of the mean population size and the rate of population (or natural) increase" (p. 80). Dynamic pool models require substantial information about the fishery presently not available and/or subject to debate, while surplus yield models require only catch and effort data. Schaefer (1970) could find no difference between Pella and Tomlinson's GENPROD (1969), where the population growth curve can vary, and the logistic for the Peruvian anchoveta, the largest commercial fishery in the world. The relation between effort and wetlands under alternative effort–catch scenarios should be investigated. We agree, but this research was well beyond the scope of this chapter, given the lack of data and the controversy surrounding growth functions.

$$\frac{dB(t)}{dtB(t)} = a\left[1 - \left(\frac{B(t)}{B^*}\right)\right] \tag{A9.3}$$

Up to this point, we have been discussing the growth in the fishery stock in the wild without intervention by man as a predator. The assumption behind logistic growth is that the environment is fixed or unchanging and that given the environment as a parameter (i.e., B^* is the upper asymptote of the biomass), fish grow or respond in a systematic manner.

Humans intervene as predators or, to be more dignified, as hunters. Let us define $F(E)$ as the rate of loss (i.e., subtraction from the rate of growth of the wild stock) of the fish due to fishing effort (E)—inputs of capital, labor, materials, and technology—or man-made mortality. Such inputs may be commercial or recreational fishing effort.

$$\frac{dB(t)}{dtB(t)} = a\left(1 - \frac{B(t)}{B^*}\right) - F(E) \tag{A9.4}$$

$F(E)$ is a general function expressing the impact of fishing effort on the biomass. Let us specify the function in the following simplified form:

$$F(E) = kE(t) \tag{A9.5}$$

From a human standpoint, fishing effort goes to harvesting fish from a growing, renewable resource. Let us assume a steady-state relation between the rate of growth in the stock and the rate of loss due to fishing effort:

$$a\left[1 - \frac{B(t)}{B^*}\right] - kE(t) = 0 \tag{A9.6}$$

Next, we hypothesize after Schaefer (1954) a fishing output function:

$$C(t) = kE(t)\,B(t) \tag{A9.7}$$

where $C(t)$ is the quantity of the stock harvested. That is, the catch depends on the amount of fishing effort and the size of the stock. The term $kE(t)$ represents the fraction of the stock removed by fishing effort. Equation A9.7 is important since it specifies two factors of production necessary to produce a harvest, fishing effort and the size of the biomass. Let us respecify Equation A9.7 by dividing each side by $kE(t)$:

$$B(t) = C(t)/kE(t) \tag{A9.7'}$$

Substituting Equation A9.7' into Equation A9.6, we obtain,

$$C(t) = kB^*\,E(t) - \left(\frac{B^*k^2}{a}\right)E^2(t) \tag{A9.8}$$

Thus, Equation A9.8 specifies a parabolic relation between catch and fishing effort. The steady-state equilibrium relationship in Equation A9.8 assumes that the present catch is a function of only the present effort. If full adjustment is made over more than one time period (i.e., if current yields are affected by current effort and also past effort levels), then a stock adjustment or Koyck (1954) model as given in Bell et al. (1973) is appropriate:

$$\underset{a}{\underline{C(t) = kB^*E(t) - (B^*k^2)\ E^2\ (t) + dC(t-1)}} \tag{A9.9}$$

Notice that B* is not only the upper asymptote of the logistic function, but is interactive with E(t) and E2(t). B* may also be considered a reflection of the carrying capacity for the fishery stock. For example, the existing level of mangrove acreage may produce only so much detritus export for estuarine-dependent species. This export may serve as a limit on the biomass for estuarine-dependent species. It should be pointed out that fishing effort can reduce the size of the stock in equilibrium, thereby requiring less carrying capacity. A destruction in mangrove acreage, where B*(1), for example, may reduce the level of B*. This deduction will in turn shift the catch–effort function downward, and catch will decline for a given level of fishing effort throughout the function.

It was hypothesized in this study that saltwater marsh (i.e., one form of wetlands) estuarine acreage is an appropriate measure of the environmental conditions impacting B*(t). Expanding human populations along the coastal area may have reduced estuarine habitat. There are six types of wetlands, one of which is salt marsh (Harris et al., 1983). To simplify the theoretical exposition, it will be assumed that salt marsh impacts B*(t). We are not neglecting or overlooking the role of mangroves and seagrasses as an input into determining B*(t), but no time-series data are available for the study area. For this exposition, we are assuming homogeneous salt marsh (i.e., quality-constant marsh acreage) and neglecting the heterogeneity of wetlands. Therefore, it is posited that:

$$B^*(t) = f[M(t-1)] \tag{A9.10}$$

where M(t – 1) = salt marsh acreage lagged one period.* In time series, the one-year lag was expected to be appropriate since increases and decreases in

---

* The following was also assumed:

$$\frac{dB(t)}{dM(t-1)} > 0; \qquad \frac{d^2B(t)}{dM^2(t-1)} < 0$$

That is, the *marginal* response of B(t) to M(t-1) should be positive, but declining, or, as marsh acreage is increased, the marginal impact on B(t) should decrease.

acreage are not expected to have an immediate effect on the catch. The actual lag structure can be investigated but is beyond the scope of this investigation (Nerlove, 1958; Koyck, 1954). An algebraic form for the relation between B(t) and M(t − 1) is hypothesized as follows:*

$$B^*(t) = g \ln M(t - 1) \qquad (A9.11)$$

Using the formulation in Equations A9.9 and A9.13, we have the following production function:

$$C(t) = b_1[\ln M(t - 1)]E(t) - b_2[\ln M(t - 1)]E^2(t) + b_3C(t - 1) \quad (A9.12)$$

where the bs are estimates of the various parameters represented by

$$b_1 = kg \qquad b_2 = k^2g/a \qquad b_3 = d \qquad (A9.13)$$

Equation A9.12 represents the fisheries wetland model which was simplified in the earlier part of the chapter where the marginal product was discussed.

The next step is to derive the marginal productivities (MP_m) of M(t − 1) using the wetland model or Equation A9.12:

$$
\begin{aligned}
MP_M &= \frac{dC(t)}{dM(t-1)} = \frac{b_1 E(t)}{M(t-1)} - \frac{b_2 E^2(t)}{M(t-1)} \\
&= \frac{1}{M(t-1)} [b_1 E(t) - b_2 E^2(t)]
\end{aligned} \qquad (A9.14)
$$

To obtain the marginal productivity functions, estimates of b_1 and b_2 must be obtained. The theoretical specification indicates that $b_1 > 0$ while $b_2 < 0$, which is explicit in Equation A9.12. Consider the MP_M. As M(t − 1) is increased, holding fishing effort constant, the MP_M declines. This decline is consistent from both an economic and biological perspective. That is, further increases in marsh acreage will have dwindling marginal productivity. Biology would dictate diminishing MP_M since there are certainly limitations on the provision of more habitat creating larger fish production at the margin while effort is constant at some level. From an economic perspective, the ratio of M to E will

---

* There are many possible functional forms between B*(t) and M(t − 1). Lynne et al. (1981) tried three other functional forms. Each failed on theoretical and/or empirical grounds for blue crab. From a theoretical point of view, the conditions found in the preceding footnote are satisfied by Equation A9.11, a diminishing returns to an increase in M(t − 1) in raising the B*(t). We have purposely kept the theoretical model simple. The hypothesized link of production to wetlands is at best a crude assertion. Certainly, this is a topic for further substantive research.

increase, indicating diminishing marginal productivity. The marginal productivity curves seem to be well behaved from both economic and biological theory. In simplified terms, these equations tell us the impact on the fishery catch of the removal or addition of an acre of wetlands. The marginal product of wetlands is higher with a relatively small acreage of wetlands and lower when wetland acreage is in relative abundance. This hypothesis would seem plausible.

To estimate Equation A9.12 by ordinary least-squares (OLS), we shall assume an error term (U) which is normal, $E(U_t) = 0$, and there is no autocorrelation. Data on catch and fishing effort for the eight estuarine-dependent species discussed above were obtained from the NMFS (U.S. Department of Commerce, 1987). Observation on salt marsh land were taken from Conroy (1979).

Table A9.1 shows the OLS results for Equation A9.12 or the basic model developed by Lynne et al. (1981). Seven of the eight species show the hypothesized signs with grouper as the only exception. Blue crab does not show statistically significant signs at the 5% level for any of the variables. Our results differ from the Lynne et al. estimated equation for two possible reasons. First, the fishing effort series included stone crab traps from 1952 to 1960. Second, Lynne et al. did not use 1975 data since fishing effort was unavailable. In terms of statistical performance, stone crabs, spiny lobsters, and shrimp have coefficients b_1 and b_2 that are statistically significant at the 1% level.

All equations, except for shrimp, exhibited a statistically insignificant (at the 5% level) intercept, which we have designated b_0. The intercept can be interpreted as picking up omitted variables since, in theory, we cannot have catch without E_t and $M(t - 1)$. As hypothesized, lagged catch is positive for seven of the eight species (except shrimp). Thus, a separate equation was estimated for shrimp without the lagged catch, which is also shown at the bottom of Table A9.1. Since the value of b_3 in Equation A9.1 is less than unity in all seven cases with the hypothesized sign, this result indicates an exponentially declining distributive lag as expressed by Koyck (1954). The R^2s are very high for all equations. Finally, the D-W statistic is not a sufficient test for autocorrelation with a lagged independent variable; therefore, Johnston (1972) indicates reliance on the Durbin-H statistic. Some autocorrelation apparently exists in the blue crab, oyster, and mullet equations where D-H values greater than one may indicate this statistical condition.

The marginal productivity of salt marsh will vary with the levels of marsh land and fishing effort, as can be seen from Expression A9.14. As is common in such analysis, the arithmetic means of salt marsh acreage and fishing effort were selected from the time series to compute the marginal productivities. The means of the data are as follows (1952 to 1976):

Table A9.1 Estimation of the Relationship between Catch and Fishing Effort and Wetlands for Selected Estuarine-Dependent Species on Florida's West Coast in the Gulf of Mexico

Species	Constant	$\ln M(t-1)E(t)$	$\ln M(t-1)E(t)$	$C(t-1)$	R	$D\text{-}W$	$D\text{-}H$	F
			Lynne et al. Model, Time Series (1952–1976)					
Blue crab	1,515,530 (0.694)	19.125 (1.318)	-0.0000854 (-0.436)	0.331 (1.693)	0.715	1.649	2.800	16.742
Stone crab	77,967 (1.448)	1.253 (4.306)	-0.00000355 (-3.974)	0.382 (2.291)	0.971	2.50	—	211.94
Spiny lobster	679 (1.732)	2.046 (3.315)	-0.00404 (-3.442)	0.337 (1.625)	0.861	1.809	—	27.71
Red snapper	-7,201.96 (-1.211)	0.520 (1.539)	-0.0000929 (-1.462)	0.650 (4.664)	0.545	2.079	-0.557	7.987
Grouper	-232.314 (-0.0231)	0.0372 (0.0651)	0.00000943 (0.0880)	0.697 (4.664)	0.536	2.107	-0.593	7.712
Oyster	-486,554 (-0.635)	372.857 (1.318)	-0.154 (-0.567)	0.484 (2.711)	0.822	1.623	1.783	30.746
Mullet	-14,036,900 (-0.334)	2,431.3 (0.388)	-1.021 (-0.348)	0.775 (5.466)	0.700	1.285	1.676	15.572
Shrimp	-22,593,700 (-2.557)	227.620 (4.441)	-0.00416 (-4.484)	-0.275 (-1.396)	0.607	1.898	-0.302	7.213
Shrimp (without lagged catch)	-24,456,800 (-2.715)	220.651 (4.195)	-0.00409 (-4.285)	—	0.553	2.475	—	9.260

Data source: Conroy (1979) and NMFS.

From Lynne et al., 1981.

Variable	Arithmetic mean
Salt marsh (acreage)	508,325
Blue crab pots	29,066
Stone crab pots	53,950
Spiny lobster traps	103,333
Red snapper handlines	2,651
Grouper handlines	2,651
Oyster tongs	528
Shrimp days	22,173
Black mullet gill nets	1,030

The interested reader can merely insert these values in Expression A9.14, along with the estimated parameters b_1 and b_2 from Table A9.1, to obtain the values shown in Table A9.2. Only the long run MP_M is shown. That is, we found that this year's catch was impacted only partially by fishing effort and salt marsh, indicating short-run responses. By including lagged catch, we are able to estimate long-run responses.

Consider, for example, the marginal product of salt marsh with respect to shrimp. A 1-acre loss in wetlands will reduce the shrimp catch by an estimated 5.68 lb using unlagged catch. All marginal productivities are expressed on an

Table A9.2 Long-Run Marginal and Value Products for Selected Species on the West Coast of Florida (Time-Series Data: Means of Variables), Lynne Model

Species	Long run			
	MPE	*MPM*	*$MVPE*	*$MVPM*
Blue crab	278.16	1.42	68.98	0.35
Stone crab	18.50	0.18	34.45	0.34
Spiny lobster	23.96	0.50	60.01	1.25
Red snapper	1,190.83	5.67	2,120.87	10.09
Grouper	3,779.42	1.07	5,208.04	1.47
Oysters	5,341.05	0.59	5,891.18	0.65
Shrimp	445.98	4.64	792.06	8.25
Shrimp (without lagged catch)	516.09	5.68	916.57	10.09
Black mullet	19,133.79	12.41	4,993.92	3.24

annual basis or annual flow of catch from the factor of production in question. This is important to remember. The marginal productivities are computed at the means of the historical series. However, we have chosen to express these marginal productivities in 1984 ex-vessel prices:*

$$MP_M \times P_{EX(1984)} = MVP_M \qquad (A9.15)$$

This equation gives us the value of the marginal product expressed in reasonably current dollars. For example, stone crab sold for $1.86 in 1984 and the marginal value product of salt marsh was $0.34 (0.182 lb × $1.86). This result means that the removal of 1 acre of salt marsh will reduce the annual value of stone crab landings by about 34 cents at the margin.

References

Batie, S. S. and C. C. Mabbs-Zeno. 1985. Opportunity costs of preserving coastal wetlands: a case of a recreational housing development. *Land Econ.* 61(1):1–9.

Bell, F. W. 1989. *Application of Wetland Valuation Theory to Florida Fisheries.* Report Number 95. Florida Sea Grant College, University of Florida, Gainesville.

Bell, F. W., E. W. Carlson, and F. V. Waugh. 1973. Production from the sea. Pages 72–91 in *Ocean Fishery Management Discussions and Research.* NOAA Technical Report NMFS CIRC-371. National Oceanic and Atmospheric Administration, Rockville, Maryland.

Beverton, R. J. H. and S. J. Holt. 1957. *On the Dynamics of Exploited Fish Populations.* Her Majesty's Stationary Office, London.

Boesch, D. F. and R. E. Turner. 1984. Dependence of fishery species on salt marsh: the role of food and refuge. *Estuaries* 7(4A):460–468.

Bromley, D. W. 1989. *Economic Interests and Institutions.* Basil Blackwell, New York.

Conroy, P. D. 1979. Economic value of the salt marsh to the Florida blue crab industry. M.S. thesis. University of Florida, Gainesville.

Costanza, R. 1980. Embodied energy and economic valuation. *Science* 210(12): 1219–1224.

Crow, J. H. and K. B. MacDonald. 1979. Wetland values: secondary productivity. Pages 289–298 in P. E. Greeson, J. R. Clark, and J. E. Clark, Eds. *Wetland Functions and Values: The State of Our Understanding.* American Water Resources Association, Minneapolis, Minnesota.

Fox, W. W., Jr. 1970. An exponential surplus-yield model for optimizing exploited fish populations. *Trans. Am. Fish. Soc.* 99:80–88.

Gosselink, J. G. and R. H. Baumann. 1980. Wetland inventories: wetland loss along the United States coast. *Z. Geomorph. N.F. Suppl. Bd.* 34:173–187.

---

* The available *final* (not preliminary) catch and value figures are for 1984 from the NMFS.

Gosselink, J. G., E. P. Odum, and R. M. Pope. 1974. *The Value of the Tidal Marsh.* LSU-SG-74-03. Center for Wetland Resources, Louisiana State University, Baton Rouge.

Graham, M. 1935. Modern theory of exploiting a fishery, and applications to North Sea trawling. *J. Conserv. Explor. Manage.* 10(3):264–274.

Harris, B. A., K. D. Haddad, K. A. Steidinger, and J. A. Huff. 1983. *Assessment of Fisheries Habitat: Charlotte Harbor and Lake Worth, Florida.* Florida Department of Natural Resources, St. Petersburg.

Heald, E. J. and W. E. Odum. 1970. The contribution of mangrove swamp to Florida fisheries. *Proc. Gulf Caribb. Fish. Inst.* 22:130–135.

Hoffman, B. A. 1986. Letter on wetland valuation techniques sent to George Spinner, Florida Department of Natural Resources, St. Petersburg.

Johnston, J. 1972. *Econometric Methods.* McGraw-Hill, New York.

Koyck, L. M. 1954. *Distributed Lags and Investment Analysis.* North-Holland, Amsterdam.

Lynne, G. D., P. Conroy, and F. J. Prochaska. 1981. Economic valuation of marsh areas for marine production processes. *J. Environ. Econ. Manage.* 8(2):175–186.

Nerlove, M. 1958. *The Dynamics of Supply.* Johns Hopkins Press, Baltimore.

Odum, E. P. 1973. A description and value assessment of South Atlantic and Gulf marshland estuaries. *Proceedings: Fish and Wildlife Values of the Estuarine Habitat.* Bureau of Sport Fish and Wildlife, Atlanta, Georgia.

Odum, H. T. 1977. Value of wetlands as domestic ecosystems. Pages 9–18 in *Proceedings of the National Wetland Protection Symposium.* U.S. Fish and Wildlife Service, Reston, Virginia.

Pella, J. V. and P. K. Tomlinson. 1969. A generalized stock production model. *Inter-Am. Trop. Tuna Comm. Bull.* 13:421–458.

Raphael, C. N. and E. Jaworski. 1979. Economic value of fish, wildlife, and recreation in Michigan's coastal wetlands. *Coastal Zone Manage. J.* 5(3):181–194.

Schaefer, M. B. 1954. Some aspects of the dynamics of population important to the management of commercial marine fishers. *Inter-Am. Trop. Tuna Comm. Bull.* 1:27–56.

Schaefer, M. B. 1970. Men, birds and anchovies in the Peru Current—dynamic interaction. *Trans. Am. Fish. Soc.* 99:461–467.

Shabman, L. A. and S. S. Batie. 1978. Economic value of natural coastal wetlands: a critique. *Coastal Zone Manage. J.* 4(3):231–247.

Tihansky, D. P. and N. F. Meade. 1976. Economic contribution of commercial fisheries in valuing U.S. estuaries. *Coastal Zone Manage. J.* 2(4):411–421.

U.S. Council on Environmental Quality. 1974. *Environmental Quality 1974.* 5th Annual Report.

U.S. Department of Commerce. 1987. Personal correspondence on time series and cross sectional data on catch and fishing effort for selected species. National Oceanic and Atmospheric Administration, National Marine Fisheries Service, Southeast Fisheries Center, Miami.

U.S. Fish and Wildlife Service. 1983. *Status and Trends of Wetlands and Deepwater Habitats in the Coterminous United States, 1950's to 1970's.* St. Petersburg, Florida.

U.S. Fish and Wildlife Service. 1984a. Florida Wetland Acreage (unpublished manuscript). St. Petersburg, Florida.

U.S. Fish and Wildlife Service. 1984b. *Wetlands of the United States: Current Status and Recent Trends*. National Wetlands Inventory. U.S. Government Printing Office, Washington, D.C.

Walker, R. A. 1974. Wetlands preservation and management: a rejoinder—economics, science and beyond. *Coastal Zone Manage. J.* 1(2):158–164.

Weintz, M. 1987. Shrimp: a troubled resource. *Fla. Sportsman*. March:26–33.

Wetland Values and Valuing Wetlands

10

Clyde F. Kiker and Gary D. Lynne

In addition to food and habitat for fish and wildlife, wetlands provide many services of importance to humans: flood protection, storm buffering, water quality maintenance, erosion control, and recreational opportunities. Approximately 50% of wetlands in the lower 48 states—and 65% in Florida—have been lost to agricultural, commercial, and residential uses (Office of Technology Assessment, 1984, p. 87).

Strong public concern has led to federal, state, and local laws, rules, and ordinances for guiding the political process pertaining to wetland conversion. The Office of Technology Assessment report (1984) on wetland use and regulation gives an overview of the extensive federal governmental involvement, and Kusler and Hamann (1985) give a similar review of state government's role. Chapter 11 summarizes the regulatory authorities of both the state and federal governments.

This chapter focuses on the heart of the regulatory function at the federal, state, and local levels—the permitting process. Requests are made for activities that can be expected to be substantially disruptive to wetland processes and as such will contribute to long-term cumulative effects far beyond the particular site of interest. However, the great majority (virtually 95% or more) of the requests related to activities involving wetlands are routine in nature and will not involve great study or public involvement. We are concerned with the process of evaluating permit requests in wetlands of special concern to special interest groups and to the public as a whole. A permitting process using accurate information and allowing for broad public participation is highly desirable and is more likely to result in acceptable decisions. The approach is presented in the spirit of what Barnard et al. (1985), coordinators of the Office of Technology Assessment study, have called the "tailoring" or adjusting of existing policies and programs "...to better focus existing protection efforts on those wetlands that are most important and to shift conversion pressures away from high valued wetlands" (p. 1053).

Many traditional economists see social governmental regulation as a necessary evil that must be occasionally tolerated. As natural resource economists, the authors view such regulation differently. The perspective is closer to one expressed by Sagoff (1988): "Social regulation reflects public values we choose collectively....[It] expresses what we believe, what we are, what we stand for as a nation..." and this includes "...a nation committed historically to appreciate and preserve a fabulous natural heritage and to pass it on reasonably undisturbed to future generations" (1988, pp. 16 and 17). We hope our approach will be useful in representing, evolving, and expressing these public values and ultimately lead to wise collective and individual choices.

Wetland Processes and Human Activities

Naturally occurring wetlands consist of intricate and interwoven sets of natural processes, nested in and interacting with other processes, and moving toward a dynamic equilibrium. Coastal salt marshes and mangroves occur in response to the ebb and flow of tides. Upland plant and animal communities affect the quality and functioning of wetlands, while wetlands may serve to improve the flora and fauna in upland areas. The resulting systems are highly interrelated, nonlinear, and dynamic.

A multitude of human activities can and do take place within wetlands. Some are nonobtrusive, in that they have little impact on natural wetland processes and functions, and others by their very nature require alteration of the wetland. Crop production, aquaculture, resource extraction, and a broad range of commercial activities give value to a site, but not necessarily for the natural processes. Many wetland functions may be disrupted and some completely lost.

The two categories of wetland use and their relationship to the value of wetland services are represented in Figure 10.1. Path A represents a holistic view with value arising from natural processes associated with the biological, ecological, physical, and hydrologic properties of the natural wetland. Along this path, any individual segment (i.e., a parcel of land) has value only because of its function in the overall wetland system. Alternatively, path B represents the isolation of a specific parcel of the overall wetland area for human activities, which generally requires modification of the land surface. In this case, the primary value of the parcel stems not from natural processes but from the economic value of the site for the human-controlled processes.

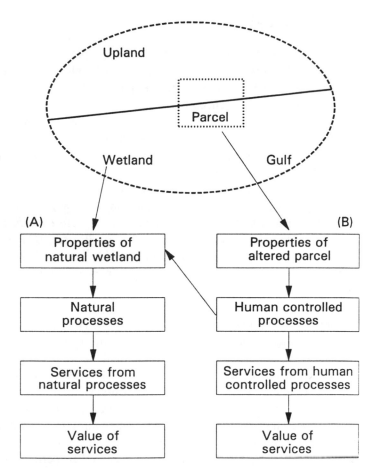

Figure 10.1 Abstract view of alternative uses of wetlands.

The human activities on such a parcel can also produce impacts on the surrounding wetlands. Drainage of a parcel, for example, can cause loss of habitat and feeding functions as well as the loss of the coastal wetland's potential for ameliorating coastal erosion and water quality problems caused from activities in nearby uplands. The impacts are identified in Figure 10.1 by the cross-linkages between the paths. Although cross-linkages may be difficult to identify and quantify, the related impacts do in fact exist and can substantially reduce the services. Impacts may be especially dramatic where conversion of many parcels causes accumulating impacts. The last unit of change may cause proportionately an even greater impact.

Regulatory Context

Natural wetland services benefit a broad range of people, whereas services from altered parcels flow primarily to the individuals who change the wetlands. Such individual actions may not necessarily be in the public interest, and conflicts can arise because of differences in values.

Institutions (laws, rules, regulations, customs, habits) evolve in response to conflict and ultimately reflect societal values. In the process, valuing and institution building take place at many levels. In the case of wetlands, a major focus has occurred at the federal level because of the responsibility for "navigable waters and waters of the United States" stemming from the interstate commerce clause of the Constitution. Doctrines originating in the early history of the United States and legislation passed by Congress in the 19th century are part of the present basis for federal involvement. In the last 30 years, Congress has passed many additional acts affecting wetlands. Florida has similarly evolved a body of law dealing with wetlands. Many local ordinances also affect wetland use.

While Congress and the state legislature have not given precise criteria for judging the merits of proposed alterations, both legislative bodies clearly intended for a broad range of participants (i.e., the public) to be involved in the decision process. Also, both recognize that an inherent aspect of property ownership is a right to reasonable private use. However, the rights and interests of the public in "navigable and other waters of the United States" and the associated wetlands are to be protected. Federal and state agencies are given responsibility for representing the public's many values through permitting processes. The General Regulatory Policy (33 CFR, Chp. II, part 320, Sec. 320.4 [a](1)) gives an overview of the factors to be considered in permitting.

> The decision whether to issue a permit will be based on an evaluation of the probable impacts, including cumulative impacts of the proposed activity and its intended use on the public interest. Evaluation of the probable impact which the proposed activity may have on the public interest requires a careful weighing of all those factors which become relevant in each particular case. The benefits which reasonably may be expected to accrue from the proposal must be balanced against its reasonably foreseeable detriments. The decision whether to authorize a proposal, and if so, the conditions under which it will be allowed to occur, are therefore determined by the outcome of this general balancing process. That decision should reflect the national concern for both protection and utilization of important resources. All factors which may be relevant to the proposal must be considered including cumulative effects thereof: among

those are conservation, economics, aesthetics, general environmental concerns, wetlands, historic properties, fish and wildlife values, flood hazards, floodplain values, land use, navigation, shore erosion and accretion, recreation, water supply and conservation, water quality, energy needs, safety, food and fiber production, mineral needs, considerations of property ownership and, in general, the needs and welfare of the people.

Clearly, benefits should be balanced against reasonably foreseeable detriments for a larger number of factors. A multidimensioned evaluation approach may be needed to facilitate the social processes that will ultimately give a balanced solution.

The Place of Values

Values concerning wetlands and specific institutions for dealing with wetland use evolve in response to societal interests and pressures. The permitting processes undertaken by various governmental agencies represent various forums for evolving wetland values. In effect, values evolve in the give and take among participants in these political processes. Given sufficient time and appropriate ways to interact in process, the wetland values held by the participants will be revealed. The societal forum within which wetland values are developed and expressed, then, can benefit from improved ways for comparing expected benefits and detriments (costs) of activities on specific parcels of wetlands.

Rolston (1985) developed an axiological model (reproduced with only slight modification in Figure 10.2) of general use in understanding both the character of the needed valuing processes and the kinds of values that will likely be revealed. The < and > signs can be read as "overridden by" and "overriding." The $\rightleftarrows$ arrows suggest "produced by"; for example, the ecosystem values underneath the line produce a situation where the social values above the line can evolve. Rolston delineates seven different (not necessarily mutually exclusive) types of value, including:

1. Naturalistic values:
    a. Ecosystemic value (Value$_{es}$) represents the good of the system of plants and animals and suggests a value of the system beyond the sum of the value of each part in the system (Chapter 8 shows one way to calculate Value$_{es}$).
    b. Organismic value (Value$_{or}$) represents what is good for an organism within an ecosystem and the value of the individual organism as a part of the ecosystem.

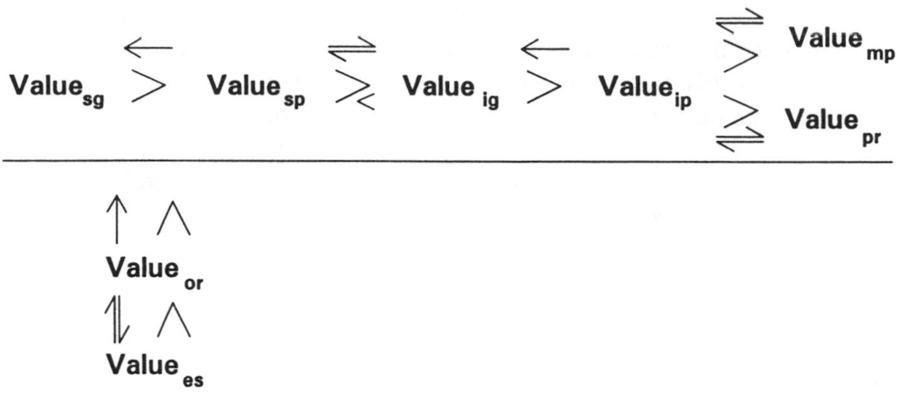

$\text{Value}_{sg} > \text{Value}_{sp} \gtrless \text{Value}_{ig} > \text{Value}_{ip}$ $\begin{cases} \text{Value}_{mp} \\ \text{Value}_{pr} \end{cases}$

Value_{or}

Value_{es}

Figure 10.2 Hierarchical wetland values. Legend: es, ecosystemic; or, organismic; sg, social good; sp, social preference; ig, individual good; ip, individual preference; mp, market price; pr, politically revealed.

Notice how Value_{or} both produces and is produced by Value_{es}, but Value_{es} overrides Value_{or}. Both override the humanistic values.

2. Humanistic values:
   a. Social good value (Value_{sg}) represents what is good for the society as a whole, comparable in idea to the Value_{es} concept but for human systems and reflecting the notion that the whole is somehow different from the sum of the parts. Value_{sg} overrides Value_{sp}, but Value_{sp} helps produce Value_{sg}.
   b. Social preference value (Value_{sp}) represents what a society may prefer, even though it may or may not be socially good. It overrides Value_{ig}, but is partially checked by Value_{ig} (small $<$).
   c. Individual good value (Value_{ig}) represents what is individually good for a person.
   d. Individual preferred value (Value_{ip}) represents what the individual prefers (wants), even though it may or may not be individually good.

All of these values serve to influence and override market values, which are revealed in the market economy:

   e. Market price value (Value_{mp}), which is produced by Value_{ip} of a large number of individuals interacting in a market economic process. (Chapter 9, on the economic value of Florida's coastal wetlands to the commercial fishery, shows one way of estimating a portion of the Value_{mp} for a wetland.)

The authors suggest another category not considered by Rolston. This value dimension is revealed in the political economy rather than the market economy:

    f. Politically revealed values ($Value_{pr}$), which are also produced by $Value_{ip}$, but the $Value_{ip}$ of individuals interacting in the political economic process. The idea is that individuals interact to generate value. Note that both $Value_{mp}$ and $Value_{pr}$ feed back to produce $Value_{ip}$; that is, both market and political values influence preferences.

Overall then, the broader naturalistic and humanistic values ultimately have an impact on both the market and political economic (valuing) processes and the actual values revealed. The extent to which such broader values get represented, however, becomes a function of the (regulatory) processes that facilitate the valuing process and wetland decisions.

The current regulatory environment does imply a broader perspective than that of the individual land owner's, which primarily focuses on $Value_{mp}$ for converting the wetland. Factors of importance to all citizens are to be included, as reflected in the broader $Value_{es}$ and $Value_{sg}$ categories. While the public may well be concerned with $Value_{mp}$, the loss of $Value_{es}$ and $Value_{sg}$ due to alteration should also be considered. The intrinsic value of the wetland system, $Value_{es}$, may be equally important as $Value_{mp}$.

Recall the link between path B and path A in Figure 10.1 which introduces the loss of ecologic and hydrologic function of the wetland and causes impacts beyond the immediate region. Clearly, humanistic values from the owner's and society's perspectives can differ. Additionally, humanistic values may differ from naturalistic values. The existence of such a wide array of values at different levels complicates the wetlands permitting process.

Shabman and Batie (1988, p. 14) have suggested that we place a heavy emphasis on $Value_{mp}$: "The wetland permit process should seek to ensure that the higher value option, e.g., wetland alteration (permit granted) versus wetland preservation (permit denied), be pursued. In most general terms, denial of a wetland alteration permit requires an analysis documenting that the benefits to human users of maintaining natural wetlands exceed the costs, measured as foregone development use values." Shabman and Batie are suggesting that $Value_{mp}$ should always be calculated and made available to the decision process. Also, those wishing to preserve wetlands to satisfy, for example, the $Value_{sg}$ category should be asked to document that $Value_{sg} > Value_{mp}$. We cannot be as sure as Shabman and Batie that $Value_{mp}$ should play a dominant role, but we do recognize that there is presently substantial pressure to use monetary values as a main measure of value in wetland decisions.

Traditionally, economic research has focused on establishing the net use (dollar) $Value_{mp}$ of sites in altered states. Considerable effort has gone into determining the $Value_{mp}$ of wetlands for agriculture, silviculture, aquaculture, and many commercial purposes. The products and services of these activities are traded in markets, as are the resources necessary to conduct the activities. The markets provide one kind of forum within which people can express willingness to pay (i.e., Value$_{mp}$) for the products and services from wetlands.

The majority of goods and services stemming directly from the wetlands, on the other hand, do not enter the marketplace as suggested by the other value categories. Another kind of forum, generally described as a political economy, is needed to express these values, Value$_{pr}$. The lack of market-established values does not mean that these wetland-provided goods and services do not have value; it means, rather, that individuals must now express such values (e.g., Value$_{es}$, Value$_{sg}$) through the political economy instead. If a forum does not exist, it becomes virtually impossible to influence the decision process with such nonmonetary values.

The problem of using monetary values is further exacerbated in that little information on the economic Value$_{mp}$ of natural wetland services exists. Shabman and Batie (1988), in a review of the literature on the socioeconomic values of wetlands, make the point clear in the conclusions: "The literature search...found relatively few articles which provide estimates of the value of wetlands. Of those articles a small number employ conceptually valid approaches to valuation....Even those (analysts) familiar with the correct applications of the concepts often report results that were, in their own view, less than satisfactory. To a large degree, this can be attributed to inadequate data and poorly documented linkages between wetland areas and wetland services in the scientific literature" (pp. 50 to 51). They also conclude that even the few conceptually correct studies have "...site-specific results (that) are not transferable beyond their case study areas" (p. 51). (It should be noted that Leitch and Ekstrom [1989] have also published a review of literature on social and economic aspects of wetland management.)

Expressing the Value$_{mp}$ associated with the altering of a specific parcel of wetlands is not presently possible in unambiguous terms. Shabman and Batie (1986, p. 11) agree "...that a policy based on a benefit–cost balancing test for wetlands permitting is technically impractical." If the permit process used a decision rule based exclusively upon monetary measures, the result would inevitably be alteration of wetlands parcels that may—if there were adequate information and a forum for consistently expressing the broader range of values outlined above—be left in a natural state.

Multiple Alternative/Multiple Attribute Evaluation

Following the spirit of the General Regulatory Policy statement cited earlier, and recognizing the existence of both the naturalistic values and the broader based humanistic values, an approach to organizing information pertaining to all types of values would clearly be helpful for the permit decision process. The intent of wetland laws and related rules seems to be anticipation and full consideration of all naturalistic and humanistic values in the decision process. To accomplish this task, information must be organized so that the consequences of particular proposed actions are clear to the decision makers, the person requesting the permit, other individuals affected, and the public as a whole.

In practice, the information collected by disciplinary specialist and field staff can be extremely technical, detailed, and voluminous. As a result, the information may be unclear to nonspecialist decision makers and the public and may result in poor understanding of the consequence of wetland change. A more constructive dialogue among the many interested parties will likely take place if information can be put forward in a way that facilitates the give and take, i.e., the market and political economic valuing processes.

The approach suggested here draws on similar efforts by the U.S. Water Resources Council to include objectives (values) beyond economic efficiency in all water and related land project evaluation processes. In the council approach, four accounts are developed: National Economic Development Account, Environmental Quality Account, Regional Development Account, and Other Social Effects Accounts (U.S. Water Resources Council, 1983, p. v). Since the wetlands evaluation process generally deals with activities proposed by individuals rather than large-scale development efforts by governmental agencies, the information for the approach developed herein is organized differently than in the council accounts and is intended to facilitate active public involvement.

By far the majority of permits requested will continue to be handled in a routine manner. Most permit requests simply do not involve activities that induce substantial change, and available information is sufficient to have decisions consistent with the broad public interest. Some requests, however, do involve activities which, if undertaken, can have substantial impacts far beyond the site. In these few cases, a process designed to use information on a range of alternatives, and including measurements of a number of attributes associated with the alternatives, will facilitate public involvement and lead to decisions consistent with public interest.

Detailed Information and Analysis

The scientific literature on wetlands suggests many perspectives from which to view wetlands and the collection of relevant information for decision making. Each scientific discipline considers the potential use of a wetland and applies broad concepts that have been identified and agreed upon by members of that discipline. Such concepts represent mental concepts of the actuality that is nature. Concepts, which are in fact analytical boundaries, are used to mentally isolate segments of nature and allow application of analytical processes. As Georgescu-Roegen (1971) has said, "no analytical boundary, no analytical processes," and one can add, "no means of measurement." Measurements are only possible where boundaries are specified. Each analyst brings different training, experience, and interests, all of which influence the conceptual boundaries being used either explicitly or implicitly (Neville, 1989). Since virtually all perspectives and boundaries are useful in the valuing process, as suggested by Figure 10.2, collecting and organizing the information from all perspectives thought to be important are desirable.

Consider a naturally occurring wetland along the west coast of Florida, for which a commercial (possibly a condominium and marina) development is being considered (Figure 10.3). From a physical science perspective (and relating to $Value_{es}$, $Value_{or}$) hydrologic, geologic, and soil analyses are needed. Hydroperiod, underlying foundation material, and soil types should be noted. The importance of the wetland to flood amelioration and water quality may be noted. Additionally, the proposed wetland-modifying activities are described. From a biological science perspective (again for $Value_{es}$, $Value_{or}$) floral, faunal, and ecological analyses may be provided. Endangered species may be

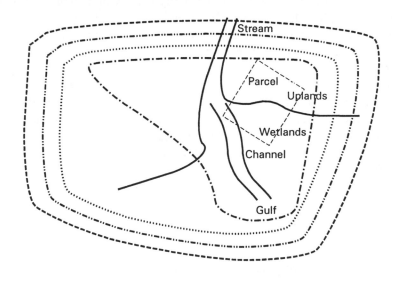

Social-Cultural-Institutional Boundary   ------- ·   Hydrologic-Geologic-Soils Boundary ··············

Economic-Financial Boundary   ■··■··■···   Floral-Faunal-Ecologic Boundary   ■·—·■·■··

Figure 10.3   Naturally occurring wetland along the west coast of Florida.

identified. Intricacies of the ecology of the wetland may de described and possible disruptions identified. From a social science standpoint, analyses of economic activities, ethics and social values, and public policy may be undertaken to identify features important to the humanistic values. Financial analysis of the proposed activities may be given, and overall economic analyses (i.e., $Value_{ip}$, $Value_{mp}$) of both the "with" proposed activities and the "without" may be provided. Broader values of society that are difficult to quantify, but can be at least qualitatively described, should also be considered ($Value_{sg}$, $Value_{sp}$). Specific local, state, and national policy aspects unique to the proposed activities and the wetland usually need to be identified. Leitch et al. (1984), Leitch and Ekstrom (1989), Adamus (1983), Adamus et al. (1987), and Roelle et al. (1987a, 1987b, 1987c) have developed conceptual frameworks useful in developing the specific kinds of detailed background information.

Summarizing Information for Decision Deliberations

The heart of the summary information system is a "multiple alternative/multiple attribute (MA/MA) array" (Kiker and Lynne, 1989). MA/MA facilitates clearly representing the many aspects of the specific wetland and the proposed alternative use(s). The emphasis is on providing information about (a) the proposal made by the individual requesting the permit, (b) the alternatives developed by the staffs of the regulatory agencies and by other interested parties, and (c) the wetland left in its original state. MA/MA helps in summarizing the detailed information compiled by the agencies' staffs, some of which may originally have come from the general public.

Table 10.1 presents an outline of a MA/MA array. The alternatives are listed across the top and attributes associated with each are in the columns. The first alternative given (Alt. 1) is the activity proposed by the individual requesting the permit. The last alternative (Alt. 4) is the option that leaves the wetland in its present (original) state. Alt. 2 and 3 give present information on potential alternatives that lie between Alt. 1 and 4, ones possibly suggested by agency staff or other interested parties. Other alternatives could easily be added. "With" and "without" comparisons can then be accomplished, with MA/MA making it easier to see what is being given up when one alternative is accepted over the others.

The listed attributes are ones identified as being important in the more detailed analysis. The attributes are organized into three categories (I through III). The three categories relate to the various groups of values. Category I attributes would be assessed by, and aid in the development of, $Value_{es}$ and $Value_{or}$. $Value_{mp}$ estimates are provided in Category II. The attributes in Category III will be assessed by, and helpful in evolving, the $Value_{sg}$, $Value_{sp}$, and

Table 10.1 MA/MA Array for Florida West Coast Wetland

	Alt. 1 (activity requested)	Alt. 2...Alt. n-1 (other alternatives)	Alt. n (original wetland)
Category I. Biological and Physical Attributes, Cardinal Measures (cm_{ij})			
1. Total acres in parcel	160	160 160	160
2. Acres of development	160	120 60	0
3. Acres of wetlands remaining	0	40 100	100
4. Feet of canals and channels	2000	1000 0	0
5. Acres of habitat for endangered species	0	0 0	0
Category II. Economic Attributes, Cardinal Measures (cm_{ij})			
a. Market-established monetary measures			
1. Value of parcel for condominiums	$\$_{1a,1}$	$\$_{1a,2}$ 0	0
2. Value of parcel for renewable harvesting	0	0 $\$_{2a,3}$	$\$_{2a,4}$
b. Nonmarket-established monetary measures			
1. Value of park	0	0 $\$_{1b,3}$	$\$_{1b,4}$
2. Storm protection value	$(\$_{2b,1})$	$(\$_{2b,2})$ $\$_{2b,3}$	$\$_{2b,4}$
Category III. Social, Cultural, and Political Attributes, Ordinal Measures			
1. Scenic attractiveness	0	0 3	4
2. Consistency with long-term community development goals	1	2 5	5

$Value_{pr}$ dimensions. $Value_{ig}$ and $Value_{ip}$ considerations will be applied in processing information for all three categories.

Note that no attempt is made to weight the numerical values associated with the attributes or to convert the items to common value terms of summing. The valuing (weighting or judging) is left to the person reading the array and ultimately those authorized to approve or disapprove the permit request.

The three categories focus on biophysical and economic attributes for which means of measurement exist, as well as attributes for which no means of cardinal measures presently exist. In explaining the three categories of information, the hypothetical example of a proposed condominium and marina development affecting a coastal wetland will be used for illustration (Figure 10.3).

Category I. Biological and Physical Attributes, Cardinal Measures

This category (pertaining primarily to $Value_{or}$ and $Value_{es}$) includes summary data on the biological and physical factors thought important. The specific attributes will stem from staff experience with the specific wetland parcel and similar ones. The measurements are the best estimates the staff analysts can provide. Gosselink et al. (1989), Roelle (1987a, 1987b, 1987c), and Greeson et al. (1978), and other chapters in this book give good overviews of biophysical factors.

A few hypothetical values are inserted for illustration. In this case, a 160-acre parcel is proposed for the condominium and marina development, 60 acres of uplands and 100 acres of wetlands. Alt. 1 is the requested development, Alt. 4 represents leaving the wetland alone. Alt. 2 and 3 were developed as alternatives by agency staff. Alt. 1 will change all the wetlands and Alt. 2 will change 60 acres. Alt. 3 and Alt. 4 will not change the wetlands. Additionally, Alt. 1 involves 2000 ft of canals and channels, Alt. 2, 1000 ft, and the others no canals or channels. None of the alternatives affect endangered species habitat. Other important biological and physical attributes of the various alternatives could be presented.

Category II. Economic Attributes, Cardinal Measures

This category represents the $Value_{mp}$ and presents the full range of benefits and costs that can be expressed in monetary terms. The values are presented in two subcategories. Subcategory IIa, market-established monetary measures, recognizes that many inputs, products, and services are traded in well-functioning markets. The underlying assumption is that the market-established values provide reasonable representations of associated benefits and costs. The category involves the full range of economic activities normally recognized as being a part of benefit–cost analyses, for example, the sale or rental of the condominium units in competitive markets. Generally, dollar value can be estimated and used in evaluation of alternatives ($\$_{1a,1}$, $\$_{1a,2}$, $\$_{2a,3}$, and $\$_{2a,4}$ in Table 10.1). Additional commercial activities associated with either developing the parcel or leaving it in wetlands would also be listed.

Subcategory IIb, nonmarket-established monetary measures, is similar to the first subcategory in that the attributes can be expressed in monetary terms, but monetary values are not established in well-functioning markets. Some of these may also reflect the slightly less quantitative $Value_{ig}$ and $Value_{ip}$ values. Recreational and flood prevention values are examples. While people are willing to pay, typically competitive markets do not exist. Monetary values, instead of stemming from markets, can be estimated using well-accepted methods (see Hufschmidt et al., 1983; Cummings et al., 1986; and Chapter 9 for examples).

Returning to the example, the parcel (upland and wetland) has potential for use as a park (Alt. 3). While there is a willingness to pay by the public for a park, there will not be a specific bid-based value for this parcel. Instead, a value will have to be estimated. Condominium development, on the other hand, requires such substantial change that no public use would exist. In addition, the required drainage would prevent storm water protection and could result in more damage. Values for such damage can similarly be estimated. As shown in Table 10.1, the storm protection (e.g., hurricane) amelioration values for Alt. 1 and 2 ($\$_{2b,1}$ and $\$_{2b,2}$) are negative, whereas Alt. 3 and 4, which retain wetland characteristics, show positive values ($\$_{2b,3}$ and $\$_{2b,4}$). Additional attributes of this type associated with each alternative can be identified and monetary values estimated.

Category III. Social, Cultural, and Political Attributes

This category is intended to allow attention to $Value_{sg}$ and $Value_{sp}$, which are represented mainly by ordinal measures and ultimately are represented mainly in $Value_{pr}$. Many of the attributes presented in ordinal scales in this section may have been identified through concerns expressed by citizens and considered in some detail by the analysts. Because of the broad range of such possible attributes, it is important that the analysts develop their rationale for inclusion of the particular attributes thoroughly in the detailed supporting reports.

One such attribute for the wetland example might be the aesthetics associated with each alternative. If the parcel were in an area recognized for its scenic properties, an ordinal measure to represent scenic attractiveness might be used. In Table 10.1, values between 0 and 4 are given. Alt. 1 and 2 have values of 0, indicating complete change of appearance, whereas Alt. 3 and 4 have values of 3 and 4, respectively, indicating scenic properties are maintained.

Additionally, in Category III, information on community intents can be included. For example, a community through its representatives may be specifying long-term plans for a wetland area, and use of an individual parcel within the area for certain activities may be incompatible with the plan. Other valued activities in the wetland area may be precluded by some alternative uses of the parcel. The ordinal values 1 through 5 represent consistency with these plans.

Agencies developing the MA/MA array would also be expected to provide complete narrative explanations for each entry in the array. The idea is that interested parties (specialists, agents, etc.) should be able to completely understand the array entries, to have sufficient information to reconstruct the numerical entries, and to locate scholarly works upon which entries are based.

Also, some forms of information useful in the decision process do not easily fit the array format. To transmit such information to decision makers and interested parties, a narrative summary introducing special issues should be provided. The intent of the narrative summary is the same for the entire MA/MA evaluation: to facilitate a constructive political dialogue in the permitting process, which leads to an appropriate valuing process that considers all the values in Figure 10.2.

Experience with MA/MA

The MA/MA evaluation methodology has been used in conjunction with a proposed freshwater wetland project in Florida. Although the Florida project was of larger scale than most wetland permit requests, a brief review gives further insights into the usefulness of the evaluation methodology.

The Green Swamp in central Florida is a major groundwater recharge area and the headwaters for four rivers. The area has been of special interest to the people of Florida because of its unique wetlands. Interest climaxed with the area being designated as an "Area of Critical State Concern" and special land use regulation put into effect (see Kiker and Lynne, 1981). A large portion has been purchased by the Southwest Florida Water Management District (SWFWMD). The public, along with many state agencies, took special interest again when a long-proposed flood-control project involving two rivers and associated wetlands was resurrected in the early 1980s. The proposed Corps of Engineers' design was seen by some as a major threat to the unique area. Others focused on the benefits of the flood control. The issue was politically quite sensitive, with many citizen groups plus the Governor's Office, the Department of Environmental Regulation, the Game and Fresh Water Commission, and other state agencies taking a keen interest in the SWFWMD staff analysis and the Governing Board decision. It was recognized that the process of considering the project's potential would have to be an open one, with opportunity for interested parties to participate, contribute, and express their values.

From the start, a traditional benefit–cost study clearly would not be at all acceptable to either the Governing Board or the many groups interested in the Green Swamp. An approach was needed that would incorporate information on the multiplicity of values in Figure 10.2, instead of only information on $Value_{mp}$ in traditional benefit–cost studies. A way was needed to facilitate the give and take of the public valuing and decision process. The methodology settled upon, while not referred to as a MA/MA evaluation, was in principle the same.

Information from primary and secondary sources was collected. Scenarios associated with six alternatives were developed in substantial detail. Ultimately three volumes were available to provide a complete portrayal of the alternatives and their associated consequences. As part of an executive report, a summary array presented the six alternatives with full information on a range of biophysical, economic, and other attributes thought to be important. (See McLeod et al. [1984] for complete documentation on the study methodology.)

The outcome of the MA/MA evaluation process is best described by quotes from letters written by some of the major participants in the decision process:

> The Green Swamp Project reports, which were prepared for the Southwest Florida Water Management District, are a good example

of how benefit–cost analysis can help political decisionmakers choose among alternatives. This study does a thorough job of assessing the alternatives and points out all the considerations upon which the final decision must be based...I am glad to see economic analysis of this quality being used to guide important decisions like this one concerning the Green Swamp.

> Letter from Lieutenant Governor Wayne Mixon
> to Walter Kolb, Office of the Governor
> January 31, 1985

Presenting a detailed description of the numerous ecological relationships and values of the Green Swamp and then attempting to quantify elements not easily quantified in the market provides a more complete picture of the long-term environmental impacts of projects and intrinsic values of our natural resources...We hope that this type of economic evaluation for water resource projects will become the accepted method rather than selectively applied project by project.

> Letter from Victoria Tschinkel,
> Secretary of the Department of Environmental Regulation*
> to Walter Kolb, Office of the Governor
> January 31, 1985

The Green Swamp Project—Economic Report is truly a state-of-the-art analysis that has been praised by state and regional agencies, local governments and public interest groups.

> Letter from Gary Kuhl, Executive Director
> of the Southwest Florida Water Management District
> to George Barson, President
> of Environmental Science and Engineering, Inc.
> April 1, 1985

Conclusion

Both public agencies and interested individuals are part of the decision process associated with wetland permits. Because of the diversified interest and wide range of values, a large number of factors will be considered in the decision process. Although the responsible agencies have substantial knowledge, which will be the basis of recommendations and decisions, information associated

---

* In 1993, the Florida Department of Environmental Regulation was combined with the Department of Natural Resources to form the Department of Environmental Protection.

with a particular permit request needs to be presented so that interested people beyond the agencies can understand the range of alternatives and consequences and be involved in the evaluation. Clear presentation of information will facilitate consensus building and usually lead to permitted activities acceptable to more of the interested parties. The authors hope that the MA/MA evaluation approach might be of use to agencies in this evolutionary process.

References

Adamus, P. R. 1983. *A Method for Wetland Functional Assessment.* Vol. 1 and 2. Report Nos. FHWA-IP-82-23 and 24. Offices of Research, Development and Technology, Federal Highway Department, U.S. Department of Transportation, Washington, D.C.

Adamus, P. R., E. J. Clairain, Jr., R. D. Smith, and R. E. Young. 1987. *Wetland Evaluation Technique (WET).* Vol. II: Methodology. Operational Draft Technical Report Y-87. U.S. Army Engineer Waterways Experiment Station, Vicksburg, Mississippi.

Barnard, W. D., C. K. Ansell, J. G. Harn, and D. Kevin. 1985. Establishing priorities for wetland management. *Water Resour. Bull.* 21(6):1049–1054.

Cummings, R. G., D. S. Brookshire, W. D. Shulze, R. C. Bishop, and K. J. Arrow, Eds. 1986. *Valuing Environmental Goods: An Assessment of the Contingent Valuation Method.* Rowman and Allanheld, Totowa, New Jersey.

Georgescu-Roegen, N. 1971. *The Entropy Law and the Economic Process.* Harvard University Press, Cambridge, Massachusetts.

Gosselink, G. G., L. C. Lee, and T. Muir, Eds. 1989. *Ecological Processes and Cumulative Impacts, As Illustrated by Bottomland Hardwood Wetland Ecosystems.* Lewis Publishing, Chelsea, Michigan.

Greeson, P. E., J. R. Clark, and J. E. Clark, Eds. 1978. *Wetland Functions and Values: The State of Our Understanding.* American Water Resources Association, Minneapolis, Minnesota.

Hufschmidt, M. M., D. E. James, A. D. Meister, B. T. Bower, and J. A. Dixon. 1983. *Environment, Natural Systems, and Development: An Economic Valuation Guide.* Johns Hopkins University Press, Baltimore.

Kiker, C. and G. D. Lynne. 1981. Areas of critical state concern: Florida's experience with the Green Swamp. *S. J. Agric. Econ.* 13(2):149–155.

Kiker, C. F. and C. G. Lynne. 1989. Can MAMA help? Multiple alternative/multiple attribute evaluation of wetland use. Pages 25–38 in J. Luzar and S. Henning, Eds. *Alternative Perspectives on Wetland Valuation and Use.* Southern Natural Resource Economics Committee Publication No. 27. Southern Rural Development Center, Mississippi State, Mississippi.

Kusler, J. and R. Hamann, Eds. 1985. *Wetland Protection: Strengthening the Role of the States.* Center for Governmental Responsibility, University of Florida, Gainesville.

Leitch, J. A. and B. L. Ekstrom. 1989. *Wetland Economics and Assessment: An Annotated Bibliography.* Garland Publishing, New York.

Leitch, J. A., K. W. Easter, and W. C. Nelson. 1984. A proposed framework for developing a multidisciplinary wetlands valuation model. *Environ. Professional* 6:117–124.

McLeod, D. S., R. A. Zwolak, G. M. Powell, and R. S. DeLotelle. 1984. *Green Swamps Project Benefit and Cost Assessment.* Vol. III. Prepared for the Southwest Florida Water Management District by Environmental Science and Engineering, Inc., Gainesville, Florida.

Neville, R. C. 1989. *Recovery of the Measure: Interpretation and Nature.* State University of New York Press, Albany.

Office of Technology Assessment. 1984. *Wetlands: Their Use and Regulation.* U.S. Congress, Washington, D.C.

Roelle, J. E., G. T. Auble, D. B. Hamilton, R. L. Johnson, and C. A. Segelquist, Eds. 1987a. *Results of a Workshop Concerning Ecological Zonation in Bottomland Hardwoods.* NEC-87-14. Prepared for U.S. Environmental Protection Agency by the National Ecological Center, Fish and Wildlife Service, U.S. Department of the Interior, Washington, D.C.

Roelle, J. E., G. T. Auble, D. B. Hamilton, G. C. Horak, R. L. Johnson, and C. A. Segelquis, Eds. 1987b. *Results of a Workshop Concerning Impacts of Various Activities on the Function of Bottomland Hardwoods.* NEC-87-15. Prepared for U.S. Environmental Protection Agency by the National Ecology Center, Fish and Wildlife Service, U.S. Department of the Interior, Washington, D.C.

Roelle, J. E., G. T. Auble, D. B. Hamilton, R. L. Johnson, and C. A. Segelquis, Eds. 1987c. *Results of a Workshop Concerning Assessment of the Functions of Bottomland Hardwoods.* NEC-87-16. Prepared for U.S. Environmental Protection by the National Ecology Center, Fish and Wildlife Service, U.S. Department of the Interior, Washington, D.C.

Rolston, H. 1985. *Philosophy Gone Wild: Essays in Environmental Ethics.* Promethus Books, Buffalo, New York.

Sagoff, M. 1988. *The Economy of the Earth: Philosophy, Law and the Environment.* Cambridge University Press, Cambridge, U.K.

Shabman, L. A. and S. S. Batie. 1986. *Mitigation Damages from Coastal Wetlands Development: Policy, Economics and Financing.* Paper prepared for Workshop on Marine Pollution and Environmental Damage Assessment, Narragansett, Rhode Island. June 5–6.

Shabman, L. A. and S. S. Batie. 1988. *Socioeconomic Values of Wetlands: Literature Review, 1970–1985.* Technical Report Y-88. Army Engineer Waterways Experiment Station, Vicksburg, Mississippi.

U.S. Army Corps of Engineers. 1986. Part 320—General Regulatory Policies. *Federal Register* 51, 219(November 13):41220–41227.

U.S. Water Resources Council. 1983. *Economic and Environmental Principles and Guidelines for Water and Related Land Resources Implementation Studies.* U.S. Government Printing Office, Washington, D.C.

Legal Protection

11

Richard Hamann and John Tucker

The Government as Proprietor

The Public Trust Doctrine

The public interest in tidelands has been legally recognized since ancient times. The Institutes of Justinian, a codification of Roman law, declared, "[b]y the law of nature these things are common to mankind—the air, running water, the sea, and consequently the shores of the sea" (Justinian). English common law vested title to the king, subject to a trust protecting the public's rights to use such areas for navigation, fishing, and other public uses (Note, 1970). This doctrine was carried over to the colonies and incorporated into the common law of the United States. Thus, when Florida became a state in 1845, it acquired title, as a sovereign, to the navigable and tidal waters of the state. The boundary of the state's proprietary interest in tidelands under the public trust doctrine extends to the line of mean high water, the average of high tides over a full lunar cycle of 19 years (Florida Coastal Mapping Act of 1974; see Maloney and Ausness, 1975; Note, 1973, pp. 610–619).

Unlike other property, submerged lands are to be held in trust by the state for the benefit of the people of Florida. Florida courts have recognized navigation, commerce, fishing, and bathing as interests protected under the public trust doctrine (*State v. Black River Phosphate Co.*, 1893, p. 648; *Broward v. Mabry*, 1909, p. 829; *White v. Hughes*, 1939, p. 453). The courts of other states have used the public trust doctrine to protect a broader range of interests, including beach access (*Borough of Neptune City v. Borough of Avon-by-the-Sea*, 54, 1972) and ecosystem integrity (*National Audubon Society v. Superior Court*, 1983; *Marks v. Whitney*, 1971). The Florida constitution expressly vests title to all land under navigable waters in the state of Florida to hold in trust for the people, except for submerged lands which have been conveyed by deed or statute (Fla. Const. art. X, § 11).

The governor and cabinet sit as the Board of Trustees of the Internal Improvement Trust Fund (Trustees), the governmental entity responsible for administering submerged lands (Fla. Stat. § 253.03). Chapters 253 and 258, F.S., delineate the powers and duties of the Trustees. Salt marshes that lie below the mean high water line and that have not been deeded away by the state are clearly owned by the state and are subject to its management policies.

History of Significant Alienations and Management of Sovereignty Submerged Lands

Florida's official policies regarding management of salt marshes and other sovereignty submerged lands have changed dramatically since 1845 (see generally Blake, 1980). Prior to the mid-1960s, the state endorsed the filling of submerged lands. Acquisition and development of submerged coastal lands was encouraged through the Riparian Acts of 1856 (Fla. Laws, 1856) and 1921 (Fla. Laws, 1921) and the Bulkhead Act of 1957 (Laws of Florida, 1957). These acts provided for title to be conveyed if the purchaser "improved and developed" (Fla. Laws, 1921) the land by filling or bulkheading. During this period, conveyances of salt marshes were made freely with little or no consideration of possible adverse environmental consequences of the loss of state ownership of these lands. As late as 1940, the Supreme Court endorsed such transactions. Deep waters have been made next to cities and attractive residential districts have replaced eyesores and unsanitary breeding places of every conceivable species of pathogenic bacteria. The potentialities of such sales nowhere approach the possibilities that they do in this state (*Caples v. Taliaferro*, 1940).

During the late 1950s and early 1960s, a few people began to voice concern that indiscriminate establishment of bulkhead lines and filling of coastal areas were detrimental to the environment and therefore to the public interest. Legislative action followed in the form of the 1967 amendments to the Butler Act, which required mandatory biological and ecological surveys prior to conveyances of sovereignty submerged lands (Laws of Florida, 1967, Ch. 67-393, § 1(2)).

During the mid-1970s, as the scientific and public understanding of wetlands values increased, the state continued to add conservation values to its management policies. In the Florida Environmental Reorganization Act of 1975, the concept of variable bulkheads was eliminated and the mean high water line was set as the bulkhead line in tidal areas for the entire state. The current statutes governing sovereignty submerged lands management retain and expand most of the provisions of the 1967 amendments.

Although the Trustees still have authority to alienate sovereignty submerged lands, current statutory and regulatory rules provide rigid criteria designed to ensure that the Trustees may only make conveyances in the public interest. Conveyances of submerged lands are practically nonexistent today except in response to owners' efforts to quiet title acquired through earlier legitimate conveyances. The Trustees are much more likely to grant a lease or easement for use of sovereign lands than to convey title to the land.

Although the state has conveyed a considerable amount of salt marsh along the Gulf coast to private parties, the precise amount is difficult to determine. Most of the conveyances of submerged lands from state to private ownership were made at a time when there was little concern over salt marsh values. The state's analysis of whether to make a sale, if expressed at all, typically exalted the virtues of increasing developable land and ridding the coast of mosquitos and completely ignored any ecological considerations. Accordingly, most of the records of conveyances do not distinguish salt marsh areas from other types of intertidal systems. Often, the land was referred to only as "submerged land." Therefore, although many of these submerged lands were undoubtedly comprised of significant areas of salt marsh, it is difficult to document the precise amount.

Current Restrictions on Alienation, Leasing, and Other Forms of Consent

The Trustees must determine the extent to which the sale of sovereignty submerged lands would interfere with the conservation of, or cause destruction of, natural resources, as well as consider "any other factors affecting the public interests" (Fla. Stat. § 253.12(2)(a)). If objections are filed, the Trustees shall withdraw the lands from sale if they determine the sale would:

a. Be contrary to the public interest;

b. Interfere with the lawful rights granted riparian owners;

c. Be, or result in, a serious impediment to navigation;

d. Interfere with the conservation of fish, marine and other wildlife, or other natural resources, including beaches and shores, to such an extent as to be contrary to the public interest; or

e. Result in the destruction of oyster beds, clam beds, or marine productivity, including, but not limited to, destruction of natural marine habitats, grass flats suitable as nursery or feeding grounds for marine life, and established marine soils suitable for producing plant growth of a type useful as nursery or feeding grounds for marine life to such an extent as to be contrary to the public interest.... (Fla. Stat. § 253.12(4)).

The Trustees' rules provide additional guidance, stating that public interest means:

> demonstrable environmental, social, and economic benefits which would accrue to the public at large as a result of a proposed action, and which would clearly exceed all demonstrable environmental, social, and economic costs of the proposed action. In determining the public interest in a request for use, sale, lease, or transfer of interest in sovereignty lands, the board shall consider the ultimate project and purpose to be served by said use, sale, lease, or transfer of lands or materials (Fla. Admin. Code § 18-21.003(38)).

Applicants for purchase of sovereignty submerged lands must conduct and submit a biological survey and ecological study and in some instances a hydrographic survey (Fla. Stat. § 253.12(7) (a)). The Trustees use these studies to help determine whether a particular sale is in the public interest (Fla. Stat. § 253.12(7)(a)). However, prior to 1988, the Florida Department of Natural Resources (DNR), the agency assigned to provide support to the Trustees, did not have sufficient personnel to conduct on-site inspections. Accordingly, most Trustee evaluations and decisions concerning management of submerged lands have been based on the biological and ecological evaluations of other state and federal agencies. In 1988, the DNR began on-site inspections of some lands. However, at the time of this writing, on-site inspections are not routine. An applicant must also apply for a fill permit, for establishment of a bulkhead line if none is established, and for dredging (Fla. Stat. § 253.12(7)(a)), or the applicant can agree to condition the sale by including in the deed a restrictive covenant prohibiting dredging or filling (Fla. Stat. § 253.12(2)(b)3).

Sales of sovereignty submerged land may not unreasonably interfere with the riparian rights of upland owners (Fla. Admin. Code § 18-21.004(3)(a)). The Trustees may only convey these lands to upland riparian owners, unless the purchaser is a state agency or county, city, or other political subdivision, in which case the purchaser must obtain the upland riparian owner's consent to the sale (Fla. Stat. § 253.12(4)(e), (5)(b)).

A number of persons filled land under the early riparian acts but never formally received title to that land. Chapter 253 addresses these persons by authorizing the Trustees to convey title to persons who filled lands prior to the enactment of the Bulkhead Act of 1957 (Fla. Stat. § 253.12(6)). The Trustees will issue upon request a disclaimer to each owner of such land (Fla. Stat. § 253.129).

In addition to regulating sales of sovereignty submerged lands, Chapter 253 also regulates uses of these lands. No person can use sovereignty submerged lands without first obtaining consent from the Trustees (Fla. Stat. § 253.77(1)). The Trustees' consent takes one of the following forms: consent of use, lease (includes standard, aquaculture, oil and gas, and dead shell mining leases), easement (public and private), management agreement, grandfather structure registration, or use agreement (Fla. Admin. Code § 18-21.005). When a person applies for a permit from some other agency to conduct an activity on sovereignty submerged lands, such as a dredge and fill permit from the Department of Environmental Regulation, that agency must notify the applicant that consent from the Trustees may also be required (Fla. Stat. § 253.77(2)). Activities on sovereignty submerged lands requiring the Trustees' consent include but are not limited to construction of docks, boat ramps, boardwalks, and pilings; dredging and filling; removal of logs, sand, or shell; and removal or planting of vegetation (Fla. Admin. Code § 18-21.003(2)).

Unlike sales of sovereignty submerged lands, which "must be in the public interest" (Fla. Admin. Code § 18-21.004(1)(a)) to be approved, activities on sovereignty submerged lands "must be not contrary to the public interest" (Fla. Admin. Code § 18-21.004(1)(a)). The less stringent showing required for activities is counterbalanced by the high degree of control the Trustees retain over most forms of consent. For example, the Trustees' rules require that all forms of approval contain whatever conditions or restrictions are necessary to protect and manage sovereignty submerged lands (Fla. Admin. Code § 18-21.004(1)(b)). The rules also dictate that only water-dependent activities are allowed on sovereignty submerged lands unless the Trustees determine it is in the public interest to allow an exception (Fla. Admin. Code § 18-21.004(1)(d)). A water-dependent activity is

> an activity which can only be conducted on, in, over, or adjacent to water areas because the activity requires direct access to the water body or sovereignty lands for transportation, recreation, energy production or transmission, or source of water, and where the use of the water or sovereignty lands is an integral part of the activity. (Fla. Admin. Code § 18-21.003(50)).

All sovereignty submerged lands are considered single-use lands and are managed primarily to maintain essentially natural conditions, to propagate fish and wildlife, and to maintain traditional recreational uses such as fishing and boating (Fla. Admin. Code § 18-21.004(2)(a)). Activities that would cause significant adverse impacts to sovereignty submerged lands are prohibited unless there is no reasonable alternative and adequate mitigation is proposed (Fla. Admin. Code § 18-21.004(2)(b)). Activities must be designed to minimize or eliminate adverse impacts on fish and wildlife habitat, with special consideration given to endangered and threatened species habitat (Fla. Admin. Code § 18-21.004(2)(i)).

The Trustees are required to consider a number of often conflicting values when deciding whether to sell or allow use of sovereignty submerged lands. Current agency attitudes and policies make sales of sovereignty submerged lands extremely rare. However, the Trustees still possess considerable discretion when deciding whether or not to allow various uses of sovereignty submerged lands.

Currently, the Trustees receive a large number of requests for construction of single-family docks. Also common are requests for consent for shoreline stabilization, multifamily projects, and commercial projects. Marinas are still commonly permitted without adequate biological assessments being conducted. Agency personnel indicate that lack of on-site inspections and insufficient enforcement capabilities continue to hinder effective management of sovereignty submerged lands (Marx, 1989).

Additional statutory and regulatory criteria govern the management of sovereignty submerged lands within Florida aquatic preserves. Aquatic preserves are sovereignty submerged lands that the Trustees have designated as "areas which have exceptional biological, aesthetic, and scientific value..." and which are "set aside forever as aquatic preserves or sanctuaries for the benefit of future generations" (Fla. Stat. § 258.36). These preserves are to be managed to "preserve, promote, and utilize indigenous life forms and habitats, including but not limited to...submerged grasses, mangroves, salt water marshes...mud flats, estuarine...and marine reptiles, game and non-game fish species...invertebrates, mammals, birds, shellfish, and mollusks" (Fla. Admin. Code § 18-20.001(3)(f)).

As with other sovereignty submerged lands, sales of land within aquatic preserves must be in the public interest (Fla. Stat. § 258.42(1)). However, the rules go beyond those for sovereignty submerged lands and require that all activities within aquatic preserves must be in the public interest (Fla. Admin. Code § 18-20.004(1)(b)). The Trustees' rules provide specific cost–benefit criteria to apply in assessing whether to allow an activity (Fla. Admin. Code § 18-20.004(2)). The criteria are to be balanced to "determine whether the social, economic and/or environmental benefits clearly exceed the costs" (Fla. Admin. Code § 18-20.004(2)). The rule requires that the following criteria be considered:

b. Benefit Categories:
1. public access (public boat ramps, boatslips, etc.);
2. provide boating and marina services (repair, pumpout, etc.);

3. improve and enhance public health, safety, welfare, and law enforcement;
4. improved public land management;
5. improve and enhance public navigation;
6. improve and enhance water quality;
7. enhancement/restoration of natural habitat and function; and
8. improve/protect endangered/threatened/unique species.

c. Costs:
1. reduced/degraded water quality;
2. reduced/degraded natural habitat and function;
3. destruction, harm or harassment of endangered or threatened species and habitat;
4. preemption of public use;
5. increasing navigational hazards and congestion;
6. reduced/degraded aesthetics; and
7. adverse cumulative impacts.

The rule also provides examples of projects considered to possess specific benefits (Fla. Admin. Code § 18-20.004(2)).

The rules direct the Division of State Lands to develop a management plan for each aquatic preserve (Fla. Admin. Code § 18-20.017). Requests for activities on sovereignty submerged lands consistent with a management plan are more likely to be approved (Fla. Admin. Code § 18-20.004(2)(a)(3)). Specific consideration is given to the quality and nature of the specific aquatic preserve, as well as to specific areas within each preserve (Fla. Admin. Code § 18-20.004(2)(a)). The rules provide varying degrees of protection for lands within the preserve depending upon the perceived quality and condition of that area. Areas within aquatic preserves are designated as Resource Protection Area 1, 2, or 3, based upon whether the area has resources of the highest quality and condition (1), resources which are in transition (2), or resources which have no significant natural resource attributes (3) (Fla. Admin. Code § 18-20.003(31-33)). Salt water marshes are designated resources of the highest quality and condition (Fla. Admin. Code § 18-20.003(31)). The Trustees are also directed to evaluate the potential cumulative impacts of an activity on the entire natural system.

Regulatory Programs

Dredging, filling, and other construction activities in salt marshes are regulated by both state and federal agencies operating under several different statutory authorities. Regulation is a vital supplement to the proprietary controls exercised by the Trustees because jurisdiction can extend to lands previously conveyed by the state and lands above mean high water or bordering nonnavigable waters. The wetlands regulatory programs of the U.S. Army Corps of Engineers (Corps), the U.S. Environmental Protection Agency (EPA), and the Florida Department of Environmental Regulation are the most comprehensive. Numerous other regulatory programs affecting the management of salt marshes are beyond the scope of this chapter. All relevant Florida programs are discussed in the state's "networked" coastal zone management program (Fla. Stat. § 380.22), including controls on discharge of pollutants (Clean Water Act; Fla. Stat. §§ 376.011-.17, .19-.21; Florida Air and Water Pollution Control Act), use of pesticides (Federal Insecticide, Fungicide, and Rodenticide Act; Fla. Stat. Chs. 388, 487), diversion and consumptive use of water (Florida Water Resources Act), siting of power plants (Florida Electrical Power Plant Siting Act), and taking of marine life (Magnuson Fisheries Conservation and Management Act of 1976; Fla. Stat. Ch. 370). Florida's complex system of land use controls is also relevant, including the Florida Land and Water Management Act of 1972, the Florida State Comprehensive Planning Act of 1972, the Florida Regional Planning Council Act, the State Comprehensive Plan, and the County and Municipal Planning and Land Development Regulation Act.

Federal Regulation

Salt marshes are directly regulated by the Corps under Section 10 of the Rivers and Harbors Appropriation Act of 1899 (RHA) (33 U.S.C. § 403) and by the Corps and EPA under Section 404 of the Clean Water Act of 1977 (33 U.S.C. § 1344; see also Ocean Dumping Act). Other federal agencies with consultative roles in the regulatory process are the U.S. Fish and Wildlife Service and the National Marine Fisheries Service. Their roles were established by the Fish and Wildlife Coordination Act, 16 U.S.C. §§ 661-667e; the Endangered Species Act, 16 U.S.C. §§ 1531-1543; the Marine Mammal Protection Act, 16 U.S.C. §§ 1361-1407; and the National Environmental Policy Act, 42 U.S.C. §§ 4321-4361 (see U.S. Congress, 1984).

Section 10 of the RHA makes it illegal to "excavate or fill, or in any manner to alter or modify the course, location, condition, or capacity of...navigable water of the United States" without authorization by the Secretary of the Army (33 U.S.C. § 403). The navigable waters subject to regulation under this act clearly include all waters subject to the ebb and flow of the tide, at least to the line of mean high water (33 C.F.R. § 322.3(a); *U.S. v. Moretti*, 1973; *U.S. v. Stoeco Homes Inc.*, 1975). Jurisdiction is not, however, necessarily limited to waters below the mean high water line or the ordinary high water mark. An activity upland of those boundaries, but which has the effect of altering the navigable waters, may be subject to Section 10 jurisdiction. For example, the construction of a canal connected to navigable waters has been held to alter the course of the waterbody, thus invoking Section 10 (*U.S. v. Sexton Cove Estates*, 1976; see also *U.S. v. Hanna*, 1983; *U.S. v. Whichard*, 1974; *U.S. v. Sunset Cove*, 1975).

Until the late 1960s, however, the Corps was reluctant to exercise this regulatory authority to protect anything other than the navigable capacity of waterways (see generally Power, 1977). Beginning in 1968, the Corps began considering the effects of dredging and filling on fish, wildlife, and other aspects of the public interest. One reason for this shift in attitude was increasing congressional pressure to implement the Fish and Wildlife Coordination Act, P.L. 91-90, 42 U.S.C.A. §§ 4331-4347, which required consideration of impacts to fish and wildlife of water development projects, including the issuance of Section 10 permits. Congress was clearly concerned with the impacts to fish and wildlife of dredging and filling in bays and estuaries (Senate Report, 1958, pp. 3446, 3448, 3450). In 1967, Congress considered authorizing the U.S. Fish and Wildlife Service to permit dredging and filling, but was convinced such dual permitting was unnecessary when the Corps agreed to begin considering fish and wildlife impacts (*Zabel v. Tabb*, 1970). The expanded review criteria were challenged and ultimately upheld in the landmark case *Zabel v. Tabb* (1970), which involved denial of a permit to dredge and fill Boca Ciega Bay (adjacent to Tampa Bay) for a trailer park. In *Zabel*, the court held that the Corps can deny a Section 10 permit for "factually substantial ecological reasons even though the project would not interfere with navigation." Although salt marshes did not appear to be involved in the case, it illustrates the increased administrative and judicial concern for tidal wetlands protection.

During the early 1970s, federal authorities began vigorously enforcing Section 10 to protect tidal wetlands in Florida from massive dredge and fill projects, primarily in the Florida Keys and Tampa Bay area (see, e.g., *U.S. v. Sexton Cove Estates*, 1976; *U.S. v. Moretti*, 1976). The success of those enforcement efforts certainly discouraged developers from attempting similar dredge and fill activities in the salt marshes further north on Florida's coast (see U.S. Congress, 1984).

The federal experience in regulating wetlands under Section 10 was well known to Congress when it began considering the legislative proposals that culminated in the Federal Water Pollution Control Act Amendments of 1972 (FWPCA) (P.L. 92-500; see Report, 1970). Under the FWPCA, the discharge of most pollutants is subject to regulation by the EPA and the states. The discharge of dredged or fill material, however, is primarily regulated by the Corps under Section 404 of the Act (33 U.S.C. § 1344, see also Ocean Dumping Act). The EPA shares authority with the Corps for administration of the Section 404 program. Permitting authority may also be delegated to the states.

Under Section 404, a permit from the Corps is required before discharging dredged or fill material into "navigable waters." One of the first cases interpreting the jurisdictional reach of this section was *U.S. v. Holland* (1974), which upheld the authority of the Corps to regulate the discharge of fill in mangrove wetlands above mean high tide adjacent to Boca Ciega Bay. Subsequent decisions have confirmed the authority of the Corps to regulate all waters subject to the ebb and flow of the tide, plus adjacent wetlands (*U.S. v. Riverside Bayview, Inc.*, 1985; *N.R.D.C. v. Callaway*, 1975).

(33 C.F.R. § 328.3(b)). A direct connection to navigable waters is not necessary for jurisdiction. A wetland is "adjacent," and thus regulated, if it is

> bordering, contiguous, or neighboring. Wetlands separated from other waters of the United States by man-made dikes or barriers, natural river berms, beach dunes and the like are "adjacent wetlands." (33 C.F.R. § 328.3(c))

Thus, all salt marshes are clearly subject to regulation, whether above or below the line of mean high water. Freshwater wetlands in the estuarine watershed are also regulated.

The decision to issue a Section 404 permit is based on application of public interest review criteria adopted by the Corps and guidelines adopted by the EPA under Section 404(b)(1) of the act. Public interest review requires evaluation of the "probable impacts, including cumulative impacts, of the proposed activity and its intended use on the public interest" (33 C.F.R. § 320.4(a); 51 Fed. Reg.). A permit will be denied if issuance is determined to be contrary to the public interest after considering and balancing all relevant factors, including:

> conservation, economics, aesthetics, general environmental concerns, wetlands, historic properties, fish and wildlife values, flood hazards, floodplain values, land use, navigation, shore erosion and accretion, recreation, water supply, energy needs, safety, food and fiber production, mineral needs, considerations of property ownership and, in general, the needs and welfare of the people. (33 C.F.R. § 320.4(a), .4(a)(3))

In addition, the Corps considers the need for the structure, the practicability of using alternatives, the extent and permanence of effects, cumulative effects, and the effects on wetlands (33 C.F.R. § 320.4(a)). According to the Corps, wetlands are "a productive and valuable public resource, the unnecessary alteration or destruction of which should be discouraged as contrary to the public interest" (33 C.F.R. § 320.4(b)). Special protection is given wetlands that "serve significant natural biological functions"; are set aside for study or as sanctuaries or refuges; are significant to "natural drainage characteristics, sedimentation patterns, salinity distribution, flushing characteristics, current patterns, or other environmental characteristics"; shield other areas from wave action, erosion, or storm forces; are valuable floodwater storage areas; are important for groundwater discharge to maintain baseflows or for recharge; which purify water; or are unique or scarce resources (33 C.F.R. § 320.4(b)).

The Section 404(b)(1) guidelines, adopted by the EPA, must also be followed by the Corps (40 C.F.R. § 230, 33 U.S.C. § 1344(b)(2)). In general, the guidelines prohibit the discharge of dredged or fill material "which will cause or contribute to significant degradation of the waters of the United States" (40 C.F.R. § 230.10(a)(3)), after considering individual and cumulative effects (40 C.F.R. § 230.1(c)). All "appropriate and practicable steps" must be taken to "minimize potential adverse impacts of the discharge on the aquatic ecosystem" (40 C.F.R. § 230.10(d)). Discharges that would violate state water quality standards (40 C.F.R. § 230.10(b)(1)) or jeopardize an endangered or threatened species (40 C.F.R. § 230.10(b)(3)) are prohibited.

If there is a practicable alternative to the proposed discharge that would have less adverse impact, then the permit must be denied (40 C.F.R. § 230.10(a)). A practicable alternative is "available and capable of being done after taking into consideration cost, existing technology, and logistics in light of overall project purposes" (40 C.F.R. § 230.10(a)(2)). The use of other property may be considered in determining the availability of practicable alternatives if it could "reasonably be obtained, utilized, expanded or managed to fulfill the basic purpose of the proposed activity" (40 C.F.R. § 230.10(a)(2)). If an activity is not "water-dependent," then "practicable alternatives...are presumed to be available" (40 C.F.R. § 230.10(a)(3)). An activity is not "water-dependent" unless it requires "access or proximity to or siting within the special aquatic site in question to fulfill its basic purpose" (40 C.F.R. § 230.10(a)(3)).

One of the first decisions to interpret the water dependency and practicable alternatives test was *Deltona Corp. v. Alexander* (1982). In *Deltona*, the Corps denied a permit to dredge and fill thousands of acres of mangrove wetlands to build residential housing adjacent to Marco Island in Collier County, Florida. One basis for denial was a determination by the Corps that the basic purpose of the project was to provide housing, which is not a water-dependent use. Because housing could be provided on upland sites as an alternative to dredging and filling valuable aquatic resources, the permit was denied. Although salt marshes were not implicated in *Deltona*, the same principles would prevent the construction of residential housing or other nonwater-dependent uses in the salt marshes of Florida's Gulf coast (see also *Hough v. Marsh*, 1982; *Shoreline*

Associates v. Marsh, 1983; *U.S. v. Weisman*, 1980; *Sierra Club v. U.S. Army Corps of Engineers*, 1985; but see *1902 Atlantic Ltd. v. Hudson*, 1983; *Friends of the Earth v. Hintz*, 1986).

Water-dependent projects, such as marinas, or projects for which there are no practicable alternatives may receive Section 404 permits, provided impacts to aquatic resources can be reduced to acceptable levels or mitigated. Mitigation is controversial but commonly used in permitting (see generally Proceedings, 1986; Leslie, 1990). The concept refers to a variety of measures used to reduce the adverse impacts of a project. The Council on Environmental Quality has adopted rules defining mitigation widely used by federal agencies (40 C.F.R. § 1508.20). These might include using an upland area instead of a wetland or redesigning a project to reduce impacts. More controversial are measures that compensate for the destruction of wetlands by constructing new wetlands or restoring degraded wetlands.

National Audubon Society v. Hartz Mountain Development Corp. (1983) illustrates the use of mitigation to make development of a salt marsh in the Hackensack Meadowlands of New Jersey permittable. The project site encompassed 406 acres of land, including 278 acres of estuarine wetlands. Hartz Mountain proposed building a large mixed-use development on the site and filling 127 acres of brackish marsh that had been degraded by nuisance plant species. Although the uses were not water dependent, the Corps determined that practicable alternatives were not available (*National Audubon Society v. Hartz Mountain Development Corp.*, 1983). As mitigation, the Corps approved a plan to enhance the quality of 63 acres of brackish marsh by replacing common reedgrass with more desirable salt marsh vegetation and replacing 88 acres of brackish marsh with freshwater wetlands. The Corps determined that the enhanced and replacement wetlands would have higher values than the existing wetlands through application of a complex point system. The opinion does not specify the point system used, but it appears to be the Habitat Evaluation Program developed by the U.S. Fish and Wildlife Service. The Corps determined that after development and mitigation, the site would provide 80% of the values it provided before development even though the acreage of wetlands would be reduced by almost half. The Corps' decision to issue the permit was upheld by the district court.

Subsequently, the EPA and the Corps entered into a joint memorandum of agreement on the use of mitigation in wetlands permitting (55 Fed. Reg.). The agreement clarifies the sequence in which compensatory mitigation may be considered. Wetland impacts must first be avoided to the maximum extent practicable. If wetland impacts are unavoidable, they must be minimized. If unavoidable adverse impacts remain after minimization, then "appropriate and practicable" compensatory mitigation will be required. The determination of what constitutes appropriate mitigation is to reflect the values and functions of the affected ecosystem, without consideration for the value of the proposed activity. Mitigation should generally provide for replacement of wetland functions on a one-to-one basis. Generally, replacement of lost acreage on a one-to-one ratio would be required. However, the agreement allows the replacement of low-value wetlands with less acreage of high-value wetlands, where the probability of mitigation success is high.

The agreement also specifies a preference for on-site, in-kind restoration as compensatory mitigation. If not practicable, other alternatives, such as the construction of a different type of wetland in a different area, may be used. The preservation of existing wetlands may serve as compensatory mitigation only in "exceptional circumstances."

Mitigation may only be required if "practicable," defined as "available and capable of being done after taking into consideration cost, existing technology, and logistics in light of overall project purposes" (40 C.F.R. § 230.3(q)). Environmentalists have expressed concern that this requirement could allow unmitigated wetland losses where restoration technology is unproven or too expensive for wetlands-destroying activities. Although the technology for restoration or creation of salt marshes appears to be relatively successful when properly designed and implemented (Lewis, 1989), it may be too expensive for some activities. It is unclear from the agreement whether an economically marginal aquaculture project, for example, would be permitted in salt marshes without mitigation that is technically possible but unaffordable for that project. As an example of a situation in which mitigation may be deemed impracticable and thus not required, a footnote to the agreement refers to areas "where there is a high proportion of land which is wetlands" (Memorandum of Agreement, 1990). Although apparently intended to create an exception for oil industry operations in Alaska, the Gulf coast of Florida certainly fits that specification. Another footnote however states, "it is important to recognize that there are circumstances where the impacts of the project are so significant that even if alternatives are not available, the discharge may not be permitted regardless of the compensatory mitigation proposed" (Memorandum of Agreement, 1990 at n. 5). Debate will continue on how to resolve competing federal policies on wetlands regulation and mitigation.

State Regulation

Dredging, filling, and construction in waters of the state are regulated by the Florida Department of Environmental Regulation (DER) under the Warren S. Henderson Wetlands Protection Act of 1984. Before passage of the Wetlands Protection Act, the DER regulated salt marshes and other wetlands under two separate statutes (Chapters 253 and 403), with inconsistent jurisdictional and permitting criteria.

Chapter 253

State regulation of dredging and filling of salt marshes began when state officials realized public interest in tidelands could not be protected solely by reform of criteria for granting permission to use sovereign submerged lands, because so much of those lands had been previously alienated. Thus, in 1957, the legislature authorized establishment of bulkhead lines and deemed any further extension of land beyond a bulkhead line to be "interference with the servitude in favor of commerce and navigation with which the navigable waters of this state are inalienably impressed" (Laws of Florida, Ch. 57-362). Bulkhead lines were intended to establish limits to land development operations that were typically accomplished by enclosing tidal wetlands or submerged lands with sheetpiling (bulkheads) and filling the area with material hydraulically dredged from adjacent submerged lands. Although permits were required for dredging and filling, the criteria for issuance were primarily concerned with the maintenance of navigable capacity and the prevention of erosion and shoaling. Environmental factors were not given much significance in permitting under Chapter 253 until the statute was substantially amended in 1967 (Laws of Florida, Ch. 67-393).

As amended, Chapter 253 required the DER to comprehensively review applications for dredge and fill permits in navigable waters (Fla. Stat. § 253.123, .124). Permitting authority was transferred to the DER in 1975 (Laws of Florida, Ch. 75-22). The DER was required to conduct an ecological study, a biological survey, and, if needed, a hydrographic survey. A permit could not be issued unless those studies showed the project "will not interfere with the conservation of fish, marine and wildlife or other natural resources, to such an extent as to be contrary to the public interest" (Fla. Stat. § 253.123(3)(d)). Applicants wanting to dredge navigation channels were required to show "the public interest which will be served by such works" (Fla. Stat. § 253.123(4)(a)).

Yonge v. Askew (1974) (see also *Shablowski v. DER*, 1979; *Albrecht v. DER*, 1978; *Hopwood v. State*, 1981) illustrates the application of Chapter 253. The applicant had applied for a permit to connect three dredged canals to the Crystal River for a 700-acre real estate development. The permit was denied by the Trustees. Review by the staff of the Trustees and other agencies identified several adverse effects that would result from the project, including the loss of wetlands that were biologically productive and filtered runoff to the river. In addition to the direct effects of dredging the canals, adverse secondary effects of the urban development would be facilitated by the canals—urban runoff, turbidity, and other pollution. The cumulative effects of this development and of subsequent developments it would portend were considered.

The court found that the applicant had the burden of showing the works were "in the public interest" and that the Trustees had a duty to consider the adverse effects of the canals and of future development in the area. Private benefits to the landowner were appropriately given no weight. A resolution of the county commissioners supporting the development as in the public interest was rejected because the relevant interest was "of all the people in the State of Florida" (*Yonge v. Askew*, 1974). The availability of alternatives—building on upland portions of the property—also supported denial of the permit.

Chapter 403

Chapter 253 was of extremely limited geographical scope, only regulating dredging and filling of navigable waters below the line of mean high water. Salt marshes or other wetlands adjacent to nonnavigable waters or above the line of mean high water were thus unprotected despite their ecological functions. Thus, state regulators turned to another statute, the Florida Air and Water Pollution Control Act (Laws of Florida, Ch. 67-436), codified in Chapter 403 of the Florida Statutes, for authority to regulate dredging and filling of additional wetlands. The responsibility for implementing the act was given to the DER in 1975 (Laws of Florida, Ch. 75-22, § 8). Under this act, the DER has authority to regulate "stationary installations" that might reasonably be expected to be sources of water "pollution," defined as "potentially harmful...alteration of the chemical, physical, biological or radiological integrity of...water..." (Fla. Stat. § 403.031(2)). Jurisdiction extends to virtually all waters in the state.

After determining that dredge and fill operations could be regulated as stationary installations, the DER had to determine the appropriate geographic jurisdiction. Although the department's jurisdiction under the act extended to virtually all waters of the state, a more limited regulatory scope was politically and administratively necessary. Therefore, the DER adopted a list of regulated waters of the state, which included all tidal waters and natural tributaries. The landward extent of jurisdiction was to include all areas dominated by certain listed species. Although the species list was short, and thus excluded many important freshwater wetlands, it included all of the significant salt marsh vegetation. Thus, after 1975, the DER had clear authority to regulate dredge and fill activities in salt marshes both above and below the line of mean high water. Above mean high water, however, the criteria for issuing permits were not nearly as broad as that used in evaluating Chapter 253 permits. Permits could only be denied for projects determined likely to violate state water quality standards. Although maintenance of the filtering effects of salt marsh vegetation was relevant, other important values and functions were not. Thus, above the line of mean high water, regulators could not consider the effects of a dredging and filling project on fish and wildlife habitat, food chain support, or aesthetics (*Environmental Confederation of Southwest Florida v. Cape Cave Corp.*, 1985).

The Warren S. Henderson Wetlands Protection Act of 1984

The Warren S. Henderson Wetlands Protection Act of 1984 addressed these problems by expanding and consolidating the wetlands regulatory program (Smallwood et al., 1985; Hamann, 1985). The geographic scope of regulation was increased by adding new indicator species, a reform that was much less significant for salt marshes than for riverine floodplain wetlands. A broad set of criteria was adopted for evaluating all permit applications, whether or not the activity would take place in navigable waters. Under the Wetlands Protection Act, applicants must provide "reasonable assurance" of their ability to meet two criteria: water quality standards and a public interest test. Generally, projects must be "not contrary to the public interest" (Fla. Stat. § 403.918(2)). However, a project that significantly degrades or is located within an Outstanding Florida Water (OFW), must be "clearly in the public interest" (Fla. Stat. § 403.918(2)). The DER is directed to consider and balance the following criteria in determining effects on the public interest.

1. Whether the project will adversely affect the public health, safety or welfare or the property of others;

2. Whether the project will adversely affect the conservation of fish and wildlife, including endangered or threatened species, or their habitats;

3. Whether the project will adversely affect navigation or the flow of water or cause harmful erosion or shoaling;

4. Whether the project will adversely affect the fishing or recreational values or marine productivity in the vicinity of the project;

5. Whether the project will be of a temporary or permanent nature;

6. Whether the project will adversely affect or will enhance significant historical and archeological resources under the provisions of § 267.01; and

7. The current condition and relative value of functions being performed by areas affected by the proposed activity. (Fla. Stat. § 403.918(2)(a))

Cumulative impacts must also be determined. In addition to the impacts of the proposed project, the DER is required to consider the impacts of existing projects and "other projects which may reasonably be expected to be located within the jurisdictional extent of waters, based upon land use restrictions and regulations" (Fla. Stat. § 403.919). If the applicant is otherwise unable to meet the permitting standards, mitigation may be used to reduce the adverse impacts of the project (Fla. Stat. § 403.918(2)(b)).

No decisions have been reported interpreting the Wetlands Protection Act with regard to salt marshes. A recent decision by the First District Court of Appeal interpreting the act as applied to a beach construction project in Key West is of significant precedential interest. The applicant in *1800 Atlantic Developers v. DER* (1989) sought a permit to fill previously eroded, privately owned submerged lands adjacent to a condominium on Key West to construct a private beach. The submerged lands were rocky tidal flats utilized by birds at low tide and fish at high tide, but were not especially productive or rare. They were, however, designated as an OFW, and the applicant was therefore required to demonstrate the development was "clearly in the public interest." After an administrative hearing, the Secretary of the DER determined that the applicant failed to meet that test.

The hearing officer had determined the project would adversely affect fish and wildlife habitat, marine productivity, and recreational values. Proposed mitigation was "ill-defined, uncertain and insufficient." Other possible, though unlikely, adverse environmental effects were increased erosion, shoaling of nearby mangroves, and smothering of off-site seagrasses. The private benefits of new beach recreational opportunities would not be shared with the public. No need for the project or lack of suitable, less destructive alternatives to protect private structures from storm forces was demonstrated. There would be no adverse cumulative effects, however, because no similar projects were existing or under construction or reasonably expected in the future. The hearing officer thus concluded the applicant had failed to provide reasonable assurance that the project was clearly in the public interest. The DER's final order adopted this conclusion and denied the permit.

The First District Court of Appeal overturned the secretary's order on several bases. First, the court determined that the DER was required to give more specific guidance to the applicant on how the project could be designed or the adverse impacts mitigated rather than denying the permit. According to the court, "absolute prohibition of dredge and filling activity...should be the rare exception in cases of extreme damage to the environment that cannot be avoided or mitigated under any circumstances" (*1800 Atlantic Developers v. DER*, 1989, p. 2607). The court thus placed on the agency the ultimate burden of designing an acceptable project and assumed the adverse effects of dredging and filling can be mitigated in almost all cases. Second, the court ruled that the DER improperly applied the public interest test by requiring some public benefit from the project. Third, the court determined no evidence existed to support a finding "that there would be some quantifiable diminution in the quality of the marine habitat attributable to this project" (*1800 Atlantic Developers v. DER*, 1989, p. 2606). Because the adverse effects of the project were not expressed in numerical form, the court assumed there were none. Since there were no adverse environmental impacts, the applicant had no burden of demonstrating a public benefit or need the project would serve that might outweigh the adverse environmental impacts. As the First District Court of Appeal interprets the Wetlands Protection Act, a project that will not "materially harm the natural environment" (*1800 Atlantic Developers v. DER*, 1989, p. 2608) can be constructed without any other showing of the public interest served by using wetlands or submerged lands.

The court's interpretation is flawed in several respects. It fails to acknowledge any distinction between the statutory standards "not contrary to the public interest" and "clearly in the public interest." Certainly the legislature intended to give greater protection to OFWs than to other waters. It also seems likely the legislature intended to incorporate existing administrative and judicial interpretations of those terms into the law. Dredge and fill projects with purely private benefits have been held in several decisions which fail an "in the public interest" test.

Perhaps the court's decision rests primarily on its belief that adverse environmental effects were not sufficiently quantified. If so, it seriously misapprehends the science of environmental impact assessment. It may not be possible for the DER to quantify the impacts to bird or fish populations of losing another relatively small rocky intertidal area. The individual effects are probably insignificant. The cumulative effects, on the other hand, are very significant. In this case, however, those effects were never considered because other similar projects were deemed unlikely. The effects of prior filling and of other dredge and fill projects were never evaluated.

The court's apparent determination that adverse environmental impacts would not occur seems also to have been influenced by a belief that mitigation would be provided by scraping down an area of uplands to create equivalent habitat somewhere in the Keys. The decision strongly supports compensatory mitigation as a device for making projects permittable. Although the court acknowledges the possibility of cases of "extreme damage to the environment that cannot be avoided or mitigated under any circumstances," it says that these should be the "rare exception" (*1800 Atlantic Developers v. DER*, 1989, p. 2607).

Because of these interpretations, the statute is inadequate to protect salt marshes and other wetlands. Some loss of salt marshes would always be allowable, unless the adverse effects could be "quantified" or could not be mitigated. Permits could only rarely be denied. In most cases, the burden would be on the agency to suggest ways of mitigating the harm to make a project permittable.

Case Studies of Submerged Lands Management and Regulation

Early efforts at management of sovereignty submerged lands encouraged development and filling of salt marshes. An analysis of early Trustees' decisions does little more than illustrate the administrative processes involved in transferring deeds or issuing leases or easements. However, as the Trustees' policies changed in the 1960s and 1970s, management decisions began to reflect some consideration of current ecological values and understanding.

Wetstone Development

In 1960, a private upland owner applied for the purchase of about 655 acres of submerged lands along the Gulf coast in Pasco County with the intent of filling the land for a residential development (Trustees, undated, a). Bulkhead lines for the area had not been established, and the applicant proposed to locate the line about 5000 ft waterward from the present shoreline. The applicant submitted a review of the proposed bulkhead line and fill project conducted by the Coastal Engineering Laboratory at the University of Florida. The review—the only documented consideration of environmental values—indicated that from a coastal engineering perspective the project would cause few or no adverse effects.

The Trustees approved the proposed bulkhead line and the sale of the submerged land. The applicant received title to the land in 1965, free and clear of any restrictions. The parcel of land extended about 4000 ft from the shoreline into the Gulf of Mexico. Apparently, the applicant did nothing with the land until 1971, when the applicant requested a dredge and fill permit so he could develop the property.

The applicant's proposal involved filling over 400 acres of salt marsh and 500 acres of other intertidal lands. Early negative feedback from the Department of Natural Resources prompted the applicant to propose a deal whereby the applicant would reconvey a portion of the submerged lands to the state if the state would guarantee dredge and fill permits for the land retained by the applicant. The Trustees rejected this offer and counteroffered that if the applicant would reconvey all of the submerged lands back to the state, the state would grant the necessary permits for dredging and filling of upland property and one main access channel from the property to the Gulf of Mexico.

After the applicant declined to accept the state's counteroffer, the Trustees requested comments from environmental agencies. The Department of Pollution Control found the proposed network of canals and elimination of existing tidal marshes would have a "definite adverse effect on water quality" (Barnes, 1971). The Department of Natural Resources, in a letter describing in detail the numerous species of vegetation and wildlife, found the proposed dredging and filling "would have extensive and permanent adverse effects on the marine biological resources" (Routa, 1971) and "no dredging or filling should be permitted in the black rush marsh, tidal streams, or bayous" (Routa, 1971). The Game and Fresh Water Fish Commission strongly objected to the filling of any marsh areas.

The Trustees informed the applicant that, based on the environmental agencies' comments, no development should take place on the land. Although no formal denial was issued, the applicant has apparently made no other attempts to develop the land.

This case vividly illustrates how the state's sovereignty submerged lands policies have changed. Although a formal dredge and fill permit was not issued in 1961 when the applicant bought the property, the sale was completed with the state's expectation that the land would be filled in the immediate future. Undoubtedly, a permit would have been issued had the applicant requested one. Ten years later, the state was unwilling to permit a similar project. Detailed ecological and biological studies by several environmental agencies indicated that filling of the salt marsh areas was unacceptable.

Florida Power Corporation Power Plant

An application by the Florida Power Corporation for acquisition of sovereignty submerged lands and dredge and fill permits illustrates how environmental concerns are balanced with apparent public need for a project. In 1964, the Florida Power Corporation applied to the Trustees for purchase of about 1033 acres of sovereignty submerged lands and to dredge and fill a portion of that property in connection with construction of a power plant near Crystal River in Citrus County (Trustees, undated, b).

The applicants had previously purchased about 3650 upland acres upon which the actual power plant was to be located. The sovereignty submerged lands the applicant sought to purchase were riparian to the applicant's uplands and covered about 5.9 mi of frontage on the Gulf of Mexico. The applicant planned to dredge two channels to provide cooling water and to allow fuel barges navigational access to the plant. The channels traversed large areas of salt marsh. The intake channel, which would also provide barge access to the plant, would extend about 8 mi into the Gulf and would require dredging of about 1.8 million yd^3 of sovereignty submerged lands. The discharge channel would extend about 2.6 mi into the Gulf and would require dredging about 350,000 yd^3 of sovereignty submerged lands. The dredged material would be placed alongside the channels to produce spoil islands. The proposal called for no dredging or altering of other salt marshes.

The Florida Board of Conservation found that the spoil from the dredging would act as "an emergent artificial fishing reef" running for "roughly six miles as a beneficial by-product of company needs for water and navigational facilities" (Woodburn, 1964). In addition, in response to company assurances that a large area of salt marshes and intertidal creeks would be left untouched, the Board of Conservation stated that "these virginal lands will largely remain inviolate from filling and other development" and "would be a conservation area although privately owned" (Woodburn, 1964). However, the board did recommend that the area remain open to sport and commercial fishing because "to deny an important segment of the coastal economy the utilization of marine resources in the subject area would not be wise conservation, good economics, or popular locally" (Woodburn, 1964).

The Trustees approved both the sale and the requests for dredge and fill. They rejected the Board of Conservation's recommendation that the public be allowed to fish within the area, stating that "[t]his invasion of the area of natural protection may defeat the protection and become a problem in the maintenance of the area by the applicant as a buffer zone around its contemplated power plant and such reservation would materially affect the appraised value..." (Trustees, 1964).

Undoubtedly, the applicant's assurances that the remaining salt marshes would not be developed played a part in the Trustees' decision. By selling the land, the Trustees gave up some power to control future uses of the land. However, future uses would be subject to the dredge and fill permitting process. The Florida Power Corporation stated that the power plant would afford a service to the state of Florida and that the sale and permits would also allow better drainage of the applicant's upland site which would benefit the entire area. Factors such as these probably influenced the Trustees.

Snow Septic Tank

In 1984, a private landowner applied to the southwest district office of the DER for a dredge and fill permit to install a septic tank in a salt marsh bordering the St. Martins River in Citrus County, Florida. Applicable county rules required a buffer of at least 50 ft between the septic tank drain field and surface water.

The project involved filling in about 4500 ft^2 of salt marsh to provide a buffer zone for installation of a septic tank to serve the applicant's residence. The application was processed pursuant to Chapter 403, Florida Statutes, which at that time only allowed the DER to consider potential water quality impacts. The DER did not assert jurisdiction under Chapter 253, Florida Statutes, which would have allowed consideration of a wider range of values, because the project probably did not fall within the scope of navigable waters. The DER issued a Letter of Intent to Deny, citing "anticipated violations of water quality standard for bacteriological quality, biological integrity, biochemical oxygen demand (BOD), dissolved oxygen, turbidity and nutrients" (*Franklin T. Snow v. State of Florida Department of Environmental Regulation*, 1984).

The landowner objected to the DER's Letter of Intent to Deny, and a series of administrative hearings followed. The hearing officer accepted as findings of fact the following DER assertions. (1) Placing fill directly on top of existing marsh vegetation would create an organic layer which would leach nutrients into adjacent surface waters and exert an oxygen demand on the water.

> [(2) T]he black rush and salt grass which presently dominate the proposed fill site perform a significant water quality function in trapping sediments, filtering runoff and assimilating nutrients....The proposed fill site also functions as a productive habitat for numerous aquatic species which comprise a portion of the estuarine food chain and ecosystem....Leaving the site in its present condition creates a public benefit to the state.

(3) The septic tank leachate would contribute to high fecal coliform densities, which would violate water quality standards (*Franklin T. Snow v. Florida Department of Environmental Regulation*, 1984).

The hearing officer initially found that the purpose of the denial of the dredge and fill permit was to preserve ecologically productive habitat, and therefore the denial amounted to a taking because it created a public benefit. He also found that the regulation of septic tanks was within the province of the Department of Health and Rehabilitative Services and not the DER. The hearing officer recommended granting the permit as long as the existing organic material and vegetation were removed and the area was filled using sand with high clay content.

The secretary of the DER remanded the recommended order back to the hearing officer, holding that the officer had no authority to decide takings issues and finding that the DER may regulate the water quality impacts of a septic tank when its installation requires a dredge and fill permit.

On remand, the hearing officer changed his reasoning but reached the same conclusion. The hearing officer found that under Chapter 253, Florida Statutes, the DER had authority to require removal of the existing vegetation as a condition of the permit, and therefore the permit should be granted with this stipulation. The hearing officer also found that if clay were installed around the septic tank, there would be no violation of applicable water quality standards.

The secretary of the DER, in the final order, again rejected portions of the hearing officer's reasoning, but adopted his ultimate recommendation that the project be permitted. The secretary of the DER found that the hearing officer's reliance on Chapter 253, Florida Statutes, was misplaced because the DER had not asserted jurisdiction or evaluated the permit application under Chapter 253. However, the DER found that the agency also had authority to impose such conditions under Chapter 403, Florida Statutes. The DER accepted the hearing officer's findings that the fill would not cause water quality violations. The permit was issued by the DER, subject to the conditions imposed by the hearing officer. The Corps promptly issued a permit two weeks later, which included no special conditions.

The Snow application illustrates how severely constrained the DER was in cases where it could only assert Chapter 403 jurisdiction before the Warren S. Huduson Wetlands Act of 1984. Essentially, the DER was limited to consideration of water quality. Values such as the preservation of wildlife and habitat could not be considered except for their potential effect on water quality.

Mikeland Properties Office Building

In 1986, Mikeland Properties, Inc. filed an application for a permit to discharge 7344 yd^3 of fill into 1.7 acres of salt marsh and 2.8 acres of adjacent upland to support construction of a six-story commercial office building and parking facility in the Tampa Bay area in Pinellas County (Florida Department of Environmental Regulation, 1986). The applicant proposed off-site mitigation for the filling consisting of (1) replacement of an existing ditch with a relocated swale, covering about 0.37 acres planted with *Spartina alterniflora*; (2) excavation of about 0.5 acres of Australian pine-covered uplands down to wetland elevation and planting with *S. alterniflora*; (3) excavation of about 1.6 acres containing 60 Brazilian pepper-covered spoil mounds down to wetland level and planting with *S. alterniflora*; and (4) dedication of the entire mitigation area, comprising 9.7 acres, and that portion of the site not used for development, comprising about 20 acres, as conservation easements (Florida Department of Environmental Regulation, 1986).

The total proposed marsh to be created covered about 2.4 acres, creating a mitigation ratio of 1.4 to 1. The total area recorded as a conservation easement covered about 30 acres. In addition, the mitigation plan called for repair of any incidental damage to adjacent wetland vegetation, a written warranty for 80%

survival for two years of all planted *S. alterniflora*, submittal of monitoring reports every six weeks for two years, and completion of the mitigation work prior to filling the salt marsh (Florida Department of Environmental Regulation, 1986).

The Biological and Water Quality Assessment Report conducted by the DER found that the "development would destroy the salt marsh habitats and could severely impact the fish populations. Disturbance by man would probably eliminate usage of this area by Wood Storks and other marsh birds" (Biological and Water Quality Assessment, 1986). However, the report indicated that the proposed mitigation would protect sufficient adjacent and nearby wildlife habitat to offset the adverse effects of the project. The report recommended that the project be permitted.

Comments from other agencies were not favorable. The EPA, aiding the Corps in its review of the project, found that development of the site would be contrary to Section 404(b)(1) guidelines because the area was functioning as an estuarine wetland (Heinen, 1986). The EPA suggested that sufficient uplands existed for the project and recommended against granting the permit. The National Marine Fisheries Service found that the project site was vegetated with black mangrove, white mangrove, and smooth cordgrass and that

> [t]hese wetlands provide food, habitat, and water quality maintenance functions that are vital for the continued production of commercially and recreationally important fishery resources in the Tampa Bay estuarine system. These resources include pink shrimp, blue crab, red drum, Spanish mackerel, red grouper, seatrout, mullet, and Gulf flounder. (Mager, 1986)

The agency stated that the proposed work was not water dependent and would "add considerably to the cumulative loss of wetlands in the Tampa Bay system which is already severely stressed by habitat alterations" (Mager, 1986). The agency found that at least 478 water development projects had been proposed for the Tampa Bay area between 1981 and 1985. A subsample of 167 of these projects proposed the alteration of more than 567 acres of wetlands.

Other agencies supported the proposal with some reservations. The Tampa Bay Regional Planning Council found that the proposal should be permitted but that all mitigation should be performed on-site (Tampa Bay Regional Planning Council, 1986). Similarly, the Board of County Commissioners, Pinellas County, Florida, was concerned that if off-site mitigation was allowed, it might become the industry standard (Cuba, 1986). In addition, the county commissioners thought that the conservation easements were meaningless because those wetland areas were already protected by wetland regulations.

Both the DER (Florida Department of Environmental Regulation, 1987) and the Corps (U.S. Army Corps of Engineers, 1988) approved the project as proposed.

DeVincenzo Permit Denial

In 1978, an applicant applied for a permit to fill about 0.9 acres of submerged tidal mud flat and adjacent mangrove fringe on a barrier island in Pinellas County, Florida. The applicant proposed depositing 13,000 yd^3 of fill and containing it behind a 460-ft seawall (In re James DeVincenzo, 1988). The proposal was revised in 1986 to fill about 0.4 acres with 5150 yd^3 of fill behind 662 ft of rip-rap revetment. The applicant proposed to mitigate the adverse impacts of the project by planting about 92 linear feet of *Spartina* sp. in the rip-rap revetment. The project site was adjacent to the Pinellas County Aquatic Preserve, an OFW.

High and mid intertidal zones at the site were dominated by black mangroves (*Avicennia germinans*) and beach carpet (*Blutaparon vermiculare*), while the lower intertidal areas were dominated by small red mangroves (*Rizophora mangle*) and smooth cordgrass (*S. alterniflora*). The majority of the area to be filled consisted of open mud flat. The DER found that a variety of marine and terrestrial organisms were utilizing the project site, including worms, snails, clams, mussels, oysters, sea squirts, shrimp, starfish, crabs, fish, snakes, mammals, birds, and other organisms.

The DER determined that the project would adversely affect marine and terrestrial organisms and plants in the project area and in the adjacent estuary. In addition, the project would adversely affect water quality by contributing to stormwater runoff and restricting flushing of channels. The DER issued a notice of intent to deny the permit based upon failure to satisfy the following statutory and regulatory criteria:

1. The applicant did not provide reasonable assurance that water quality standards would not be violated, as required by Chapter 403.918(1), Florida Statutes, and Chapter 17-3, Florida Administrative Code.

2. The applicant did not provide reasonable assurance that the project is not contrary to the public interest as specified in Chapter 403.918(2), Florida Statutes.

3. The applicant did not provide assurances that the project would not violate standards for mangrove alterations as required by Rule 17-27.050, Florida Administrative Code.

4. Development of the general area around the project site has resulted in "cumulative loss[es] of mangrove-cordgrass fringe communities and other

functioning estuarine zones" which "has been directly reflected in a recent decline in numbers of sport and commercial fishes" (In re James DeVincenzo, 1988). The cumulative losses increase the need to protect the project site.

5. The applicant's mitigation proposal was inadequate because:

   1) The applicant did not provide the Department with reasonable assurance that the mitigation proposed would ameliorate the negative impacts to fish and wildlife and their habitats that would result from the proposed project. The installation of 92 linear feet of *Spartina* sp [sic] in a rip-rap revetment can not replace the functions performed by the existing mangrove forest, cordgrass fringe, and associated mudflat. The decrease in the diversity of the habitats available would result in a overall decrease in species diversity. A net loss in food resources, and refugia for larval fishes and roosting and nesting habitat for wading birds would result. Further, the applicant did not provide the Department with assurances of the success of the mitigation plan that was proposed. These assurances would include but not be limited to the source, size and spacing of plant material, methods for handling and installation of plants, the methods and frequency for monitoring and a contingency plan in the event of failure.

   2) The applicant did not provide the Department with reasonable assurance that the mitigation proposed will cause a net improvement of the water quality standards specified in this document.

   3) The applicant did not propose measures to ameliorate the cumulative impacts anticipated by the proposed project. (In re James DeVincenzo, 1988 at 10, 11)

6. The applicant failed to propose or consider reasonable project modifications or alternatives.

Conclusions

Salt marshes receive substantial protection under current laws and regulations. Those lands the state holds under the public trust doctrine receive the greatest protection. The governor and cabinet have considerable discretion to withhold permission to use state-owned submerged lands. State and federal regulatory programs restrict the development of salt marshes and other wetlands that were never part of the public trust or have passed into private ownership.

Under none of these programs, however, are salt marshes inviolate. Salt marshes are sometimes sacrificed for more important public interest considerations. Today, such destruction is usually limited and mitigation is required.

Mitigation, in theory, can eliminate adverse effects. If a salt marsh in one location is destroyed but replaced with another salt marsh of equal or greater value, the environment suffers no net loss of salt marsh. It may even gain in the long run. Much of the regulatory activity involving salt marshes involves application of this theory, as illustrated by the case studies described above. But the gap between theory and reality may be very large.

We were unable to physically investigate the sites of our case studies to evaluate the mitigation projects. Several recent investigations, however, bring into question the use of mitigation in current regulatory programs. In a review of eleven mitigation sites in Manatee and Sarasota counties, including five marine wetlands, one investigator found problems with all of the projects as regulated by the DER (Crewz, 1990). Ten of the sites had hydrological problems. Eight of the sites were improperly designed or were not constructed as designed. Plant mortality and failures to plant the correct species or the correct densities, or to replant when necessary, were observed at seven sites. Most sites had uncorrected invasions of exotic species. Nine of the sites lacked the required monitoring reports. The DER had almost uniformly failed to require corrective action.

Other investigations are revealing similar results. The South Florida District of the DER reviewed ten dredge and fill projects with mitigation (Blair and Dentzau, 1989). Only five had been completed as required in the permit, and only one of those was successful. Four of the projects were not completed and one was not even attempted.

Southwest Florida Aquatic Preserve staff initiated another review of 43 dredge and fill projects permitted by the DER adjacent to aquatic preserves in southwest Florida (Beever, 1990). The mitigation had not even been attempted on 42% (18) of these projects. Where attempted, planting was initially done correctly, but the plants failed at 72% (18) of those sites. Native vegetation subsequently restored five of those sites.

The creation and restoration of tidal wetlands is possible, but successful mitigation requires skill and commitment (Lewis, 1989). Projects must be carefully designed and constructed. Even then, failures occur that must be corrected. Maintenance, monitoring, and the willingness to invest additional resources to achieve successful mitigation are essential. Most mitigation, however, is conducted as a condition for receiving dredge and fill permits. Once the permitted activity has occurred, the permittee often has little interest in continuing to monitor and maintain a mitigation project, much less undertake corrective action. Some permittees will not implement the mitigation plan at all, if possible. The vigorous enforcement of mitigation conditions is necessary.

References

1800 Atlantic Developers v. DER. 552 So. 2d 946, 14 F.L.W. 2604 (Fla. 1st D.C.A. 1989), *reviewing Florida Keys Citizens Coalition v. 1800 Atlantic Developers,* D.O.A.H. Case No. 86-1216, Final Order, 8 F.A.L.R. 5564 (1986), *review denied Department of Environmental Regulation v. 1800 Atlantic Developers,* 562 So. 2d 345 (Fla. 1990).

1902 Atlantic Ltd. v. Hudson. 1983. 574 F. Supp. 1381.

33 C.F.R. §§ 320, 322, and 328.

33 U.S.C. §§ 403, 1344, and 1401.

40 C.F.R. § 230.

51 Fed. Reg. 41223. 1986.

54 Fed. Reg. 51319. 1989.

55 Fed. Reg. 9210. 1990.

Albrecht v. DER. 1978. 353 So. 2d 883.

Ansbacher, S. and Q. J. Knetsch. 1989. The public trust doctrine and sovereignty lands in Florida: a legal and historical analysis. *J. Land Use Environ. Law* 4:337–375.

Barnes, B. A. 1971. Letter from the Bureau of Permitting, Department of Pollution Control, to John DuBose, Board of Trustees of the Internal Improvement Trust Fund.

Beever, J., III. 1990. Memorandum from the Resource Management and Research Coordinator, Southwest Florida Aquatic Preserves, Department of Natural Resources, State of Florida, to Hank Smith, Environmental Specialist III, Bureau of State Lands Management, Division of State Lands.

Biological and Water Quality Assessment. 1983. Permit Application Appraisal. Florida Department of Environmental Regulation file no. 52-49189.

Biological and Water Quality Assessment. 1985. Permit Application Appraisal. Florida Department of Environmental Regulation file no. 150977219.

Biological and Water Quality Assessment. 1986. Permit Application Appraisal. Florida Department of Environmental Regulation file no. 521198323.

Biological and Water Quality Assessment. 1989a. Permit Application Appraisal. Florida Department of Environmental Regulation file no. 521546349.

Biological and Water Quality Assessment. 1989b. Permit Application Appraisal. Florida Department of Environmental Regulation file no. 520491899.

Blair, L. and M. Dentzau. 1989. State of Florida, Department of Environmental Regulation. Interoffice Memorandum to Janet Llewellyn et. al.

Blake, N. 1980. *Land into Water, Water into Land.* University Presses of Florida, Tallahassee.

Borough of Neptune City v. Borough of Avon-by-the-Sea. 1972. 294 A.2d 47.

Brantley, Col. R. 1985. Letter from the Executive Director, Florida Game and Fresh Water Fish Commission, to Victoria Tschinkel, Department of Environmental Regulation.

Broward v. Mabry. 1909. 50 So. 826.

Caples v. Taliaferro. 1940. 144 Fla. 1; 197 So. 861.

Clean Water Act. 1989. 33 U.S.C.A. §§ 1251-1387.

County and Municipal Planning and Land Development Regulation Act. 1989. Fla. Stat. §§ 163.3161-.3243.

Crewz, D. 1990. *Habitat-Mitigation Evaluations for Manatee-Sarasota Counties.* Mid-Project Summary: Projects 1-11, Report to Manasota 88.

Cuba, T. 1986. Letter from the Chief, Environmental Management Division, Department of Environmental Management, Board of County Commissioners, Pinellas County, Florida, to Sheila Turner, I.C. and R. Coordinator, Tampa Bay Regional Planning Council.

Deltona Corp. v. Alexander. 1982. 682 F.2d 888.

Endangered Species Act. 16 U.S.C. §§ 1531-1543.

Environmental Confederation of Southwest Florida v. Cape Cave Corp. 1985. Final Order 8 F.A.L.R. 317; Recommended Order 8 F.A.L.R. 343.

Federal Insecticide, Fungicide, and Rodenticide Act. 1989. 7 U.S.C.A. §§ 136-136y.

Federal Water Pollution Control Act. P.L. 92-500, codified at 33 U.S.C. §§ 1251-1376.

Fish and Wildlife Coordination Act. 16 U.S.C. §§ 661-667e.

Fish and Wildlife Coordination Act. P.L. 91-90; 42 U.S.C.A. §§ 4331-4347.

Fla. Admin. Code. 1987. Rule 17-12.340(2).

Fla. Admin. Code. 1987. §§ 17-4, 18-20, and 18-21.

Fla. Const. 1970. Art. X, § 11.

Fla. Laws. 1856. Ch. 791.

Fla. Laws. 1921. Ch. 8537.

Fla. Laws. 1957. Ch. 57-362.

Fla. Laws. 1967. Ch. 67-393, § 1(2).

Fla. Stat. 1989. §§ 253, 258, 370, 380, 388, 403, and 487.

Florida Air and Water Pollution Control Act. 1989. Fla. Stat. §§ 403.011-.4153.

Florida Coastal Mapping Act of 1974. 1989. Fla. Stat. §§ 177.25-.40.

Florida Department of Environmental Regulation permit application no. 150977219. 1984.

Florida Department of Environmental Regulation. 1985. Permit no. 150977219.

Florida Department of Environmental Regulation. 1986. Permit application no. 521198323.

Florida Department of Environmental Regulation. 1987. Permit no. 521198323.

Florida Department of Environmental Regulation. 1989a. File no. 521546349.

Florida Department of Environmental Regulation. 1989b. File no. 520491899.

Florida Electrical Power Plant Siting Act. 1989. Fla. Stat. §§ 403.501-.517.

Florida Keys Citizens Coalition v. 1800 Atlantic Developers. 1989. 552 So. 2d 946.

Florida Land and Water Management Act of 1972. 1989. Fla. Stat. §§ 380.012-.10.

Florida Regional Planning Council Act. 1989. Fla. Stat. §§ 186.501-.513.

Florida State Comprehensive Planning Act of 1972. 1989. Fla. Stat. §§ 186.001-.031.

Florida Water Resources Act. 1989. Fla. Stat. Ch. 373.

Franklin T. Snow v. State of Florida Department of Environmental Regulation. 1984. Recommended Order, D.O.A.H. case no. 84-2836. Exceptions to Recommended Order. D.O.A.H. case no. 84-2836; O.G.C. file no. 84-0567. 1985.

Friends of the Earth v. Hintz. 1986. 800 F.2d 822.

Frye, O. E., Jr. 1981. Letter from the Director, Florida Game and Fresh Water Fish Commission, to Joel Kuperberg, Director, Trustees of the Internal Improvement Trust Fund.

Hamann, R. 1985. The evolution of Florida wetlands regulation. In J. Kusler and R. Hamann, Eds. *Wetland Protection: Strengthening the Role of the States.* State Wetland Managers, Zorna, New York.

Hartman, B. 1989. Letter from the Director, Office of Environmental Services, Florida Game and Fresh Water Fish Commission, to Dale Twachtmann, Secretary, Department of Environmental Regulation.

Hayes v. Bowman. 1957. 91 So.2d 795.

Heinen, E. T. 1986. Letter from the Chief, Marine and Estuarine Branch, Water Management Division, Region IV, U.S. Environmental Protection Agency, to Col. Charles Myers, III, District Engineer, Jacksonville District, U.S. Army Corps of Engineers.

Hoogland, R. 1986. Letter from the Assistant Regional Director, Habitat Conservation Division, Southeast Regional Office, National Marine Fisheries Service, to Col. Charles Myers, III, District Engineer, Jacksonville District, U.S. Army Corps of Engineers.

Hopwood v. State. 1981. 402 So. 2d 1296.

Hough v. Marsh. 1982. 557 F. Supp. 74.

In re James DeVincenzo. 1988. Department of Environmental Regulation. File no. 520095553.

Justinian. 1987. *Institutes I*, 2.1.1.

Laws of Florida. 1957. Ch. 57-362. Codified at Fla. Stat. § 252.122.

Laws of Florida. 1967. Chs. 67-393, 436.

Laws of Florida. 1975. Ch. 75-22.

Leslie, M. 1990. Mitigation Policy. *Issues in Wetlands Protection: Background Papers Prepared for the National Wetlands Policy Forum.* The Conservation Foundation, Washington, D.C.

Lewis, R., III. 1989. Creation and restoration of coastal plain wetlands in Florida. Pages 73–101 in J. Kusler and M. Kentula. Eds. *Wetland Creation and Restoration: The Status of the Science.* Vol. I: Regional Reviews. EPA/600/3-89/038a,b. Environmental Research Laboratory, Corvallis, Oregon.

Mager, A., Jr. 1986. Letter from the Acting Assistant Regional Director, Habitat Conservation Division, Southeast Regional Office, National Marine Fisheries Service, to Col. Charles Myers, III, District Engineer, Jacksonville District, U.S. Army Corps of Engineers.

Mager, A., Jr. 1988. Letter from the Acting Assistant Regional Director, Habitat Conservation Division, Southeast Regional Office, National Marine Fisheries Service, to Col. Robert L. Herndon, District Engineer, Jacksonville District, U.S. Army Corps of Engineers.

Mager, A., Jr. 1989. Letter from the Acting Assistant Regional Director, Habitat Conservation Division, Southeast Regional Office, National Marine Fisheries, to Col. Robert L. Herndon, District Engineer, Jacksonville District, U.S. Army Corps of Engineers.

Magnuson Fisheries Conservation and Management Act of 1976. 1989. 16 U.S.C.A. §§ 1801-1882.

Maloney, F. and R. Ausness. 1975. The use and legal significance of the mean high water line in coastal boundary mapping. *N.C. L. Rev.* 53:185.

Maloney, F., S. Plager, and F. Baldwin, 1968. *Water Law and Administration: The Florida Experience.* University of Florida Presses, Gainesville.

Marine Mammal Protection Act. 16 U.S.C. §§ 1361-1407.

Marks v. Whitney. 1971. 491 P.2d 374.

Martin v. Waddell. 1842. 41 U.S. 367.

Marx, J. 1989. Interview. Environmental Administrator, Bureau of Submerged Lands Management, Division of State Lands, Florida Department of Natural Resources.

Memorandum for Record. 1987. Approved by Nancy Schwall, Chief, Gulf Coast Permits Section, Jacksonville District, U.S. Army Corps of Engineers.

Memorandum of Agreement Between the Environmental Protection Agency and the Department of the Army. 1990. The Determination of Mitigation Under the Clean Water Act, Section 404(b)(e), Guidelines, fn. 7. Published at 55 Fed. Reg. 9210.

Myer, J. 1989. Project Manager, Letter from the Intergovernmental Coordination and Review, Tampa Bay Regional Planning Council, to Constance Bersok, Florida Department of Environmental Regulation.

National Audubon Society v. Hartz Mountain Development Corp. 1983. 14 Envt'l. Law Reptr. 20724.

National Audubon Society v. Superior Court. 1983. 33 Cal.3d 419; 658 P.2d 709.

National Environmental Policy Act. 42 U.S.C. §§ 4321-4361.

Note. 1970. The public trust in tidal areas: a sometime submerged traditional doctrine. *Yale L. J.* 70:762–789.

Note. 1973. Florida's sovereignty submerged lands: what are they, who owns them and where is the boundary? *F.S.U. L. Rev.* 1:596.

N.R.D.C. v. Callaway. 1975. 392 F. Supp. 685.

Ocean Dumping Act. P.L. 92-532, codified at 33 U.S.C. §§ 1401-1445.

Phillips Petroleum Co. v. Mississippi. 1988. 108 S. Ct. 791; 98 L. Ed. 2d 877.

Pollard's Lessee v. Hagan. 1845. 44 U.S. 212.

Pollutant Spill Prevention and Control Act. 1989. Fla. Stat. §§ 376.011-.17 and .19-.21.

Power, G. 1977. The fox in the chicken coop: the regulatory program of the U.S. Army Corps of Engineers. *Va. L. Rev.* 63:503.

Proceedings. 1986. National Wetland Symposium: Mitigation of Impacts and Losses. New Orleans, Louisiana.

Report. 1970. *Our Waters and Wetlands: How the Corps of Engineers Can Prevent Their "Destruction and Pollution."* H.R. Rep. No. 91-917, 91st Cong., 2d Sess.

Routa, R. A. 1971. Letter from the Chief, Survey and Management, Department of Natural Resources, to Stanley C. Burnside, Clerk, Board of County Commissioners, Pasco County, Florida.

Senate Report. 1958. Number 1981. 85 Cong. 2d Sess. *U.S. Code Cong. & Admin. News* 1958:3446, 3448, 3450.

Shablowski v. DER. 1979. 370 So. 2d 50.

Shively v. Bowlby. 1894. 152 U.S. 1.

Shoreline Associates v. Marsh. 1983. 555 F. Supp. 169; aff'd, 725 F.2d 677 (4th Cir. 1984).

Sierra Club v. U.S. Army Corps of Engineers. 1985. 772 F.2d 1043.

Smallwood, M., S. Alderman, and M. Dix. 1985. The Warren S. Henderson Wetlands Protection Act of 1984: a primer. *Fla. St. Univ. J. Environ. Land Use Law* 1: 211–270.

State Comprehensive Plan. 1989. Fla. Stat. §§ 187.101-.201.

State v. Black River Phosphate Co. 1893. 13 So. 646.

Tampa Bay Regional Planning Council. 1986. Agenda Item No. 3. Approved by the Clearinghouse Review Committee.

Tampa Bay Regional Planning Council. 1989. Agenda Item no. 8.C.2.

Troxel, J. 1986. Letter from the Acting Field Supervisor, Division of Ecological Services, Panama City, Fla., Fish and Wildlife Service, U.S. Department of the Interior, to District Engineer, Jacksonville District, U.S. Army Corps of Engineers.

Trustees of the Internal Improvement Trust Fund. Undated, a. File no. 919-51-253.12.

Trustees of the Internal Improvement Trust Fund. Undated, b. File no. 1450-09-253.12.

Trustees of the Internal Improvement Trust Fund file no. 1450-09-253.12 minutes (Feb. 4, 1964).

U.S. Army Corps of Engineers. 1986. Permit no. 86IPT-20272.

U.S. Army Corps of Engineers. 1988. Permit no. 86IPE-20483.

U.S. Congress. 1984. *Wetlands: Their Use and Regulation.* Office of Technology Assessment. OTA-0-206.

U.S. v. Hanna. 1983. 639 F.2d 194; 19 E.R.C. 1068.

U.S. v. Holland. 1974. 373 F. Supp. 665.

U.S. v. Moretti. 1973. 478 F.2d 418.

U.S. v. Riverside Bayview, Inc. 1985. 106 S. Ct. 455.

U.S. v. Sexton Cove Estates. 1976. 526 F.2d 1293.

U.S. v. Stoeco Homes, Inc. 1973. 498 F.2d 297; *cert denied*, 420 U.S. 927.

U.S. v. Sunset Cove. 1973. 5 E.R.C. 1023; *remanded on other grounds*, 514 F2.d 1080.

U.S. v. Tull. 1985. 769 F. 2d 182..

U.S. v. Weisman. 1980. 489 F. Supp. 1332.

U.S. v. Whichard. 1974. 4 E.L.R. 20,819.

Wesley, D. 1986. Letter from the Field Supervisor, Endangered Species Field Stations, Jacksonville, Fla., Fish and Wildlife Service, to John Adams, Chief, Regulatory Division, Jacksonville District, U.S. Army Corps of Engineers.

White v. Hughes. 1939. 190 So. 446.

Woodburn, K. D. 1964. Letter from a biologist, Division of Salt Water Fisheries, Florida Board of Conservation, to Van Ferguson, Director, Trustees of the Internal Improvement Trust Fund.

Yonge v. Askew. 1974. 293 So. 2d 395.

Zabel v. Tabb. 1970. 430 F.2d 199; *cert denied*, 401 U. S. 910. 1971.

Management

12

Kenneth D. Haddad and Edwin A. Joyce, Jr.

The authors of the preceding chapters have shown that much is known about the physical, chemical, biological, and ecological aspects of Gulf coast intertidal marshes. We understand their importance to fisheries production, sediment stabilization, erosion control, and pollutant filtration. Still, we need to learn and understand more in order to ensure long-term, effective marsh management.

Most marine resource management strategies and actions in Florida have been reactive rather than preventive or corrective and have been oriented to single species. As technical data on the status and trends of our coastal and marine resources have become available, the inadequacy of this approach over the long term has become evident. Habitat has been lost, species abundances have declined, polluted waters have reduced areas of safe shellfish harvest, and fisheries have declined. This realization has stimulated the evolution of an ecosystem approach to resource management.

An ecosystem is a community of animals and plants and their physical environment, which interact as an ecological unit. For example, the estuarine ecosystem could be defined by watersheds, which would include not only the estuary but also the land and waters that drain into the estuary. The Gulf coast salt marshes should then be considered as a functioning unit within a specific estuary and its watershed and should be managed accordingly. An ecosystem approach is used when the components and processes of an ecosystem—such as species interactions, communities, community interactions, and cumulative environmental impacts (both natural and man induced)—are understood and managed in a manner that benefits the whole.

A fully conceptualized and implemented ecosystem approach to resource management has not been realized and is long term and evolutionary. However, the process can be initiated by evaluating and integrating management decisions within a defined ecosystem to reduce cumulative impacts. Northwest Florida would benefit from a commitment to develop and implement this approach.

Status of Marsh Management

Historically, marsh management has been driven by the socioeconomic needs of an area, and little attention has been paid to the scientific knowledge of marsh systems. Early marsh management practices in Florida were characterized by draining, dredging, and filling marshes. Marsh management has evolved to include a much more complex approach that attempts to balance the preservation and maintenance of marshes with the pressures of population growth. Managers, however, are still not taking full advantage of the advances in scientific understanding. Environmental warnings often go unheeded by Florida's resource managers until a problem is in its advanced stages and requires drastic remedial action.

A good, contemporary definition of marsh management is elusive. Actions taken in the name of marsh management have ranged from completely destroying to totally preserving marshes. Florida's Gulf coast marshes have experienced the full range of management options. Fortunately, the current perspective on salt marsh management, supported by rules and regulations, is to maintain or preserve marshes as a standard practice and to grant permits to destroy or alter marshes only when it is in the public interest (see Chapter 11). Although the salt marshes in northwest Florida are protected by layers of regulations, we must continue to actively manage these marshes in order to prevent large-scale losses, which often are the products of the accumulation of small losses resulting from what appear to be minor resource management decisions or from unregulated activities and indirect impacts. Active management, as opposed to "leave it alone" passive management, is needed to ensure maintenance and preservation. Contemporary marsh management is a complex and active process that includes:

1. Identifying the role (e.g., ecological, socioeconomic, recreational) of the marsh in a given geographically designated ecosystem. This role could vary by estuary or region depending, for example, on the health and areal distribution of the marsh, contribution by the marsh complex to maintenance of the ecosystem, and socioeconomic considerations.

2. Developing resource management plans customized to the designated ecosystem or region and in which management of the marshes is placed within the context of an overall management strategy for that ecosystem.

3. Encouraging the complex of management and regulatory authorities to network and communicate so that an accumulation of seemingly unrelated decisions does not impact the integrity of the ecosystem.

4. Implementing the management plan with active on-site monitoring. A management plan must include active on-site monitoring to ensure that the plan is meeting its objectives and that it is continually updated with new and better information and strategies.

Marsh management, as defined above, is not a simple process and has never been fully achieved because of the complex planning and implementation required. Northwest Florida salt marshes could benefit from such management. Otherwise, mistakes made in other areas of the state could be repeated.

Northwest Florida and Big Bend Intertidal Marshes

Salt marshes and submerged aquatic vegetation (e.g., seagrasses) are the significant marine habitat components of the ecosystem in northwest Florida. Knowing the areal extent and distribution of intertidal marshes is an important first step in managing them. An ecosystem approach necessitates a knowledge of the landscape; that knowledge also puts the significance of salt marshes to this low-energy coast into perspective.

The National Oceanic and Atmospheric Administration (NOAA) has compiled estimates of the distribution and abundance of coastal wetlands of the U.S. (Field et al., 1991) from U.S. Fish and Wildlife Service, National Wetlands Inventory maps (Tiner, 1984). Table 12.1 summarizes the abundance of

Table 12.1 **Acreage of Intertidal Salt Marshes within Each County of Florida's Northern Gulf Coast (Field et al., 1991)**

County	Salt marsh acreage	County	Salt marsh acreage
Bay	8,200	Levy	35,400
Citrus	27,400	Okaloosa	300
Dixie	20,300	Pasco	4,600
Escambia	1,300	Santa Rosa	6,500
Franklin	18,400	Taylor	21,800
Gulf	2,800	Wakulla	18,600
Hernando	9,200	Walton	2,700
Jefferson	3,800		
		Total acreage	181,300

salt marshes within each northern Gulf coast county. The salt marshes primarily
include *Juncus* but also include other herbaceous, salt-tolerant plants, such as
Spartina, *Batis*, and *Salicornia*. Levy County has the greatest abundance of salt
marsh, whereas Okaloosa County has the least.

The southern portion of the area, often referred to as the Big Bend (Pasco,
Taylor, Citrus, Dixie, Wakulla, Franklin, Levy, Hernando, and Jefferson counties) includes the greatest abundance of salt marshes. Salt marsh habitat is
clearly a major component of marine wetlands of the northern Gulf coast,
comprising approximately 181,000 acres. In fact, this acreage represents 50%
of the total salt marshes in Florida and 70% of the total along the Florida Gulf
coast (Field et al., 1991).

Fortunately, the intertidal marshes of this region are located some distance
from the urban centers of Florida. Thus, most of these resources are still intact
and presumably functioning. However, pressures associated with the population
growth that started farther south are moving into the area, and the natural
systems could be eliminated or drastically altered if they are not carefully
managed. Figure 12.1 shows the type of development that has occurred in the
salt marshes of the Big Bend area. Salt marsh habitat is impacted when development occurs directly on the waterfront along many areas of the Gulf or
adjacent estuaries. Due to the characteristically low relief of the area and the
resultant abundance of salt marshes, waterfront properties cannot be developed
without dredging and filling the marshes.

Charlotte Harbor Management: An Ecosystem Perspective

Analysis of habitat changes in Charlotte Harbor, located in Lee and Charlotte
counties in southwest Florida, provides a perspective on how strides in managing estuaries have been made. The analysis also shows that estuarine management must not be confined to the estuary itself but should be broadened to the
ecosystem. The Charlotte Harbor estuary has been managed since the 1970s,
prior to large population increases, and as a result, wetlands buffers along the
shoreline, to a large extent, have been maintained.

Haddad and Hoffman (1986) assessed land use alterations in the Charlotte
Harbor region between the 1940s and 1980s. Some of the results of this analysis are presented in Table 12.2. Four observations can be made that are relevant
to the enhancement of long-term management of marine resources of northwest
Florida:

1. Mangrove acreage actually increased by 10%, a gain of 5107 acres. In
   the late 1960s, the Charlotte Harbor area was the focus of effective state,

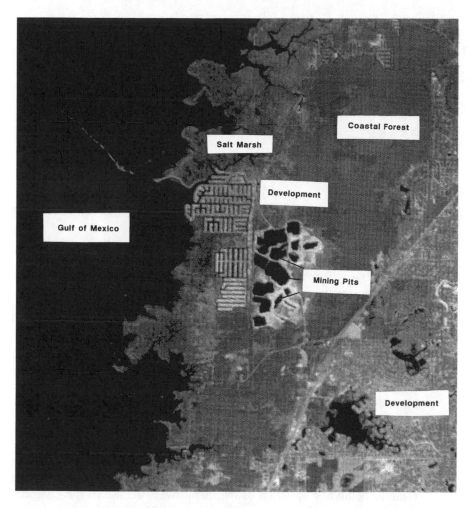

Figure 12.1   Landsat satellite image depicting the community of Hernando Beach, Florida.

regional, and local planning. This planning provided for the acquisition, through purchases, mitigation, and donation, of a buffer zone of wetlands fringing the harbor. This marsh and mangrove habitat is now maintained as a functioning part of the estuarine system. Because mangroves were protected, very few trees were removed during property development. The harbor was not being managed to increase mangrove habitat, but to protect it. Factors such as recruitment to mud flats, sea level rise, and altered freshwater flow may explain the mangrove in-

Table 12.2   Historic and Recent Acreage in Charlotte Harbor for Various Categories of Land Use and Vegetation

Land use or vegetation category	1945	1982	% change
Urban	3,710	96,105	+2490
Rangeland	106,219	20,704	−81
Salt marsh	7,251	3,547	−51
Seagrasses	82,959	58,495	−29
Mangroves	51,524	56,631	+10

From Haddad and Hoffman, 1986.

creases, but without the management strategy, loss would have far exceeded any gains.

2. While the fringing shoreline of the harbor was being protected and managed, the uplands were undergoing massive urban conversion. Urban area, for example, increased by 2490% while undeveloped rangeland decreased by 81%. This development caused large-scale modification of the estuarine drainage area surrounding Charlotte Harbor and the introduction of extensive canals (Figure 12.2) and drainage systems emptying into the estuary.

3. Salt marshes covered a much smaller acreage than mangroves and incurred a considerable loss (51%), amounting to thousands of acres. In Charlotte Harbor, salt marsh is typically found in the lower salinity regimes behind or upstream of the mangroves because of the competitiveness of mangroves in more saline waters. This stratification put salt marsh within the reach of development while maintaining the fringing mangroves. Additional loss occurred when mangroves encroached upon the salt marsh and may be attributed to the modified drainage system that reduced freshwater sheet flow into the marshes. In addition, sea level within this region rose approximately 1.2 mm/year, or 4.8 cm, between 1940 and 1980 (National Research Council, 1987). In a low-relief salt marsh and mangrove system, this type of rise could have created conditions (e.g., salinity, water depth) that favored mangroves.

4. The apparent success in the management of mangroves is tempered by a 29% decline in seagrass, resulting in a loss of 24,464 acres. Although some loss was due to direct mechanical impact from dredging and fill-

Figure 12.2 Satellite image of Cape Coral, Florida, depicting canals and upland alteration (dark area is water, light area is development, marsh is intermediate shade). Image licensed by EOSAT Corporation.

ing, much of the loss was simply conversion to bare bottom. Losses often occurred in deeper areas, indicating that lower light levels within the water column could be influencing growth patterns of the seagrasses. Increases in nutrients, which promote phytoplankton growth, and increases in resuspendable fine organic and clay sediments may explain reduced water clarity, but this has not been documented. Suspected contributors to seagrass loss include sediment resuspension from dredging, nutrient and sediment loading from the extensive canals that serve as sinks until a large rainstorm occurs, a drainage system that bypasses the potential filtration capabilities of the existing marshes and mangroves, and loss of the salt marshes and uplands that act as natural filters of water entering the harbor.

We have learned from the Charlotte Harbor experience that an entire estuary cannot be managed by protecting only one component; all elements of the estuary, and the watershed that influences it, must be included in a management strategy. The development of a wetlands buffer zone in Charlotte Harbor has been a success for mangroves, but the loss of seagrass suggests a failure in managing the entire area as a system. The long-term degradation of the Charlotte Harbor estuary is a reality, but we would not realize it if our measure of success was based only on the success of mangroves.

Good management of the salt marshes of the Florida Gulf coast does not ensure that the ecosystem of which they are part will not be degraded by changes in land and water uses. As development pressure in the region increases, this must be kept in mind, and the Charlotte Harbor scenario must not be repeated.

Ecosystem Management: Fisheries Implications

When managing an ecosystem, perturbations to component habitats and the cumulative effects of those perturbations need to be measured. For example, loss of wetlands on a case-by-case basis may be considered to have minor impact. However, when one considers that loss of wetlands causes loss of fisheries habitat and loss of fish, the full significance of the depletion of wetlands can be realized. The following study provides a perspective on habitat loss and the potential impact to fisheries. Like the Charlotte Harbor example, it elucidates important considerations for long-term management of our northern Gulf marshes.

Tampa Bay is one of the most impacted estuaries in Florida. Lewis et al. (1985) estimated that 44% of the mangroves and salt marshes and 81% of the seagrasses have been lost in the Tampa Bay area since the late 1800s. Haddad (1989) estimated that Tampa Bay lost over 50% of its seagrasses and 30% of its salt marshes between 1950 and 1982. During the same period, the region experienced a 192% increase in urban area and a 78% increase in agricultural lands.

Significant loss of fishery habitats has occurred in the Tampa Bay area as a result of dredging and filling, with degradation of water quality being implicated in much of the seagrass loss (Lewis et al., 1985). Figures 12.3 and 12.4, which show Boca Ciega Bay in 1950 and 1982, depict the most dramatic example of the impact of development in Tampa Bay. Over 90% of the fisheries habitat was lost to massive fill areas for development of residential and commercial properties.

Because the Tampa Bay area has experienced such large increases in urban and agricultural development, which contribute to pollutant loading and decreases in ecologically important natural habitats such as the salt marsh, we should not be surprised if changes occur to the species that depend on the estuary.

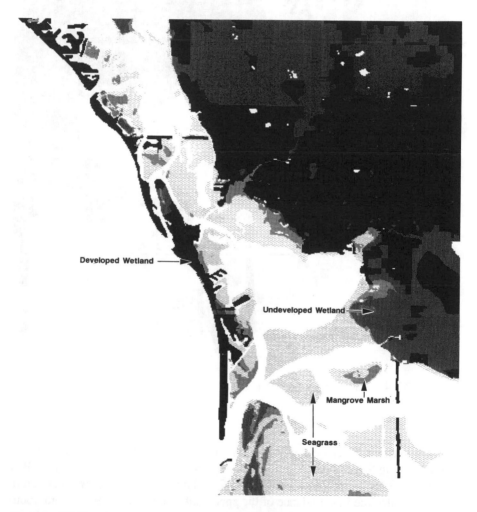

Figure 12.3   1950 distribution of marsh/mangrove and seagrass habitats in Boca Ciega Bay, Florida.

Seagrass

Developed Upland

Mangrove/Marsh

Figure 12.4 1982 distribution of marsh/mangrove and seagrass habitats in Boca Ciega Bay, Florida.

One way to measure change is by analyzing trends in commercial fish landings. Landings of spotted seatrout (*Cynoscion nebulosus*) in Tampa Bay since the 1950s are presented in Figure 12.5. Trends in catch show an apparent significant decline in abundance of the species. Conclusive scientific data documenting the apparent seatrout decline do not exist; however, this species is

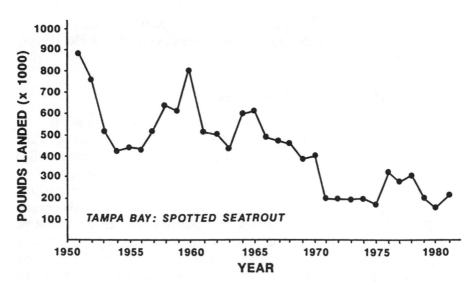

Figure 12.5 Landings of spotted seatrout in Tampa Bay. (After Haddad, 1989).

highly estuarine dependent, and any changes in the estuary may impact the population. This assumption of lowered abundance is supported by existing knowledge of juvenile spotted seatrout in the Tampa Bay system. McMichael and Peters (1989) found that seagrass meadows in Tampa Bay appear to be an important nursery ground for juvenile seatrout. Of the juveniles collected in their study, 78% were found in seagrass, even though less than 40% of the collections were made in that habitat. Seatrout are nonmigratory, spending their entire life cycle in a given estuary, and, thus, the Tampa Bay population is likely to reproduce and support itself with minimal external recruitment. Although numerous factors control the spotted seatrout population, a loss of 50% of the seagrasses in Tampa Bay since the 1950s could be expected to affect landings.

Even though the decline of spotted seatrout may be linked to seagrass loss, the concurrent loss of salt marsh and mangrove has likely exacerbated the loss of seagrass by the loss of their water filtration and nutrient assimilation functions. This may have contributed to the water quality degradation of the estuary and subsequent seagrass loss.

Although this Tampa Bay study focused on seagrass and fisheries, rather than salt marsh and fisheries, it is an example of what may happen when fisheries habitat is degraded. Fortunately, we have no such examples in northwest Florida relative to salt marsh. However, Turner (1977) has correlated the extent of the marsh/water interface and freshwater flow with shrimp production in Louisiana. This implies that a reduction of marsh in northwest Florida may impact shrimp production.

Habitat and Fisheries Management

Spotted seatrout are representative of the many commercial and recreational species caught in northern Gulf estuaries. Over 70% of the commercial and recreational species caught in Florida utilize the estuaries during some portion of their life cycles (Haddad and Hoffman, 1986). Significant declines in the Tampa Bay fisheries habitat have occurred, and concurrent significant declines in spotted seatrout landings have been observed. Although a conclusive cause-and-effect relationship between habitat loss and fisheries decline cannot be developed, the correlative aspects of the information cannot be ignored by managers. This may be considered an unbiased correlation by the objective scientist, and, in fact, it is the scientist's job to provide factual and valid scientific information to the manager and to ensure that the manager is aware of the scientific value and limitations of the information. However, the resource manager's job is to evaluate all of the information available and to make the best decision possible. Management decisions and strategies are often based on a combination of science, socioeconomic factors, and politics. Ideally, the decisions and strategies are conservative toward the protection of the resource, but in reality that is not always the case.

From a resource manager's perspective, the conclusion that loss of fisheries habitat impacts the potential population of fishery species is valid. It is also fair to conclude that a primary socioeconomic reason for maintaining the health and abundance of the salt marshes of the northern Florida Gulf coast is for long-term maintenance of the fishery resources. Because salt marshes are the primary intertidal fisheries habitat in this region, their management and maintenance are critical.

Current Salt Marsh Management Practices

All of the impacts to salt marshes and the various components of other Florida estuaries (e.g., Tampa Bay and Charlotte Harbor) have also occurred on the northern Florida Gulf coast, but on a smaller scale. The opportunity exists to take into account what has been learned in those other areas. A number of marsh management practices that either are occurring or may occur in the future can have detrimental or beneficial impacts, depending on how they are applied. They include the designation of natural resource management areas, marsh restoration activities, mosquito control techniques, and aquaculture.

Special Management Areas

Resource managers generally accept the premise that preservation is the most economical and effective form of wetlands management. Florida has recognized the need for special management of marine, estuarine, and freshwater habitats by developing the Aquatic Preserve Program within the Florida Department of Environmental Protection, Division of Marine Resources. Aquatic preserves are designed for managing certain submerged lands owned by the state. Salt marshes are typically included within aquatic preserves because they occupy the intertidal zone below the mean high water boundary of state sovereignty lands.

For northwest Florida, the St. Joseph Bay Aquatic Preserve Management Plan details the authority and purpose of aquatic preserves through Florida Statute (F.S.) and Florida Administrative Code (F.A.C.) (Florida Department of Natural Resources, 1987).

> Chapter 18-20, F.A.C., addresses the aquatic preserves specifically and derives its authority from Sections 258.35, 258.36, 258.37 and 258.38, F.S. The general rules in Chapter 18-20, F.A.C., are supplemental to the rules in Chapter 18-21, F.A.C., in the regulation of activities in the aquatic preserves. The intent of this chapter is found in Section 18-20.001, F.A.C., which states:
>
> (1) All sovereignty lands within a preserve shall be managed primarily for the maintenance of essentially natural conditions, the propagation of fish and wildlife, and public recreation, including hunting and fishing where deemed appropriate by the board of the managing agency.
>
> (2) The aquatic preserves which are described in Section 258.39, 258.391, 258.392 and 258.393, F.S., Chapter 85-345, Laws of Florida, and in Section 18-20.02, F.A.C., were established for the purpose of being preserved in an essentially natural or existing condition so that their aesthetic, biological and scientific values may endure for the enjoyment of future generations.
>
> (3) The preserves shall be administered and managed in accordance with the following goals:
>
> > (a) to preserve, protect, and enhance these exceptional areas of sovereignty submerged lands by reasonable regulation of human activity within the preserves through the development and implementation of a comprehensive management program;

(b) to protect and enhance the waters of the preserves so that the public may continue to enjoy the traditional recreational uses of those waters such as swimming, boating, and fishing;

(c) to coordinate with federal, state, and local agencies to aid in carrying out the intent of the Legislature in creating the preserves;

(d) to use applicable federal, state, and local management programs, which are compatible with the intent and provisions of the act and these rules, and to assist in managing the preserves;

(e) to encourage the protection, enhancement, or restoration of the biological, aesthetic, or scientific values of the preserves, including but not limited to the modification of existing manmade conditions toward their natural condition, and discourage activities which would degrade the aesthetic, biological, or scientific values, or the quality, or utility of a preserve, when reviewing applications, or when developing and implementing management plans for the preserves;

(f) to preserve, promote, and utilize indigenous life forms and habitats, including but not limited to: sponges, soft coral, hard corals, submerged grasses, mangroves, salt water marshes, fresh water marshes, mudflats, estuarine, aquatic and marine reptiles, game and non-game fish species, estuarine, aquatic, and marine invertebrates, estuarine, aquatic, and marine mammals, birds, shellfish and mollusks;

(g) to acquire additional title interests in lands wherever such acquisitions would serve to protect or enhance the biological, aesthetic, or scientific values of the preserve;

(h) to maintain those beneficial hydrologic and biologic functions, the benefits of which accrue to the public at large.

The above authority and purpose make it clear that aquatic preserves are to preserve, protect, and enhance the included sovereign submerged lands, but the authority does not rule out reasonable regulation and utilization of the resources. This is a case of resource managers being given a tool that can be subject to broad interpretation during its use. The success of the Aquatic Preserve Program can be measured only in time.

TABLE 12.3   Salt Marsh Acreage in Aquatic Preserves of the Northern
Florida Gulf Coast

Aquatic preserve	Salt marsh acreage[a]
Alligator Harbor	302
Apalachicola Bay	7,177
Big Bend seagrasses	85,070
St. Joseph Bay	290
St. Andrews Bay	147
St. Martins marsh	9,202
Ft. Pickens	N/A
Rock Bayou	N/A
Total acreage	102,188

[a] N/A = not available.

The Aquatic Preserve Program preserves and protects salt marshes in a large portion of the northern Florida Gulf coast. There are eight aquatic preserves in the region which include over 102,000 acres of salt marshes (Table 12.3) and represent over 56% of the total salt marsh acreage of the region.

Federal, state, and local parks and reserves also provide protection to salt marshes. Additional special-area management in the region includes the designation of Areas of Critical State Concern (Florida Statute 380) and the designation by water management districts of Priority Water Bodies through authority of the Surface Water Improvement and Management Act. Both of these designations favorably affect the preservation, restoration, and enhancement of salt marshes.

Special management areas provide the tools to manage and preserve a large portion of the salt marshes. The challenge to the resource manager is to integrate all of these tools in order to facilitate effective management from an ecosystem perspective.

Marsh Restoration, Enhancement, Creation, and Mitigation

One important component of marsh management, the planting of salt marsh vegetation, can be approached from several different management perspectives. It is important to define terms when discussing salt marsh restoration, creation, enhancement, and mitigation. Lewis (1990) proposes the following standard definitions:

Restoration: returned from a disturbed or totally altered condition to a previously existing natural or altered condition by some action of man; refers to the return to a pre-existing condition.

Creation: the conversion of a persistent non-wetlands area into wetlands through some activity of man.

Enhancement: the increase in one or more values of all or a portion of existing wetlands by man's activities, often with the accompanying decline in other wetlands values; the intentional alteration of existing wetlands to provide conditions that did not previously exist and that, by consensus, increase one or more values.

Mitigation: the actual restoration, creation, or enhancement of wetlands to compensate for permitted wetlands losses.

Marsh planting projects are not new to the northern Florida Gulf coast. They have generally been coupled with mitigation of permitted habitat destruction or small, independently sponsored projects. In recent years, the public has become more aware that marsh planting may be a viable approach for enhancing the functions of our estuaries, and public monies have been directed to salt marsh planting projects. However, no systematic approach has been taken to monitor and evaluate the results of salt marsh planting nor have any comprehensive plans been developed for a systematic approach to marsh planting (Crewz and Lewis, 1991). The presence of a healthy, growing, planted marsh plant does not mean that the intended habitat functions or that other optimal functional aspects of a marsh are being performed.

The Florida Department of Environmental Protection, water management districts, and others have been conducting and sponsoring marsh planting projects designed to make significant contributions toward our understanding of the dynamics of resource recovery. Success is not guaranteed in restoring, creating, or enhancing natural vegetated habitats (Crewz and Lewis, 1991). Factors controlling planting success may be site specific and may vary with planting-stock sources, handling, and physical setting.

The principal purpose of salt marsh restoration and creation programs should be to approximate the complex functions of a natural habitat, which presumably begins with revegetation. An understanding of the utilization of planted marshes by fisheries organisms, the function of associated flora and fauna, and the optimization of techniques to engineer marsh planting will provide a perspective on the value of planted versus natural environments. This information is essential to the long-term management of our coastal resources and marine fisheries.

From a management perspective, it is critically important to consider all aspects of salt marsh planting for this region and not initiate expensive large-scale or many small-scale planting efforts based on the vague and unsupported assumption that any plant in the ground will enhance the estuary. To make large-scale marsh restoration, enhancement, creation, or mitigation more effective as management tools, planting techniques, factors affecting survival of plantings, and functions of the habitat must be understood. Unfortunately, funds are more easily made available for planting, and the needed applied research is often overlooked and underfunded.

Assuming we can understand and optimize successful marsh planting techniques, the horticultural and functional success of individual plantings must be balanced with an understanding of the ecological function of salt marshes in an estuary, and that function varies among estuaries. These factors are often overlooked by those seeking solutions to complex environmental problems. It may be easy and convenient to plant a salt marsh in an estuary, regardless of whether the estuary will function any better as an ecosystem. Proper management of the system, on the other hand, may lead to the commitment of resources to reducing nutrient inputs as a first priority rather than to planting marshes.

Mosquito Control

One of the earliest and undoubtedly one of the most extensive and damaging marsh management activities on Florida's Gulf coast was initiated for the purpose of mosquito control. Termed "source control," the most damaging methods were ditching and impounding. Both methods eliminated habitat necessary for mosquito reproduction. Currently, these source-control techniques are not frequently used in this region, but as the population grows, control of mosquito populations will become more of an issue.

Ditching is the most widespread method of source control. Ditching simply drains the small tidal pools and puddles of a marsh that are required for the 5-day aquatic larval stage of the saltwater mosquito. Thousands of miles of dragline ditches were dug, crisscrossing virtually every major tidal marsh and wetlands area in much of Florida. The effects were often devastating; not only were ditches created, but the spoil taken out of each ditch by the dragline became a dike that impeded the sheet flow and changed the drainage patterns responsible for creating the marsh in the first place. The rotary ditcher, a later invention, has been hailed as the solution to the problems caused by ditching, because it does not leave a spoil dike. A rotary ditch may be as much as 4 ft deep and 5 ft across, but the spoil is thrown out over the surface of the marsh in a fine layer seldom more than 0.75 in. deep. This is certainly an improvement over the dikes that are several feet high, but in some instances, even this fine layer of spoil may be enough to cause changes in the marsh. Regardless of the technique, ditching for mosquito control has a negative effect on the marsh.

Whereas ditching eliminates the water necessary for larval development, impounding eliminates the dry or damp surface required by the female for egg deposition. The major salt marsh mosquito does not lay its eggs on or in the water; it lays them on the dry or damp surfaces of low areas. The eggs remain there until rain or tide inundates them, and hatching takes place shortly thereafter. In some areas of Florida, entire marsh areas, some many hundreds of acres in size, were dammed and flooded. This technique reduces biological production of the marsh system: marsh vegetation and mangroves die from the effects of constant submersion, and the estuarine-dependent larval and juvenile forms of marine species no longer have unlimited access to the area they need for survival.

Impounding has not been a major form of mosquito control on the Gulf coast. Some impounding of marsh habitat has occurred in northwest Florida, but not for mosquito management. Impounded marshes are maintained for migrating waterfowl in the St. Marks Wildlife Refuge.

Both impoundments and ditches are frequently sprayed or otherwise treated with a wide variety of pesticides. In the past, massive amounts of DDT and Paris Green were used, and even today, mosquito control districts use some of the most toxic organophosphates developed by man. Such treatments have led to striking contradictions in resource management. For example, Florida has one of the strongest oil spill regulations in the country, yet in one year, one mosquito control district sprayed its wetlands in one county with over 150,000 gal of diesel fuel, which was just the carrier for the pesticides. Recent regulations and guidelines requiring that certain standards be met prior to spraying are helping to relieve some of these damaging practices.

As the population of northwest Florida increases, there will be pressure to control mosquitoes at a much greater level than currently exists. Careful planning and design of mosquito control programs balanced with maintenance of the ecosystem will be necessary.

Aquaculture

Aquaculture of marine species has seldom been undertaken in Florida, particularly in the natural environment. Oysters and clams have been successfully cultured for decades, and the culture of these species can be expected to increase in the future. In addition, culture of marine fishes and other marine invertebrates can be expected to increase as the economics and technologies improve. Placement of aquaculture activities in the natural environment will become a larger issue than it is now as the requests for use of public submerged lands increase. Florida aquaculture laws were initially developed to address a large shrimp culture operation in a northern Gulf coast bay. Some 2500 acres of marsh and open water was isolated from the rest of the bay by over 3 mi of barrier nets. Further, approximately 600 acres of salt marsh was diked and impounded for intensive culture. After some 15 years of operation, this enterprise was abandoned. Several breaches were made to the impoundment dikes, restoring tidal exchange. The marsh appears to be re-establishing itself as an integral component of the bay system. Any form of aquaculture that reduces the natural functions of important marine habitat should not again be an option for northwest Florida.

The Future of Salt Marsh Management

The salt marshes of northwest Florida are abundant, covering over 180,000 acres. Unfortunately, this seemingly high number can be misinterpreted by specific interests. There is often the temptation to look at this number and conclude that salt marsh loss can be accrued with little measurable impact on such a vast amount of wetlands. The perceived natural value of a small piece of marsh is then so diminished that the management process is affected. At this point, science, education, misuse of information, lack of information, and politics all blend together to form the final management strategy, often at the expense of the resource.

We can no longer target for management a single species or single habitat, such as salt marsh, without understanding its role in the system and managing for that role. Unfortunately, the financial resources necessary to adequately quantify the ecological roles of species and habitats that occur in our estuaries have not been provided in the past, and it is unlikely that they will be provided in the near future. A basic conflict exists between the science and the politics of resource management; management requires short-term responses to immediate problems that require long-term research for adequate response. Long-term research often does not provide the short-term information frequently perceived as necessary in budgetary allocation issues. The result has been a disdain by politicians, interest groups, and some managers toward the gathering of information necessary to make effective decisions or monitor the effectiveness of decisions (e.g., spend money and effort to plant marshes, and do not spend any money to find out if they live, die, or contribute to the fishery or the estuary). A shift in priorities and attitudes by political, resource, and fiscal managers is required to remedy this conflict.

Resource managers are beginning to explore the concepts of ecosystem management, and they realize that an ecosystem approach to management is complex and requires data collection, environmental monitoring, and continuing research. New techniques for information management, analyses, and reporting are needed in order to fully benefit from this more holistic approach. In fact, effective information management, analyses, and utilization must form the foundation on which to build an ecosystem approach to resource management. One of the greatest failures in resource management has been the inability to effectively synthesize and translate scientific information into management information. This is the fault of both scientists, who often lack the ability or interest to provide any information beyond their specific expertise, and managers, who pay little attention to developing an organized process for receiving the information required to meet their needs. The result is that managers have become accustomed to making decisions without adequate information. This cannot be tolerated if an ecosystem approach to management is to be successful. New tools are available which, if properly implemented, could facilitate more informed marsh management. Based on present trends in information technology, it appears that computer networking and Geographic Information System (GIS) technologies will become the foundation for processing resource management information. Networking will bring information from a variety of sources, and GIS will bring a new and powerful tool that can integrate and analyze disparate resource data and display that information in a form easily interpreted by managers. Careful attention must be given to implementation of these tools, or ecosystem management will not be successful.

Conclusions

Effective management of northwest Florida salt marshes will provide long-term benefits to the natural resources and the economic value of the region. Several management activities and issues have been discussed in this chapter, but the coverage has certainly not been comprehensive. Issues such as thermal effluent, freshwater flow reductions, water circulation changes, and pollution impacts have not been addressed, although they are important. The intent has not been to expound on every issue but rather to provide a perspective on marsh management within the context of the ecosystem. Because the salt marshes on the northern Gulf coast of Florida have been minimally impacted relative to habitats in other areas of the state, an ecosystem approach to their management is the only viable way to assure that salt marshes are successful in their role in maintaining a functioning estuary. Otherwise, the estuaries of the region that have apparently healthy ecosystem components (i.e., salt marshes) might become so degraded or altered that the existence of a single constituent is inconsequential. The northern Florida Gulf coast is certainly a candidate for developing and implementing new strategies in resource management.

References

Crewz, D. W. and R. R. Lewis, III. 1991. *An Evaluation of Historical Attempts to Establish Emergent Vegetation in Marine Wetlands in Florida.* Florida Sea Grant College, University of Florida, Gainesville.

Field, I., A. Reyer, P. Genovese, and B. Shearer. 1991. *Coastal Wetlands of the United States.* National Oceanic and Atmospheric Administration, Strategic Assessment Branch, Rockville, Maryland. 59 pp.

Florida Department of Natural Resources, 1987. *St. Joseph Bay Aquatic Preserve Management Plan, 1987.* Florida Department of Natural Resources, Tallahassee. 127 pp.

Haddad, K. D. 1989. Fisheries and habitat trends in Tampa Bay. Pages 113–128 in *Tampa and Sarasota Bays: Issues, Resources, Status, and Management.* NOAA Estuary of the Month. Seminar Series No. 11. U.S. Department of Commerce, Washington, D.C.

Haddad, K. D. and B. A. Hoffman. 1986. Charlotte Harbor habitat assessment. Pages 175–192 in E. D. Estevez, J. Miller, J. Morris, and R. Hammon, Eds. *Proceedings of the Conference: Managing Cumulative Effect in Florida Wetlands.* New College Environmental Studies Program Publication No. 37. Omni Press, Madison, Wisconsin. 326 pp.

Lewis, R. L. 1990. Wetland restoration/creation/enhancement terminology: suggestions for standardization. Pages 417–422 in J. A. Kusler and M. E. Kentula, Eds. *Wetland Creation and Restoration: The Status of the Science.* Island Press, Washington, D.C.

Lewis, R. R., III, M. J. Durako, M. D. Moffler, and R. C. Phillips. 1985. Seagrass meadows of Tampa Bay. Pages 210–246 in S. Treat, J. Simon, R. Lewis, III, and R. Whitman, Jr., Eds. *Proceedings of the Tampa Bay Area Scientific Information Symposium.* Florida Sea Grant College Report No. 65. University of Florida, Gainesville.

McMichael, R., Jr. and K. Peters. 1989. Early life history of spotted seatrout *Cynoscion nebulosus* (Pisces: Sciaenidae), in Tampa Bay, Florida. *Estuaries* 12(2):98–110.

National Research Council. 1987. *Responding to Changes in Sea Level: Engineering Implication.* National Academy Press, Washington, D.C.

Tiner, R. W., Jr. 1984. *Wetlands of the United States: Current Status and Recent Trends.* National Wetlands Inventory. U.S. Fish and Wildlife Service, Washington, D.C. 59 pp.

Turner, R. E. 1977. Intertidal vegetation and commercial yield of Penaeid shrimp. *Trans. Am. Fish. Soc.* 106(5) 411–416.

Terrestrial Vertebrates of Florida's Gulf Coast Tidal Marshes

A

Michael D. Hubbard and C.S. Gidden

There are a small number of terrestrial vertebrate species which either depend directly upon the tidal salt marshes of Florida's northern Gulf coast for habitat or food or are common visitors to the marshes. This group includes approximately 2 species of amphibians, 9 species of reptiles, 15 species of birds, and 12 species of mammals. These species are discussed briefly here, along with notes on their distribution, biology, and relationships with the marsh.

The terrestrial vertebrate species which depend directly upon the tidal salt marshes of Florida's northern Gulf coast for habitat or food are not numerous, although there are several species that are greatly dependent upon this ecosystem for survival.

A larger number of terrestrial vertebrate species are common as incidental visitors or passersby in these marshes. The number of vertebrate species feeding or living in the marshes increases in an inland direction as the marshes gradually change from tidal salt marshes to brackish and then freshwater tidal marshes.

Because of the generally dense nature of the saline/brackish marsh vegetation in this area (black needlerush [*Juncus roemerianus* Scheele] and smooth cordgrass [*Spartina alterniflora* Loisel] in higher saline areas and sawgrass [*Cladium jamaicense* Crantz.] and cattail [*Typha domingensis* Pers.] in lower saline portions), feeding by most marsh and wading birds is not possible, although an apparently abundant food supply is present. Activities such as burning or deposition of mats of dead vegetation by storm tides, which result in openings in the dense marsh, allow feeding by a wide variety of herons, egrets, ibis, and larger shorebirds. Feeding in these species is generally restricted to tidal creeks, shallow water, and mud flats outside the marsh proper at low tide. In addition, a wide variety of marsh and water birds, shorebirds, gulls, terns, and allied species depend on sand barrens and salt spikegrass (*Distichlis spicata* [L.] Greene) flats for roosting, loafing, and feeding sites at high tide. Wolfe et al. (1988) report that more than 60 species of birds use habitats within Florida Panhandle salt marshes in some manner.

The more common terrestrial vertebrate species which may be found in Florida's northern Gulf coast tidal marshes are listed below. Further information about most of these species can be found in the useful books by Longstreet et al. (1931), Carr and Goin (1955), Golley (1962), Stevenson (1976), and Ashton and Ashton (1988) and references therein. Because of the fluctuating situation in regard to nomenclature of North American vertebrates, scientific and common names listed here generally follow the use of Stevenson (1976) for the amphibians, reptiles, and mammals and that of the American Ornithologists' Union (1983) for the birds.

Little is known about the biology of most animals in the tidal salt marsh ecosystem and their specific adaptations to the saline conditions found there.

List of Terrestrial Vertebrates of Northern Florida Gulf Coast Tidal Marshes

Amphibians

Green Tree Frog
Hyla cinerea (Schneider)
The green tree frog feeds and lives in tidal brackish and freshwater marshes of Florida's Gulf coast.

Leopard Frog
Rana pipiens Schreber
The southern leopard frog feeds and breeds in tidal brackish and freshwater marshes of Florida's Gulf coast.

Reptiles

Alligator
Alligator mississippiensis (Daudin)
Alligators often pass through the tidal salt and brackish marshes and the freshwater marshes on their way from one body of water to another. They frequently feed in the tidal channels of the tidal salt marshes and sun along the banks.

Alabama Red-Bellied Turtle
Chrysemys alabamensis Baur
Stout (1984) reported the Alabama red-bellied turtle as occurring in the tidal salt marshes of western Florida.

Diamondback Terrapin
Malaclemys terrapin (Schoepff)
The two northern Florida subspecies of the diamondback terrapin, the Florida diamondback terrapin (*Malaclemys terrapin macrospilota* [W. P. Hay]) and the Mississippi diamondback terrapin (*Malaclemys terrapin pileata* [Weed]) nest on the beaches of the tidal salt marshes and are often found feeding in the tidal channels and flooded marshes at high tide.

Florida Cooter
Chrysemys floridana floridana (Le Conte)
Stout (1984) reported the Florida cooter as occurring in the tidal salt marshes of western Florida.

Banded Water Snake
Nerodia fasciata pictiventris (Cope)
The banded water snake feeds principally on fish and amphibians and lives in tidal brackish and freshwater marshes of Florida's Gulf coast. It is extremely pugnacious and, although nonpoisonous, can bite savagely.

Cottonmouth Moccasin
Agkistrodon piscivorus conanti Gloyd
The venomous cottonmouth moccasin feeds and lives in the tidal brackish and freshwater marshes of Florida's Gulf coast. It eats almost any animal small enough for it to swallow.

Gulf Salt Marsh Snake
Nerodia fasciata clarkii (Baird & Girard)
The gulf salt marsh snake (Clark's water snake) feeds and lives in tidal brackish and freshwater marshes of Florida's Gulf coast. This snake seldom enters freshwater habitats and can be considered the only true saline/brackish water tidal marsh snake along the Gulf coast. It feeds primarily on small fish, occasionally taking crabs or other small invertebrates.

Rough Green Snake
Opheodrys aestivus (L.)
The rough green snake feeds and lives in tidal brackish and freshwater marshes of Florida's Gulf coast. It feeds on soft-bodied invertebrates.

Green Anole (Carolina Anole)
Anolis carolinensis Voigt
The green anole feeds and lives in tidal brackish and freshwater marshes of Florida's Gulf coast.

Birds

Black-Crowned Night Heron
Nycticorax nycticorax (L.)
The black-crowned night heron has been observed nesting in tidal marsh on Smith and Palmetto Islands off Spring Creek, Wakulla County, Florida. Active nests have been located in stands of black needlerush (*Juncus roemerianus*) and rank growths of salt meadow cordgrass (*Spartina patens*).

Green-Backed Heron
Butorides striatus (L.)
The green-backed heron, one of the smallest of Florida's herons (slightly larger than the least bittern), feeds and nests in tidal brackish and freshwater marshes of Florida's Gulf coast. Its nest is made wholly of long, dried sticks and may hold as many as six eggs. It feeds upon small fish, tadpoles, aquatic and terrestrial insects, and sometimes even small mammals and snakes.

Least Bittern
Ixobrychus exilis (Gmelin)
The least bittern, the smallest of Florida's herons, feeds and nests in tidal brackish and freshwater marshes of Florida's Gulf coast. It is especially common in the summer.

Snowy Egret
Egretta thula (Molina)
The snowy egret has been observed nesting in black needlerush (*Juncus roemerianus*) on Smith and Palmetto Islands off Spring Creek, Wakulla County, Florida. It is a highly gregarious bird, especially at mating time, and makes a flimsy nest in trees, often with several nesting close together. It feeds upon small terrestrial and aquatic animals.

Tricolored Heron (Louisiana Heron)
Egretta tricolor (Müller)
The tricolored heron has been observed nesting in tidal marsh on Smith and Palmetto Islands off Spring Creek, Florida. It feeds mostly in shallow waters, although grasshoppers also form an important part of its diet.

Northern Harrier
Circus cyaneus (L.)
The northern harrier (marsh hawk) is frequently observed soaring low over tidal marshes during the fall and winter in search of small mammals and other prey.

Clapper Rail
Rallus longirostris Boddaert
A shy bird, more often heard than seen, the clapper rail feeds and nests in the tidal salt marshes. Its eggs are often preyed upon by the fish crow (*Corvus ossifragus*).

Sora Rail
Porzana carolina (L.)
The sora rail feeds in tidal brackish and freshwater marshes of Florida's Gulf coast during the winter.

Virginia Rail
Rallus limicola Vieillot
The Virginia rail feeds in tidal brackish and freshwater marshes of Florida's Gulf coast during the winter. It is also possible that it nests in northern Florida as well (J. C. Ogden, personal communication).

Laughing Gull
Larus atricilla L.
The laughing gull has been seen nesting in salt spikegrass (*Distichlis spicata*) on Smith and Palmetto Islands off Spring Creek, Wakulla County, Florida. They are uncommon in northern Florida during the winter.

Fish Crow
Corvus ossifragus Wilson
The fish crow often feeds in the tidal marshes of Florida's northern Gulf coast.

Marsh Wren
Cistothorus palustris (Wilson)
The marsh wren, a fairly abundant but seldom-seen bird, feeds and nests in the tidal salt marshes.

Seaside Sparrow
Ammodramus maritimus (Wilson)
The seaside sparrow frequently nests in black needlerush (*Juncus roemerianus*) and feeds on spiders, grasshoppers, and beetles. It also feeds on some small bivalves, gastropods, and small crabs in the tidal salt marshes.

Sharp-Tailed Sparrow
Ammodramus caudacutus (Gmelin)
The sharp-tailed sparrow feeds in the tidal salt marsh along the Gulf coast in winter. A rather secretive bird, it can be found feeding in smooth cordgrass (*Spartina alterniflora*) fringes on the seaward side of barrier beaches and the borders of sand barrens near the coast.

Red-Winged Blackbird
Agelaius phoeniceus (L.)
Red-winged blackbirds often pass through and sometimes feed in the tidal salt marshes, usually on the seeds of grasses and weeds. In winter they feed in great conspicuous flocks.

Mammals

Nine-Banded Armadillo
Dasypus novemcinctus L.
The nine-banded armadillo is a common occasional visitor in the upper tidal salt marshes. It is usually nocturnal and feeds mostly on insects and worms.

Cotton Rat
Sigmodon hispidus Say & Ord
One of the more common mammals of northern Florida, the cotton rat is often quite abundant at the upper edges of the tidal salt marshes, living primarily in the nearby pine flatwoods.

Marsh Rabbit
Sylvilagus palustris (Bachman)
The marsh rabbit feeds and breeds at the upper edges of the tidal salt marshes.

Long-Tailed Weasel
Mustela frenata Lichtenstein
Stout (1984) reported the long-tailed weasel as feeding in the tidal salt marshes of the Gulf coast.

Mink
Mustella vison Schreber
Stout (1984) reported the mink as feeding in the tidal salt marshes of the Gulf coast. It preys on small animals (Golley, 1962).

Nutria
Myocastor coypus (Molina)
Stout (1984) reported the nutria as a common resident of tidal salt marshes of the Gulf coast.

Otter
Lutra canadensis (Schreber)
Stout (1984) reported the otter as feeding in the tidal salt marshes of the Gulf coast. It consumes mostly fish and crustaceans (Golley, 1962).

Raccoon
Procyon lotor (L.)
The raccoon is quite often found feeding in the tidal salt marshes, often on shellfish.

White-Tailed Deer
Odocoileus virginianus Zimmermann
The white-tailed deer often crosses the upper tidal salt marshes while traveling between foraging habitats.

Bobcat
Lynx rufus (Schreber)
The bobcat is sometimes an incidental visitor to the tidal salt marshes, feeding on small mammals and on birds which are loafing or feeding on sand barrens.

Feral Pig
Sus scrofa L.
Feral pigs, descended from the European wild boar, are often seen crossing and rooting in the upper tidal salt marshes in the night, early morning, and late afternoon, especially in the salt barrens area, where they often feed on fiddler crabs. They give birth to young on higher ground above the marshes.

Man
Homo sapiens L.
Man is, without doubt, the animal that has had the greatest impact on Florida's Gulf coast tidal marshes. Burning, diking, draining, and construction, along with man's many other activities, have had a tremendous influence on the tidal marshes and will certainly continue to do so unless drastic measures are taken to avert these activities.

Acknowledgments

We thank John C. Ogden of Everglades National Park, Homestead, Florida, and Kerrie M. Swadling (University of Tasmania) and John Gibson (ANARE) for critical comments on the manuscript.

References

American Ornithologists' Union. 1983. *Check-List of North American Birds.* 6th Ed. xxix + 877 pp.

Ashton, R. E., Jr. and P. S. Ashton. 1988. *Handbook of Reptiles and Amphibians of Florida. Part One. The Snakes.* 2nd Ed. Windward Publishing, Miami. 176 pp.

Carr, A. and C. J. Goin. 1955. *Guide to the Reptiles, Amphibians, and Fresh-Water Fishes of Florida.* University of Florida Press, Gainesville.

Golley, F. B. 1962. *Mammals of Georgia: A Study of Their Distribution and Functional Role in the Ecosystem.* University of Georgia Press, Athens.

Longstreet, R. J., R. W. Williams, H. L. Stoddard, F. M. Weston, L. R. Lovelace, S. W. Partridge, and A. H. Howell. 1931. Florida birds. Biographies of selected species of birds and compiled list of all species occurring in Florida. *Q. Bull. Dept. Agric. Fla.* 41(3):1–189, A1–A9.

Stevenson, H. M. 1976. *Vertebrates of Florida.* University Presses of Florida, Gainesville. 607 pp.

Stout, J. P. 1984. *The Ecology of Irregularly Flooded Salt Marshes of the Northeastern Gulf of Mexico: A Community Profile.* Biological Report 85(7.1):i–xiv, 1–98. U.S. Department of the Interior, Fish and Wildlife Service, Washington, D.C.

Wolfe, S. H., J. A. Reidenauer, and D. B. Means. 1988. *An Ecological Characterization of the Florida Panhandle.* Biological Report 88(12):i–xix, 1–277. U.S. Department of the Interior, Fish and Wildlife Service, Washington, D.C.

The Aquatic Insects of Northern Florida's Gulf Coast Tidal Marshes

B

Michael D. Hubbard

Many species of insects associated with Florida's Gulf coast tidal marshes are dependent upon the water for development. A number of these (especially species of Diptera, such as mosquitoes, sandflies, horseflies, and deerflies) are important nuisance species to both man and animals. These aquatic insects form an important part of the tidal salt marsh ecosystem.

Anyone who has spent any time at all in or around northern Florida's Gulf coast tidal marshes is acutely aware that insects are associated with the marshes. Mosquitoes, sandflies or "no-see-ums," dog flies, deer flies, yellow flies, and horse flies are extremely common and make their presence known to the intruder in an unmistakable way.

These ubiquitous biting inhabitants of the tidal marshes, all of which belong to the insect order Diptera (the true flies), along with other insects (such as dragonflies), are often dependent on water for a developmental habitat. The immature stages of most of these insects are spent in the water or mud substrate of the marsh until they are ready to emerge as adults. After this transformation, they feed on and interact with man and other animals and reproduce, continuing the cycle with their aquatic offspring.

This appendix deals only with those insect inhabitants of Florida's Gulf coast tidal marshes that are usually thought of as truly "aquatic"; *viz.*, those directly dependent upon water for completion of the immature stages of their life cycles. The terrestrial insect fauna are addressed in Chapter 7.

©St. Lucie Press CCC 1-57444-026-8 1/97/$100/$.50

Aquatic Insects

Sandflies

Diptera: Ceratopogonidae

Sandflies are fierce biters and often make a visit to the tidal marshes an excruciatingly painful and unpleasant experience. The females of most species of the common sandfly genus *Culicoides* depend for maturation of their eggs on blood meals which they take by biting vertebrates. Male *Culicoides* (and females of *C. bermudensis*) do not take blood meals; hence they do not bite. Males, and sometimes females, feed on nectar of flowering plants. Larvae of *Culicoides* are not strictly aquatic, but they cannot develop without water. They feed primarily on living food (motile algae and small invertebrates). The larvae of few species can exist at a depth of more than a few centimeters below the air–water interface.

Sandflies can be vectors of several diseases (especially protozoal and viral), affecting birds and mammals (Blanton and Wirth, 1979). It is not known if the sandflies of northern Florida's Gulf coast tidal marshes are important disease vectors.

In the tidal salt marshes of northern Florida's Gulf coast, which are characterized by the growth of black needlerush (*Juncus roemerianus* Scheele), *C. mississippiensis* is the predominant sandfly, and there are lesser numbers of *C. furens*, *C. melleus*, and *C. bermudensis* (Blanton and Wirth, 1979).

Culicoides bermudensis Williams
The sandfly *C. bermudensis* breeds in salt marshes associated with salt spikegrass (*Distichlis spicata* [L.] Greene), in salt marsh pools, and in mud covered with mangroves. Because the females of *C. bermudensis* do not take blood meals, they are not biting pests.

Culicoides furens (Poey)
The sandfly *C. furens* commonly breeds in mangrove swamps and in those northern Florida's Gulf coast tidal salt marshes dominated by smooth cordgrass (*Spartina alterniflora* [Loisel]). The larvae probably feed primarily on living food.

Culicoides melleus (Coquillett)
Females of the sandfly *C. melleus* oviposit on the upper intertidal portion of the beaches. The nymphs are predators on protozoa, motile algae, and small invertebrates.

Culicoides mississippiensis Hoffman
The sandfly *C. mississippiensis* breeds in northern Florida's Gulf coast tidal salt marshes.

Other Sandflies

Rey and McCoy (1986) reported the presence of four species of the sandfly genus *Dasyhelea*, and one species each of the sandfly genera *Forcipomyia*, *Atrichopogon*, *Bezzia*, and *Stilobezzia* in the St. Marks National Wildlife Refuge salt marshes.

Mosquitoes

Diptera: Culicidae

The two most important and dominant salt marsh mosquitoes found along the more than 2000-km coastline of Florida are the black salt marsh mosquito (*Aedes taeniorhynchus* [Wiedemann]) and the eastern salt marsh mosquito (*A. sollicitans* [Walker]) (Nayar, 1985), which at one time made many areas of Florida virtually uninhabitable because of their density and fierce biting habits. The two species are extremely common in northern Florida's Gulf coast tidal marshes.

These two species of mosquito have been shown to have the potential to be good to excellent vectors of human encephalitis and the dog heartworm (*Dirofilaria immitis*) and may be important in the spread of these diseases (King et al., 1960).

Aedes (Ochlerotatus) sollicitans (Walker)

The larvae and adults of the mosquito *A. sollicitans* may be found any time during the year. The adults are strong fliers and often migrate in large numbers to communities many miles from the saltwater marshes in which they breed. The females are persistent biters and will attack any time during the day or night. The adults rest in the vegetation during the daytime and will attack anyone invading their haunts, even in full sunlight (Carpenter and LaCasse, 1955).

The females lay their eggs (which can withstand long periods of dryness) on the mud of the moist marshes, and the eggs hatch when flooded by high tides or rains (Matheson, 1944; King et al., 1960).

Aedes (Ochlerotatus) taeniorhynchus (Wiedemann)

The larvae and adults of the mosquito *A. taeniorhynchus* may be found any time during the year in the extreme south, but populations are usually heavier following high tides or heavy rains during the summer and early fall. The most prolific breeding in the northern Florida salt marshes occurs in temporary ponds and holes in areas dominated by the salt marsh plants *Batis* and *Distichlis* (King et al., 1960). Like *A. sollicitans*, the females rest in the vegetation during the daytime, are fierce and persistent biters, and will attack any time during the night or day, even in full sunlight. The adults of this species are strong fliers and may be an extreme nuisance many miles from the saltwater marshes in which they breed (Carpenter and LaCasse, 1955).

Other Mosquitoes
Rey and McCoy (1986) also reported the mosquito species *Anopheles atropos* Dyar & Knab, *Anopheles bradleyi* King, and *Culex salinarius* Coquillett as not uncommon in the St. Marks National Wildlife Refuge salt marshes, although these mosquitoes are not considered to be important nuisance species.

Other Diptera

Blind Midges (Diptera: Chironomidae)
The larvae of the blind midge family Chironomidae pass their life in the water. Several species of chironomid midges have been collected or reared from northern Florida's Gulf coast tidal marshes (J. Epler, personal communication). *Apedilum elachistus* Townes, *Dicrotendipes lobus* (Beck), and *Goeldichironomus devineyae* (Beck) breed in pools in the tidal marsh. *Cunio marshalli* Stone and Wirth and *Chironomus* sp. also breed in the marsh. *Dicrotendipes modestus* (Say), *Goeldichironomus holoprasinus* (Goeldi), and *Polypedilum* sp. have been collected in the marshes, but whether or not they breed there is unknown. These blind midges do not bite, but sometimes may become a nuisance because of the large numbers in which they can emerge.

Crane Flies (Diptera: Tipulidae)
Tipulidae, the largest family of Diptera, usually have aquatic larvae which directly depend on atmospheric oxygen and cannot venture far from the water surface. Usually, the larvae leave the water to pupate in nearby soil or litter (Byers, 1978). Rey and McCoy (1986) reported the tipulid species *Limonia floridana* (Osten Sacken) and another undetermined species as uncommon in the St. Marks National Wildlife Refuge salt marshes. Crane flies do not bite and are not considered a nuisance to humans or animals.

Horse Flies and Deer Flies (Diptera: Tabanidae)
Horse flies and deer flies are often painful biters and can be a severe nuisance to vertebrates (including man) in the marsh, especially in late summer and early fall. Adult females lay egg-masses on vegetation close to moist soil suitable for larval development. Principally daylight feeders, the females of most species require a blood meal for development of eggs (Axtell, 1976; Jones and Anthony, 1964). Tabanid larvae are commonly found in areas dominated by *Spartina* and relatively few are found in *Juncus* areas.

Rey and McCoy (1986) reported the tabanid species *Tabanus nigrovittatus* Macquart as common and *Chrysops fuliginosus* Wiedemann and *Chrysops atlanticus* Pechuman as uncommon in the St. Marks National Wildlife Refuge salt marshes. The larvae of *T. nigrovittatus* tend to congregate in areas of the marsh reached only by high tide. *C. atlanticus* may not actually breed in the salt marsh, but in nearby fresh water (Jones and Anthony, 1964).

Other Aquatic Insects

Many of the insect orders traditionally thought of as aquatic are poorly adapted to and rarely found in heavily saline habitats and are usually lacking from the salt marsh fauna. In particular, the Ephemeroptera (mayflies), the Trichoptera (caddisflies), and the Plecoptera (stoneflies), almost all species of which require water for development of the immature stages, do not occur in northern Florida's Gulf coast tidal marshes (Berner and Pescador, 1988; Leader, 1976; Stark and Gaufin, 1979).

Except for the seaside dragonlet (*Erythrodiplax berenice* [Drury], Dunkle (1989) does not report any dragonflies (Odonata), another water-dependent group, breeding in the northern Florida Gulf coast tidal marshes, although dragonfly adults of several species often perch on *Juncus* stems and apparently feed on small insects flying over the marsh. Except for the seaside dragonlet, the only true saltwater dragonfly in North America, these dragonflies probably breed in nearby brackish and freshwater habitats and enter the marshes only for feeding.

Other groups of insects which are often referred to as "aquatic" (e.g., the order Megaloptera, some groups of Diptera, and some members of the orders Collembola, Coleoptera, Hemiptera, Hymenoptera, Lepidoptera, and Neuroptera) are not dealt with here because of an unfortunate lack of knowledge of this fauna in northern Florida's Gulf coast tidal salt marshes. Many of these groups probably do not occur in the tidal salt marshes. Further general information on salt marsh and marine insects is provided by Cheng (1976).

Effects of Aquatic Insect Control on the Salt Marsh

Attempts—many successful—to control the nuisance aquatic insects in tidal salt marshes have induced important changes to salt marshes in many cases and have the potential for even more. In many places, attempts to control populations of mosquitoes and sandflies have involved extensive modification of the tidal salt marsh environment. Ditching and diking are two important methods often used to control these insects. Both of these methods have been used for some time and result in a major change in the salt marsh and the function of its associated ecosystem. Habitat control has proved of little use in control of tabanids in the salt marsh. The extensive use of pesticides in and around salt marshes for control of mosquitoes, sandflies, and tabanids has resulted in the introduction of a great deal of toxic material into the ecosystem, with a generally unknown effect on nontarget organisms.

Acknowledgments

I thank John H. Epler for the benefit of his knowledge of the Chironomidae, and Kerrie M. Swadling (University of Tasmania) and John Gibson (ANARE) for critical comments on the manuscript.

References

Axtell, R. C. 1976. Coastal horse flies and deer flies (Diptera: Tabanidae). Pages 415–445 in L. Cheng, Ed. *Marine Insects*. North-Holland, Amsterdam.

Berner, L. and M. L. Pescador. 1988. *The Mayflies of Florida*. Rev. Ed. University Presses of Florida (Florida A&M University Press, Tallahassee and University of Florida Press, Gainesville). 415 pp.

Blanton, F. S. and W. W. Wirth. 1979. The sand flies (*Culicoides*) of Florida (Diptera: Ceratopogonidae). *Arthropods of Florida and Neighboring Land Areas* 10:i–xv, 1–204.

Byers, G. W. 1978. Tipulidae. Pages 285–310 in R. W. Merritt and K. W. Cummins, Eds. *An Introduction to the Aquatic Insects of North America*. Kendall/Hunt, Dubuque, Iowa.

Carpenter, S. J. and W. J. LaCasse. 1955. *Mosquitoes of North America (North of Mexico)*. University of California Press, Berkeley. 360 pp.

Cheng, L. 1976. *Marine Insects*. North-Holland, Amsterdam. 581 pp.

Dunkle, S. W. 1989. *Dragonflies of the Florida Peninsula, Bermuda and the Bahamas*. Scientific Publishers, Gainesville, Florida. 154 pages.

Jones, C. M. and D. W. Anthony. 1964. *The Tabanidae (Diptera) of Florida*. Agricultural Research Service Technical Bulletin No. 1295:1–85. U.S. Department of Agriculture, Washington, D.C.

King, W. V., G. H. Bradley, C. N. Smith, and W. C. McDuffie. 1960. *A Handbook of the Mosquitoes of the Southeastern United States*. Agriculture Handbook 173: 1–188. U.S. Department of Agriculture, Washington, D.C.

Leader, J. P. 1976. Marine caddisflies (Trichoptera: Philanisidae). Pages 291–302 in L. Cheng, Ed. *Marine Insects*. North-Holland, Amsterdam.

Matheson, R. 1944. *Handbook of the Mosquitoes of North America*. 2nd Ed. Comstock Publishing, Ithaca, New York. 314 pp.

Nayar, J. K. 1985. Bionomics and physiology of *Aedes taeniorhynchus* and *Aedes sollicitans*, the salt marsh mosquitoes of Florida. *Fla. Agric. Exp. Sta. Bull.* 852: 1–148.

Rey, J. R. and E. D. McCoy. 1986. Terrestrial arthropods of northwest Florida salt marshes: Diptera (Insecta). *Fla. Entomol.* 69:197–205.

Stark, B. P. and A. R. Gaufin. 1979. The stoneflies (Plecoptera) of Florida. *Trans. Am. Entomol. Soc.* 104:391–433.

INDEX

347